NUMERICAL ANALYSIS
with
ELEMENTS *of* COMPUTER SCIENCE

NUMERICAL ANALYSIS
with
ELEMENTS *of* COMPUTER SCIENCE

For BSc (General Course) Students

A N Das MSc (Gold Medallist), PhD
Associate Professor in Mathematics
Alipurduar College, Jalpaiguri, West Bengal
Ex-Lecturer in Mathematics, St Joseph's College, Darjeeling

New Central Book Agency (P) Ltd
LONDON
HYDERABAD ERNAKULAM BHUBANESWAR
NEW DELHI KOLKATA PUNE GUWAHATI

NCBA

REGD OFFICE
8/1 Chintamoni Das Lane, Kolkata 700 009, India
email: ncbapvtltd@eth.net

OVERSEAS
NCBA (UK) Ltd, 149 Park Avenue North, Northampton, NN3 2HY, UK
email: ncbauk@yahoo.com

EXPORT
NCBA Exports Pvt. Ltd, 212 Shahpur Jat, New Delhi 110 049
email: ncbaexp@ncbapvtltd.com

BRANCHES

212 Shahpur Jat
New Delhi 110 049
email: ncbadel@ncbapvtltd.com

Shop Nos. 3 & 4, Vinayak Towers
681/B Budhwar Peth, Appa Balwant Chowk
Pune 411 002
email: ncbapune@ncbapvtltd.com

Shop No. 15, Bharati Towers, Ground Floor
Forest Park, **Bhubaneswar** 751 001
email: ncbabub@ncbapvtltd.com

House No. 3-1-315, 1st Floor
Nimboliadda, Kachiguda, **Hyderabad** 500 027
email: ncbahydb@ncbapvtltd.com

GSS Shopping Complex, 1st Floor
Opposite College Post Office, Convent Road
Ernakulam 682 035
email: ncbaernk@ncbapvtltd.com

Radhamayee Complex (Second Floor)
Opp. SBI, Panbazar, Jaswanta Road
Guwahati 781 001
email: ncbaguwahati@ncbapvtltd.com

NUMERICAL ANALYSIS WITH ELEMENTS OF COMPUTER SCIENCE • A N Das

First Published: July 2012
Reprinted: June 2013

PUBLISHER AND TYPESETTER
New Central Book Agency (P) Ltd
8/1 Chintamoni Das Lane, Kolkata 700 009

PRINTER
New Central Book Agency (P) Ltd
Web-Offset Division, Dhulagarh, Sankrail, Howrah

COVER DESIGNER
Soumen Paul

COVER PRINTER
Techno Process, Kolkata

TECHNICAL EDITOR
Dipan Roy

PROJECT TEAM
Suman Dutta, Koushik Samanta and Prabhat Jas

ISBN 978 81 7381 755 7

Price ₹ 235.00

CONTENTS

ELEMENTS OF COMPUTER SCIENCE

PREFACE

Efforts have been made in this book to make the treatment logical and simple—throughout the book the need of the students have been kept in view. To make it more helpful to the students, quite a few short questions have been added at the end of the book.

I take the opportunity to express my deep debt of gratitude to several authors whose books have been consulted during the preparation of this treatise.

I shall be grateful to the readers for pointing out any errors and omissions that, in spite of all care, might have crept in. Suggestions for improvement of the book will be highly appreciated.

Finally, I would like to thank Mr Amitabha Sen, Director, New Central Book Agency (P) Ltd, for his interest and tireless toil in publishing the book.

Good Friday
6 April 2012

A N Das

NUMERICAL ANALYSIS

Chapter 1

NUMERICAL APPROXIMATIONS

1.1 INTRODUCTION

The aim of numerical analysis is to solve complex numerical problems using only simple operations of arithmetic, to develop and evaluate method of computing numerical results from given data. While solving problems in numerical analysis one generally uses a polynomial for interpolation or one starts with some initial data to get an approximate result, which is corrected up to two or three decimal places. So the error in a computed result occurs due to the error in given data or due to the machine used (as each machine has fixed length) or both. In the next section we will discuss about errors.

1.2 ERROR

In many cases one has to work with an approximate number, which represents the exact number to certain degree of accuracy. For example, 0.1111 is an approximate numbers approximating the exact number 1/9. Obviously, an exact number may have many approximate numbers of different degree of accuracies.

Absolute Error: If the number x^* is an approximation to the exact number x, then numerical value of the difference of the exact value and approximated value, i.e., $E_A = |x - x^*|$ is called absolute error in x^*.

Relative Error: A relatively better measure of error is given by the relative error which is defined by

$$E_R = E_A / x = |x - x^*| / x.$$

From this it is clear that the relative error is a dimensionless quantity.

Percentage Error: When the relative error expressed in percentage, it is known as percentage error, thus percentage error, $E_p = 100 \times E_R$.

1.3 SIGNIFICANT DIGITS AND ROUNDING-OFF

Significant digits are those, which are used to express a number. For example, the numbers 2.137, 0.5347, 0.6046 each contains four significant digits but the number 0.0049 has two significant digits since in this case zeros preceding the non-zero digit serve for fixing the decimal point only. As a human or machine can handle only a limited number of digits to express a number, one has to discard all but a proposed number of digits of a given number. This process of rejection of unwanted digits of a given number is known as rounding-off the number. To round-off a number to *n*-significant digits, discard all digits to the right of the *n*th digit, and if the $(n + 1)$th digit is

(a) less than 5; leave the *n*th digit unchanged,

(b) greater than 5; increase the *n*th digit by 1,

(c) exactly 5; increase the *n*th digit by 1 if it is odd otherwise leave it unchanged.

For example, the numbers 2.5684, 11.681, 0.185382, 1.24359, 2.13853 when rounded-off to four significant digits yield 2.568, 11.68, 0.1854, 1.244, and 2.138. The simplest way of reducing the number of significant digits in the representation of a number is merely to ignore the unwanted digits. This procedure is known as chopping. For example, π chopped to four decimal places is 3.1415, but it is 3.1416 when rounded.

If a number x is rounded to q decimal places and this rounded number is denoted by x^*, then we say that x^* is corrected up to q decimal places and we write, $x = x^* + \delta\cdot 10^{-q}$ $(|\delta| \le 1/2)$.

1.4 LOSS OF SIGNIFICANCE DUE TO SUBTRACTION

One of the sources of error in numerical computations is loss of significant digits. During subtraction of two nearly equal numbers, a serious error may occur due to loss of significant digits. Suppose we have to calculate the number $z = x - y$. If x^*, y^*, the approximate values of x, y respectively, are corrected up to r significant digits, then $z^* = x^* - y^*$, the approximate value of z may be corrected fewer than r significant digits.

For example, suppose $x^* = 53.469$, $y^* = 52.883$ are the approximate values of x, y corrected to 5 significant digits. Then $z^* = x^* - y^* = 0.586$, i.e., their difference is corrected to 3 significant digits only.

Such loss of significant digits is often avoidable and the error thus caused can be reduced by

(i) taking the numbers as accurate as possible, and

(ii) by transforming the expression whose values are to be evaluated.

For example, evaluation of $1 - \cos x$ for small values of x is to be made using its equivalent formulae $\dfrac{\sin^2 x}{1+\cos x}$ or $\dfrac{x^2}{2} - \dfrac{x^4}{24}$.

1.5 FLOATING POINT REPRESENTATION OF REAL NUMBERS

Floating point representation of real numbers have an additional convenience in scientific calculations. In this form numbers are described by three parts, a sign, a mantissa and an exponent with sign. An *n*-digit floating point number in base A has the form

$$\pm (0.b_1 b_2 b_3 \cdots)_A A^e,$$

where $(0.b_1b_2b_3\cdots)_A$ is A-fraction called mantissa and e is an integer called exponent. For example, the number 0.4932 might appear as $+0.4932 \times 10^0$. This number can also be written as other ways like $+4.932 \times 10^{-1}$ or $+0.04932 \times 10^1$ but the standard practice is to write the number in normalized form which is the case when $b_1 \neq 0$ or all $b_i = 0$. In this case the normalized form is $+0.4932 \times 10^0$.

1.6 ADDITION AND SUBTRACTION

The mantissas are added and subtracted; (after shifting the mantissa and increasing the exponent of the smaller number, if necessary, to make the exponents agree) the final result is obtained by rounding (after shifting the mantissa and adjusting the exponent, if necessary). Thus:

$$4.23 \times 10^2 + 2.46 \times 10^2 = 6.69 \times 10^2$$

$$4.27 \times 10^1 + 2.71 \times 10^{-1} = 4.27 \times 10^1 + 0.0271 \times 10^1 = 4.2971 \times 10^1 \approx 4.30 \times 10^1$$

$$3.65 \times 10^1 - 2.78 \times 10^{-1} = 3.65 \times 10^1 - 0.0278 \times 10^1 = 3.6222 \times 10^1 \approx 3.62 \times 10^1.$$

1.7 MULTIPLICATION AND DIVISION

The exponents are added and mantissas are multiplied; the final results are obtained by rounding (after shifting the mantissa right and increasing the exponent by 1, if necessary). Thus:

$$4.23 \times 10^2 \times 2.46 \times 10^2 = 10.4058 \times 10^4 \approx 10.41 \times 10^4$$

$$4.27 \times 10^1 \times 2.71 \times 10^{-1} = 11.5717 \times 10^0 \approx 1.16 \times 10^1.$$

In case of division the exponents are subtracted and the mantissas are divided; the final results are obtained by rounding (after shifting the mantissa left and increasing the exponent by 1, if necessary).

Thus, $4.23 \times 10^2 \div 2.46 \times 10^2 = 1.7195 \times 10^0 \approx 1.72 \times 10^0$

$$4.27 \times 10^1 \div 2.71 \times 10^{-1} = 1.5756 \times 10^2 \approx 1.58 \times 10^2.$$

The order of evaluation is determined in a standard way and the result of each operation is a normalized floating point number. Thus

$$4.23 \times 10^2 \div 2.46 \times 10^2 + 4.27 \times 10^1 \div 2.71 \times 10^{-1} - 3.27 \times 10^1 \div 3.71 \times 10^{-1}$$

$$= 1.7195 \times 10^0 + 1.5756 \times 10^2 - 0.8814 \times 10^2$$

$$= 0.0172 \times 10^2 + 1.5756 \times 10^2 - 0.8814 \times 10^2$$

$$= 0.7114 \times 10^2 \approx 0.71 \times 10^2.$$

The peculiarities of floating point arithmetic lead to some unexpected and unfortunate consequences.

(i) Addition or subtraction of a small number (but not zero) may have no effect, e.g.,

$$4.27 \times 10^2 + 2.71 \times 10^{-1} = 4.27 \times 10^2 + 0.00271 \times 10^2$$

$$= 4.27271 \times 10^2 \approx 4.27 \times 10^2.$$

(ii) The result $a.(1/a)$ may not be equal to 1, e.g.,

If $a = 3.00 \times 10^0$, then $1/a = 3.33 \times 10^{-1}$ and $a.(1/a) = 9.99 \times 10^{-1}$

(iii) The result $a + (b + c)$ may not be equal to $(a + b) + c$, e.g.,

if $a = 6.31 \times 10^1$, $b = 4.24 \times 10^0$, $c = 2.47 \times 10^{-1}$

$$\begin{aligned} a + (b + c) &= 6.31 \times 10^1 + (4.24 \times 10^0 + 2.47 \times 10^{-1}) \\ &= 6.31 \times 10^1 + (4.24 \times 10^0 + 0.247 \times 10^0) \\ &= 6.31 \times 10^1 + 4.49 \times 10^0 \\ &= 6.31 \times 10^1 + 0.449 \times 10^1 \\ &= 6.759 \times 10^1 = 6.76 \times 10^1 \end{aligned}$$

and

$$\begin{aligned} (a + b) + c &= (6.31 \times 10^1 + 4.24 \times 10^0) + 2.47 \times 10^{-1} \\ &= (6.31 \times 10^1 + 0.424 \times 10^1) + 2.47 \times 10^{-1} \\ &= 6.73 \times 10^1 + 0.0247 \times 10^1 \\ &= 6.75 \times 10^1. \end{aligned}$$

1.8 ERROR PROPAGATION AND GENERATION

We know that a number is to be represented by a finite number of digits. It is expected that the result of any arithmetic calculation involving a set of numbers will have an implicit error relating to the error in the original numbers. We say that the initial errors propagate through the computation. In addition, errors may be generated at each step in the algorithm and we speak of the total cumulative error at any step as the accumulated error. In extended calculations it is likely that many round off error will be made. Each of these plays the role of an input error for the remainder calculation, and each has an impact on the eventual output.

Algorithms for which the accumulated error is limited, so that a useful result is generated, are called *stable algorithm.* Unfortunately, there are cases where the accumulative error is devastating and the solution is overwhelmed by errors. Needless to say that such algorithms are called *unstable*.

Consider two numbers $x = x^* + e_1$ and $y = y^* + e_2$.

Then $x \pm y = x^* \pm y^* + e_1 \pm e_2$

so that $e \approx (x \pm y) - (x^* \pm y^*) = e_1 \pm e_2$

and hence $|e| \leq |e_1| + |e_2|$, i.e., max $|e| = |e_1| + |e_2|$, i.e., the magnitude of the propagated error is not more than the sum of initial absolute errors.

We also see that $xy = x^*y^* + x^*e_2 + y^*e_1 + e_1e_2$

so that

$$\left|\frac{xy - x^*y^*}{x^*y^*}\right| \leq \left|\frac{e_2}{y^*}\right| + \left|\frac{e_1}{x^*}\right| + \left|\frac{e_1e_2}{x^*y^*}\right|$$

and so $$\text{Max}\left|\frac{xy - x^*y^*}{x^*y^*}\right| \leq \left|\frac{e_2}{y^*}\right| + \left|\frac{e_1}{x^*}\right|,$$ assuming $\left|\frac{e_1e_2}{x^*y^*}\right|$ is small.

1.9 A GENERAL FORMULA

In this section, we derive a general formula for the error committed in using a certain formula. Let $u = f(x_1, x_2, \cdots, x_n)$ be a function of several variables. If the error in each x_i be δx_i then the error δu in u is given by

$$\delta u = f(x_1 + \delta x_1, x_2 + \delta x_2, \cdots, x_n + \delta x_n) - f(x_1, x_2, \cdots, x_n)$$

$$\delta u = \sum_{i=1}^{n} \frac{\partial f}{\partial x_i} \delta x_i$$

(after expanding in Taylor's series and neglecting the terms $(\delta x_i)^2$ and higher powers).

The formula for relative error is

$$\frac{\delta u}{u} = \sum_{i=1}^{n} \frac{\partial f}{\partial x_i} \frac{\delta x_i}{u}.$$

1.10 INHERENT ERRORS

As one starts with an approximate data and as every machine has a fixed length, the result of most calculations are inexact. In a numerical calculation all the initial data or results of subsequent steps are rounded off causing rounded off errors. These errors can be minimised by using better data and by using computing machines of higher precision. The round off errors can also be reduced by carrying calculations with more significant figures at each step and after the last operation it is rounded off. If a number X is rounded to N decimal places, then

$$\text{absolute error} \le 0.5 \times 10^{-N}.$$

For example, if a number X = 0.43 is corrected up to 2 decimal places, then maximum absolute error is 0.005 and the relative error is given by 0.005/0.43 = 0.01163.

1.11 TRUNCATION ERROR

In numerical calculation we have often to use an infinite series of x, instead of a function $f(x)$, after truncating it at a certain stage. The error thus caused is called truncation error. Obviously, the study of such error is associated with the problem of convergence. Retaining more terms in the series we can reduce truncation error; this of course involve extra work.

Examples

1. Round off the following numbers to 3 places of decimals:
   (a) 6.23105 (b) 0.34679 (c) 2.6782 (d) 0.0228.
   Solution: (a) 6.231 (b) 0.347 (c) 2.678 (d) 0.023.
2. Round off the following numbers to four significant digits:
   (a) 6.23105 (b) 0.34679 (c) 2.6783 (d) 0.022827 (e) 0.500678.
   Solution: (a) 6.231 (b) 0.3468 (c) 2.678 (d) 0.02283 (e) 0.5007.
3. Calculate $\sqrt{3.2} - \sqrt{3}$ correct to 4 decimal places:
   Solution: Since $\sqrt{3.2} = 1.788854$ and $\sqrt{3} = 1.732051$,
   $$\sqrt{3.2} - \sqrt{3} = 0.056803 \approx 0.0568.$$

4. Three approximations of 1/9 are 0.1,0.11, and 0.12. Which of these three is the best approximation?

Solution: We have $|1/9 - 0.1| = 1/90,$

$|1/9 - 0.11| = 1/900,$

$|1/9 - 0.12| = 2/225.$

It follows that 0.11 is the best approximation.

5. Evaluate the sum S = $\sqrt{3} + \sqrt{6} + \sqrt{7}$, correct to 4 significant digits and find its absolute and relative errors.

Solution: We have $\sqrt{3} = 1.732$, $\sqrt{6} = 2.449$, $\sqrt{7} = 2.646$.

So S = 6.827 and the absolute error = 0.0005 +0.0005 + 0.0005 = 0.0015. The total absolute error shows that the sum is correct to 3 significant figures only. Hence we take S = 6.83 and then relative error is = 0.0015/6.83 = 0.0002.

6. Find the sum of the following approximate numbers if the numbers are correct to the last digits:

2.14, 6.348, 3.9156, 1.002, 3.167.

Solution: Required sum = 2.14 + 6.348 + 3.9156+ 1.002 + 3.167

= 16.5726

= 16.57 (the result is rounded to two decimal places which is equal to the minimum number of digits after decimal places in the given numbers).

7. If $u = x_1x_2 + x_3x_2 + x_1x_3$, find the relative error in the computation of u at $x_1 = 2.104$, $x_2 = 1.935$ and $x_3 = 0.845$.

Solution: Here x_1, x_2, x_3 are given to 3 decimal places, so absolute error in x_1, x_2, x_3 are 0.0005.

Now relative error in u $= \frac{\partial u}{\partial x_1}\frac{\delta x_1}{u} + \frac{\partial u}{\partial x_2}\frac{\delta x_2}{u} + \frac{\partial u}{\partial x_3}\frac{\delta x_3}{u}$

$$= \frac{1}{u}[(x_2 + x_3)\delta x_1 + (x_1 + x_3)\delta x_2 + (x_2 + x_1)\delta x_3]$$

$$= \frac{0.0005 \times 2 \times 4.884}{7.4842} = 0.000653 = 6.53 \times 10^{-5}.$$

EXERCISE

1. Round off the following numbers to 3 places of decimals:
   (a) 46.2356 (b) 0.72679 (c) 12.00678 (d) 0.0258 (e) 0.34654 (f) 1.2355.
2. Round off the following numbers to four significant digits:
   (a) 46.2356 (b) 0.72679 (c) 12.00678 (d) 0.025845 (e) 0.34654 (f) 1.2355.
3. Calculate $\sqrt{3.2} - \sqrt{3}$, correct to 3 significant figures.
4. Calculate $\sqrt{2.1} - \sqrt{2}$, correct to 3 decimal places.
5. Calculate the value of $\sqrt{102} - \sqrt{101}$, correct to 4 significant figure.
6. Three approximations of 1/3 are 0.3,0.33, 0.34. Which of these three is the best approximation?

7. If 2/3 is correct up to five significant digits, then find absolute, relative and percentage errors.
8. If 5/6 is correct up to five significant digits, then find absolute, relative and percentage errors.
9. If we take = 3.14 instead of = 22/7, then find absolute, relative and percentage errors.
10. The floating numbers are given in a decimal computer with a four digit normalised mantissa:

    $a = 0.3425 \times 10^4$ $\quad b = 0.4176 \times 10^{-3}$ $\quad c = 0.2538 \times 10^{-1}$

    Calculate the following operations, assuming symmetric rounding:

    (i) $a + b + c$, (ii) $a - b - c$, (iii) a/c, (iv) ab/c, (v) $b/(ac)$.
11. Determine the greatest value of absolute error in the following calculations:

    $a > 3.12 + 0.147 - 2.13325$,

    $b > 1.328 \times 0.27$,

    $c > 0.628/3.2174$.
12. If $u = x_1x_2$, find the relative error in the computation of at $x_1 = 2.341$, $x_2 = 1.653$.
13. Find the sum of the following approximate numbers if the numbers are correct to the last digits: 2.014, 16.0348, 0.09156, 4.702, and 6.5167
14. Show that the result $a + (b + c)$ is not equal to $(a + b) + c$, if $a = 6.31 \times 10^1$, $b = 4.24 \times 10^0$, $c = 2.47 \times 10^{-1}$.

ANSWERS

1. (a) 46.236 (b) 0.727 (c) 12.007 (d) 0.026 (e) 0.346 (f) 1.236
2. (a) 46.24 (b) 0.7268 (c) 12.01 (d) 0.02584 (e) 0.3465 (f) 1.236
3. 0.0568
4. 0.0349
5. 0.04963
6. 0.33
7. 0.000003, 0.000004, 0.0004%
8. 0.000003, 0.000004, 0.0004%
9. 0.00286, 0.00091, 0.091%
13. 29.359

Chapter 2

OPERATOR

2.1 INTRODUCTION

If S denotes the class of all functions defined in $(-\infty, \infty)$, and for every function $f(x)$ in S, we can find a definite function $g(x)$ in S by a definite rule, then this correspondence is called *operator* and we write as $Af(x) = g(x)$. Such operator is said to be linear if for any two functions $f(x)$, $g(x)$ in S and for any two constants c, d

$$A\{cf(x) + dg(x)\} = cA\{f(x)\} + dA\{g(x)\}.$$

The sum and product of two operators A, B on S are defined by

$$(A + B) f(x) = Af(x) + Bf(x) \text{ and } (AB) f(x) = A[Bf(x)]$$

for every $f(x)$ in S.

If $y = f(x)$ be a function, then for $x = a, a + h, a + 2h, a + 3h, \cdots$ the values of y are $f(a)$, $f(a + h), f(a + 2h), f(a + 3h), \cdots$. The forward difference operator is defined as

$$\Delta f(a) = f(a + h) - f(a).$$

$$\begin{aligned}
\text{Now, } \quad \Delta^2 f(a) &= \Delta[\Delta f(a)] \\
&= \Delta[f(a + h) - f(a)] \\
&= \Delta f(a + h) - \Delta f(a) \\
&= f(a + 2h) - f(a + h) - f(a + h) + f(a) \\
&= f(a + 2h) - 2f(a + h) + f(a) \\
\text{and } \quad \Delta^3 f(a) &= \Delta[\Delta^2 f(a)] \\
&= \Delta[f(a + 2h) - 2f(a + h) + f(a)] \\
&= \Delta f(a + 2h) - 2\Delta f(a + h) + \Delta f(a) \\
&= f(a + 3h) - f(a+2h) - 2[f(a + 2h) - f(a + h)] + f(a + h) - f(a) \\
&= f(a + 3h) - 3f(a + 2h) + 3f(a + h) - f(a) \text{ and so on.}
\end{aligned}$$

In general

$\Delta^n f(a) = f(a + nh) - {}^nc_1 f\{a + (n-1)h\} + {}^nc_2 f\{a + (n-2)h\} - {}^nc_3 f\{a + (n-3)h\} + \cdots + (-1)^n f(a)$. If r, s are two non negative integers then

$$\Delta^{r+s} f(a) = \Delta^r [\Delta^s f(a)] = \Delta^s [\Delta^r f(a)].$$

2.2 FORWARD DIFFERENCE TABLE

x	y	Δy	$\Delta^2 y$	$\Delta^3 y$	$\Delta^4 y$	$\Delta^5 y$	$\Delta^6 y$	$\Delta^7 y$
x_0	y_0							
		Δy_0						
x_1	y_1		$\Delta^2 y_0$					
		Δy_1		$\Delta^3 y_0$				
x_2	y_2		$\Delta^2 y_1$		$\Delta^4 y_0$			
		Δy_2		$\Delta^3 y_1$		$\Delta^5 y_0$		
x_3	y_3		$\Delta^2 y_2$		$\Delta^4 y_1$		$\Delta^6 y_0$	
		Δy_3		$\Delta^3 y_2$		$\Delta^5 y_1$		$\Delta^7 y_0$
x_4	y_4		$\Delta^2 y_3$		$\Delta^4 y_2$		$\Delta^6 y_1$	
		Δy_4		$\Delta^3 y_3$		$\Delta^5 y_2$		
x_5	y_5		$\Delta^2 y_4$		$\Delta^4 y_3$			
		Δy_5		$\Delta^3 y_4$				
x_6	y_6		$\Delta^2 y_5$					
		Δy_6						
x_7	y_7							

The backward difference operator ∇ is defined by $\nabla y_k = y_k - y_{k-1}$. In a similar way, we can define the higher order differences, i.e.,

$$\nabla^2 y_k = \nabla y_k - \nabla y_{k-1} = y_k - 2y_{k-1} + y_{k-2},$$

$$\nabla^3 y_k = \nabla^2 y_k - \nabla^2 y_{k-1} = y_k - 3y_{k-1} + 3y_{k-2} - y_{k-3} \text{ and so on.}$$

In general $\quad \nabla^n y_k = y_k - {}^nc_1 y_{k-1} + {}^nc_2 y_{k-2} - {}^nc_3 y_{k-3} + \cdots + (-1)^n y_{k-n}.$

2.3 BACKWARD DIFFERENCE TABLE

x	y	∇y	$\nabla^2 y$	$\nabla^3 y$	$\nabla^4 y$	$\nabla^5 y$	$\nabla^6 y$	$\nabla^7 y$
x_0	y_0							
		∇y_1						
x_1	y_1		$\nabla^2 y_2$					
		∇y_2		$\nabla^3 y_3$				
x_2	y_2		$\nabla^2 y_3$		$\nabla^4 y_4$			
		∇y_3		$\nabla^3 y_4$		$\nabla^5 y_5$		
x_3	y_3		$\nabla^2 y_4$		$\nabla^4 y_5$		$\nabla^6 y_6$	
		∇y_4		$\nabla^3 y_5$		$\nabla^5 y_6$		$\nabla^7 y_7$
x_4	y_4		$\nabla^2 y_5$		$\nabla^4 y_6$		$\nabla^6 y_7$	
		∇y_5		$\nabla^3 y_6$		$\nabla^5 y_7$		
x_5	y_5		$\nabla^2 y_6$		$\nabla^4 y_7$			
		∇y_6		$\nabla^3 y_7$				
x_6	y_6		$\nabla^2 y_7$					
		∇y_7						
x_7	y_7							

Another operator E

Shifting operator E is given by $Ef(a) = f(a + h)$,

so $\Delta f(a) = f(a + h) - f(a) \Rightarrow \Delta f(a) = Ef(a) - f(a)$

$$\Rightarrow [1+\Delta] f(a) = Ef(a) \Rightarrow 1+\Delta = E \Rightarrow \Delta = E - 1.$$

$\Delta^n = (E - 1)^n$ for all n.

Again, $E^2 f(a) = Ef(a + h) = f(a + 2h)$, in general $E^n f(a) = f(a + nh)$.

Also $Ef(a - h) = f(a) \Rightarrow E^{-1} f(a) = f(a - h)$.

The specific operators are as follows

(1) The operator Δ, called *forward difference operator*, is defined as $\Delta y_k = y_{k+1} - y_k$,

(2) the operator E, called *shifting operator*, is defined as $Ey_k = y_{k+1}$.

(3) the backward difference operator ∇ is defined by $\nabla y_k = y_k - y_{k-1}$.

(4) the central difference operator δ is defined as $\delta = E^{1/2} - E^{-1/2}$.

(5) the operator μ, called *averaging operator*, is defined as $\mu = (E^{1/2} + E^{-1/2})/2$.

(6) the differentiation operator $D = d/dx$ is related to the operator E as $E \equiv e^{hD}$.

From the above definitions of operators, it is clear that the operators are related. In the next section we will discuss how the operators are related.

2.4 RELATION OF OPERATORS

Since $\Delta y_k = y_{k+1} - y_k$ and $Ey_k = y_{k+1}$,

$$\Delta y_k = Ey_k - y_k \Rightarrow \Delta y_k + y_k = Ey_k \Rightarrow (\Delta + 1)y_k = Ey_k,$$

as this is true for all k, $\Delta + 1 = E$ (2.1.1)

Again, from the results

$$\nabla y_k = y_k - y_{k-1} \Rightarrow y_{k-1} = y_k - \nabla y_k \Rightarrow y_{k-1} = E^{-1} y_k = (1 - \nabla) y_k,$$

which implies $E^{-1} = (1 - \nabla)$ (2.1.2)

Also $E(\nabla y_k) = E(y_k - y_{k-1}) = E(y_k) - E(y_{k-1}) = y_{k+1} - y_k = \Delta y_k$,

which shows that $E\nabla = \Delta$. (2.1.3)

Now, $\Delta E(y_k) = \Delta(y_{k+1}) = y_{k+2} - y_{k+1} = E(y_{k+1}) - E(y_k) = E(y_{k+1} - y_k) = E(\Delta y_k)$,

so $\Delta E = E\Delta$. (2.1.4)

The central difference operator is defined as $\delta = E^{1/2} - E^{-1/2}$,

so $\delta y_{1/2} = y_1 - y_0 = \Delta y_0 = \nabla y_1$ and so on and also $\delta E^{1/2} = E - 1 = \Delta$. (2.1.5)

The averaging operator μ is given by $\mu = (E^{1/2} + E^{-1/2})/2$, so that

$$\mu y_{1/2} = (y_1 + y_0)/2.$$

Also $\delta^2 = (E^{1/2} - E^{-1/2})^2 = (E^{1/2} + E^{-1/2})^2 - 4 = 4(\mu^2 + 1)$ (2.1.6)

and $E^{1/2} = (E^{1/2} + E^{-1/2})/2 + (E^{1/2} - E^{-1/2})/2 = \mu + \delta/2$ (2.1.7)

$E^{-1/2} = (E^{1/2} + E^{-1/2})/2 - (E^{1/2} - E^{-1/2})/2 = \mu - \delta/2$. (2.1.8)

Examples

1. If $f(x) = x^2$, then show that $\Delta^n f(x) = 0$ for $n \geq 3$.

Solution: If $f(x) = x^2$, then $\Delta f(x) = f(x + h) - f(x) = (x + h)^2 - x^2 = 2xh + h^2$.

and $\Delta^2 f(x) = f(x+2h) - 2f(x+h) + f(x) = (x+2h)^2 - 2(x+h)^2 + x^2 = 2h^2 =$ constant, so $\Delta^3 f(x) = 0$.

i.e., $\Delta^r f(x) = 0$, for $r \geq 3$.

2. If $f(x) = a^x$, then show that $\Delta^n f(x) = 2^x$ for all n, if $a = 2$, $h = 1$.

Solution: If $f(x) = a^x$, then $\Delta f(x) = f(x+h) - f(x) = a^{x+h} - a^x = a^x(a^h - 1)$ and $\Delta^2 f(x) = (a^h - 1)\Delta a^x = a^x(a^h - 1)^2$, proceeding in this way we get

$$\Delta^n f(x) = a^x(a^h - 1)^n \text{ for all } n.$$

If $a = 2$, $h = 1$, then $\Delta^n f(x) = 2^x$ for all n.

3. Determine (a) $\Delta[u_x v_x]$, (b) $\Delta[u_x / v_x]$

Solution:

(a) $\Delta[u_x v_x] = u_{x+1} v_{x+1} - u_x v_x = u_{x+1} v_{x+1} - u_{x+1} v_x + u_{x+1} v_x - u_x v_x = u_{x+1} \Delta v_x + v_x \Delta u_x$.

(b) $\Delta[u_x / v_x] = u_{x+1} / v_{x+1} - u_x / v_x = [u_{x+1} v_x - v_{x+1} u_x] / v_x v_{x+1} = [v_x \Delta u_x - u_x \Delta v_x] / v_x v_{x+1}$.

4. Prove that $\nabla\Delta = \Delta\nabla = \Delta - \nabla = \delta^2$.

Solution: We have

$$\begin{aligned}\nabla\Delta y_k &= \nabla(y_{k+1} - y_k) \\ &= y_{k+1} - y_k - y_k + y_{k-1} \\ &= \Delta y_k - \Delta y_{k-1} \\ &= \Delta\nabla y_k.\end{aligned}$$

Again,

$$\begin{aligned} y_{k+1} - y_k - y_k + y_{k-1} &= \Delta y_k - \nabla y_k \\ \Rightarrow (\Delta - \nabla) y_k &= y_{k+1} - y_k - y_k + y_{k-1} \\ &= E y_k - 2 y_k + E^{-1} y_k \\ &= (E - 2 + E^{-1}) y_k \\ &= \delta^2 y_k.\end{aligned}$$

$\therefore \Delta - \nabla = \delta^2$ **(Proved)**.

5. Prove that $\Delta^k y_0 = \sum_{i=0}^{k} (-1)^i \binom{k}{i} E^{k-i} y_0$

Solution: We know that $\Delta = E - 1$,

$$\text{so } \Delta^k = (E-1)^k = \left[E^k - \binom{k}{1} E^{k-1} + \binom{k}{2} E^{k-2} + \cdots (-1)^k\right] = \sum_{i=0}^{k} (-1)^i \binom{k}{i} E^{k-i}.$$

So, $\Delta^k y_0 = \sum_{i=0}^{k} (-1)^i \binom{k}{i} E^{k-i} y_0$ **(Proved)**.

6. Prove that $y_k = \sum_{i=0}^{k} \binom{k}{i} \Delta^i y_0$.

Solution: Since $E = 1 + \Delta$, $\therefore E^k = (1+\Delta)^k = \sum_{i=0}^{k} \binom{k}{i} \Delta^i \Rightarrow y_k = E^k y_0 = \sum_{i=0}^{k} \binom{k}{i} \Delta^i y_0$

(Proved).

7. Establish a relation between E and D.

Solution: Using Taylor's theorem, we get

$$Ey_k = y_{k+1} = y_k + hy_k' + \frac{1}{2!}h^2 y_k'' + \frac{1}{3!}h^3 y_k''' + \cdots + \frac{1}{n!}h^n y_k^{(n)} + \cdots$$

$$= y_k + hDy_k + \frac{1}{2!}h^2 D^2 y_k + \frac{1}{3!}h^3 D^3 y_k + \cdots + \frac{1}{n!}h^n D^n y_k + \cdots \text{ where } D = d/dx$$

$$= [1 + hD + \frac{1}{2!}h^2 D^2 + \frac{1}{3!}h^3 D^3 + \cdots + \frac{1}{n!}h^n D^n + \cdots]\, y_k = e^{hD} y_k.$$

So, $E = e^{hD}$.

8. For small value of h, prove that $\Delta^{n+1} f(x_0) \approx h^{n+1} f^{n+1}(x_0)$.

Solution: From Differential Calculus we know that $\underset{h\to 0}{lt} \dfrac{f(x_0+h) - f(x_0)}{h} = f'(x_0)$

$$\Rightarrow \underset{h\to 0}{lt} \frac{\Delta f(x_0)}{h} = f'(x_0)$$

$$\Rightarrow \Delta f(x_0) \approx hf'(x_0), \text{ for small values of } h.$$

Again $\underset{h\to 0}{lt} \dfrac{f'(x_0+h) - f'(x_0)}{h} = f''(x_0)$

$$\Rightarrow \underset{h\to 0}{lt} \frac{\Delta f(x_0+h)/h - \Delta f(x_0)/h}{h} = f''(x_0)$$

$$\Rightarrow \underset{h\to 0}{lt} \frac{\Delta f(x_0+h) - \Delta f(x_0)}{h^2} = f''(x_0)$$

$$\Rightarrow \Delta^2 f(x_0) \approx h^2 f''(x_0), \text{ for small values of } h.$$

Proceeding in this way we get $\Delta^{n+1} f(x_0) \approx h^{n+1} f^{n+1}(x_0)$.

9. Prove that $u_1 + u_2 + \cdots + u_n = {}^nC_1 u_1 + {}^nC_2 \Delta u_1 + \cdots + \Delta^{n-1} u_1$

Solution: L.H.S. $= u_1 + u_2 + \cdots + u_n = u_1 + Eu_1 + \cdots + E^{n-1} u_1$

$$= (1 + E + E^2 + \cdots + E^{n-1}) u_1$$

$$= \frac{E^n - 1}{E - 1} u_1 = \frac{(1+\Delta)^n - 1}{\Delta} u_1$$

$$= \left[{}^nC_1 + {}^nC_2 \Delta + \cdots + {}^nC_{n-1} \Delta^{n-2} + \Delta^{n-1}\right] u_1$$

$$= {}^nC_1 u_1 + {}^nC_2 \Delta u_1 + \cdots {}^nC_{n-1} \Delta^{n-2} u_1 + \Delta^{n-1} u_1$$

= R.H.S. (**Proved**).

10. Prove that $u_0 - u_1 + u_2 - u_3 + \cdots = \frac{1}{2}u_0 - \frac{1}{4}\Delta u_0 + \frac{1}{8}\Delta^2 u_0 - \frac{1}{16}\Delta^3 u_0 + \cdots$.

Solution: L.H.S. $= u_0 - u_1 + u_2 - u_3 + \cdots$

$$= u_0 - Eu_0 + E^2 u_0 - E^3 u_0 + \cdots = (1 - E + E^2 - E^3 + \cdots)\, u_0$$

$$= (1 + E)^{-1} u_0 = (2 + \Delta)^{-1} u_0$$

$$= \frac{1}{2}\left[1+\frac{\Delta}{2}\right]^{-1} u_0 = \frac{1}{2}\left[1-\frac{\Delta}{2}+\frac{\Delta^2}{4}+\frac{\Delta^3}{8}+\cdots\right]u_0$$

$$= \frac{1}{2}\left[u_0 - \frac{\Delta u_0}{2}+\frac{\Delta^2 u_0}{4}+\frac{\Delta^3 u_0}{8}+\cdots\right].$$

11. Prove that $u_1 x + u_2 x^2 + u_3 x^3 + \cdots = \frac{x}{1-x}u_1 + \frac{x^2}{(1-x)^2}\Delta u_1 + \frac{x^3}{(1-x)^3}\Delta^2 u_1 + \cdots$.

Solution: L.H.S. $= xu_1 + x^2 E u_1 + x^3 E^2 u_1 + \cdots$

$$= x[1 + xE + x^2E^2 + \cdots]u_1$$

$$= x(1 - xE)^{-1}u_1$$

$$= x[(1-x) - x\Delta]^{-1}u_1$$

$$= \frac{x}{1-x}\left[1+\frac{x}{1-x}\Delta+\frac{x^2}{(1-x)^2}\Delta^2+\cdots\right]u_1$$

$$= \frac{x}{1-x}u_1 + \frac{x^2}{(1-x)^2}\Delta u_1 + \frac{x^3}{(1-x)^3}\Delta^2 u_1 + \cdots$$

= R.H.S. (**Proved**).

12. Show that $\sum_{x=0}^{\infty} u_{2x} = \frac{1}{2}\sum_{x=0}^{\infty} u_x + \frac{1}{4}\left(1-\frac{\Delta}{2}+\frac{\Delta^2}{4}-\ldots\right)u_0$

Solution: R.H.S. $= \frac{1}{2}\sum_{x=0}^{\infty} u_x + \frac{1}{4}\left(1-\frac{\Delta}{2}+\frac{\Delta^2}{4}-\ldots\right)u_0$

$$= \frac{1}{2}(u_0 + u_1 + u_2 + \ldots) + \frac{1}{4}\left(1+\frac{\Delta}{2}\right)^{-1}u_0$$

$$= \frac{1}{2}(1 + E + E^2 + \ldots)u_0 + \frac{1}{2}(2+\Delta)^{-1}u_0$$

$$= \frac{1}{2}(1-E)^{-1}u_0 + \frac{1}{2}(1+E)^{-1}u_0$$

$$= (1 + E^2 + E^4 + \ldots)u_0$$

$$= \sum_{x=0}^{\infty} u_{2x}$$

= L.H.S. (**Proved**).

EXERCISE

1. Define the forward difference operator, backward difference operator and shifting operator.
2. Establish the relations among the forward difference operator, backward difference operator and shifting operator.

3. **Prove that** i. $\nabla = \delta E^{-1/2} = 1 - E^{-1}$

ii. $2\delta\mu = \Delta(1 + E^{-1})$

iii. $E \approx e^{hD}$

iv. $\Delta(Cy_k) = C\Delta(y_k)$

v. $\nabla\Delta = \Delta\nabla = \Delta - \nabla = \delta^2$

vi. $\delta = \Delta E^{-1/2} = E^{1/2}\nabla$

vii. $(E + a)(\Delta + b) = (\Delta + b)(E + a)$

viii. $\Delta/\nabla - \nabla/\Delta = \nabla + \Delta$

ix. $(E + 1)\delta = 2(E - 1)\mu$

x. $\Delta\sqrt{f_x} = \dfrac{\Delta f_x}{\sqrt{f_x} + \sqrt{f_{x+1}}}$

xi. $\Delta^n y_k = \nabla^n y_{k+n}$

4. **Show that** $u_0 + \dfrac{u_1 x}{1!} + \dfrac{u_2 x^2}{2!} + \cdots = e^x\left[u_0 + x\Delta u_0 + \dfrac{x^2}{2!}\Delta^2 u_0 + \cdots\right].$

5. **Show that for the function** $u_x = ax^2 + bx + c$,

$$u_{2n} - \binom{n}{1} 2u_{2n-1} + \binom{n}{2} 2^2 u_{2n-2} - \cdots + (-2)^n u_n = (-1)^n (c - 2an).$$

6. **Define** ***k*****th order difference of a function** ***f*****(*****x*****). Prove that**

$$\Delta^k f(x) = \sum_{i=0}^{\infty} (-1)^i \binom{k}{i} f[x + (k - i)h], \text{ where } h \text{ is the step length.}$$

Chapter 3

INTERPOLATION

Interpolation is a very useful technique in numerical analysis with several utilities. In our daily life we are sometimes confronted with the problem of finding the value of a function for some value in the given interval with the help of a given set of observations. Simply, it is a method of finding the value of the function for a given set of values when the values of the function are known for a set of values.

3.1 DEFINITION

The method of obtaining the value of a function for any intermediate value of the argument when the values of the function are known for a set of values of the arguments is known as *interpolation*.

Mathematically, if the values of the function $y = f(x)$ at $x = a, a + h, \cdots, a + nh$ be known then finding the value of the function at $x = b$, where $a < b < a + nh$ is known as *interpolation*. If x lies outside the above said range, then the corresponding process is called *extrapolation*.

In this section we shall discuss various interpolation formulae for both equi-spaced and non-equi-spaced arguments.

3.2 POLYNOMIAL INTERPOLATION

Let a function $f(x)$ be such that the analytical form of the function is unknown, but the values of the function at $n + 1$ distinct values of x: $x_0, x_1, x_2, \ldots, x_n$ are known. We also assume that $f(x)$ is a smooth function in R. Our problem is to find the value of the function at $x \in (x_0, x_n)$. As the explicit form of the function is unknown, the precise value of the function cannot be obtained. In such case, it is required to find a simpler function

$\varphi(x)$ such that $f(x)$ and $\varphi(x)$ agree at $x = x_i$, $0 \le i \le n$. As polynomials are very easy to use, we generally interpolate $f(x)$ by a polynomial $\varphi(x)$ and such interpolation is called *polynomial interpolation.* As a justification for the approximation of a function by a polynomial, we state here, without proof, a theorem due to Weierstrass [1885] : if a function $f(x)$ is continuous in (x_0, x_n), then for given any $\varepsilon > 0$, there exists a polynomial $P(x)$ such that $| f(x) - P(x) | < \varepsilon$ for all $x \in (x_0, x_n)$.

In approximating a function by a polynomial naturally the following questions arise:

1. What should be the error in polynomial approximation?
2. How many polynomials can interpolate a function at specified points?

3.3 ERROR AND UNIQUENESS IN POLYNOMIAL INTERPOLATION

Let the function $f(x)$ be continuous and $(n + 1)$ times differentiable in an interval I containing the points x_i, $0 \le i \le n$ and let the nth degree polynomial $\varphi_n(x)$ approximates the function $f(x)$ in I, i.e., $f(x_i) = \varphi_n(x_i)$, for $0 \le i \le n$. Since the difference is zero at the points of collocation, we can assume,

$$f(x) - \varphi_n(x) = C\Pi(x) = C(x - x_0)\,(x - x_1) \cdots (x - x_n).$$

Now consider the function

$$F(x) = f(x) - \varphi_n(x) - C\Pi(x),$$

obviously $F(x)$ vanishes at x_i, for $0 \le i \le n$. If we choose a new argument $x = x_{n+1}$ and

$$C = \frac{f(x_{n+1}) - \phi_n(x_{n+1})}{\Pi(x_{n+1})},$$

then $F(x_{n+1})$ will also be zero. So, $F(x) = 0$ has $n+2$ zeros. By Rolle's theorem, $F'(x) = 0$ has $n + 1$ zeros between those of $F(x) = 0$, while $F''(x) = 0$ has n zeros between those of $F'(x) = 0$. Proceeding in this way we can show that $F^{n+1}(x) = 0$ has one zero, say $x = \xi$, in the interval (x_0, x_n). Now

$$F^{n+1}(x) = f^{n+1}(x) - 0 - C(n + 1)!$$

$$\Rightarrow \quad F^{n+1}(\xi) = 0 = f^{n+1}(\xi) - C(n + 1)!$$

$$\Rightarrow C = \frac{f^{n+1}(\xi)}{(n+1)!}.$$

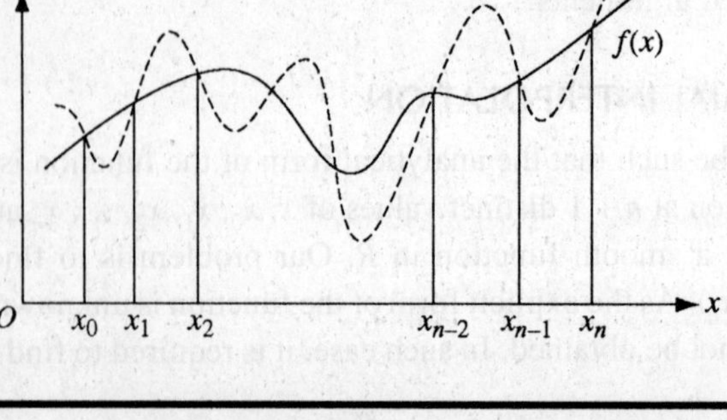

Fig.3.1 Interpolation of a function $f(x)$ by a polynomial $P(x)$

This may be substituted to get

$$f(x_{n+1}) - \varphi_n(x_{n+1}) = \Pi(x_{n+1})\frac{f^{n+1}(\xi)}{(n+1)!}.$$

Since x_{n+1} can be any argument between x_0 and x_n except for x_i, $0 \le i \le n$, and since this result is also true for $x = x_i$, $0 \le i \le n$, we can write

$$f(x) - \varphi_n(x) = \Pi(x)\frac{f^{n+1}(\xi)}{(n+1)!}. \tag{3.1}$$

Uniqueness: If possible, there exists two nth degree polynomials $\varphi_1(x)$, $\varphi_2(x)$ which interpolate the function $f(x)$ at specified arguments $x = x_i$, $0 \le i \le n$. Then $p(x) = \varphi_1(x) - \varphi_2(x)$ would be of degree n or less and has $n + 1$ zeros, which is impossible unless $p(x)$ vanishes identically. So there exists at most one polynomial, which can interpolate a function at specified arguments.

3.4 NEWTON'S FORWARD DIFFERENCE POLYNOMIAL FOR EQUISPACED POINTS

Let a set $n + 1$ points (x_i, y_i), $0 \le i \le n$ be given. We have to find an nth degree polynomial $\varphi_n(x)$ such that $\varphi_n(x_i) = y_i$ for $0 \le i \le n$.

Let us take $\varphi_n(x) = a_0 + a_1(x-x_0) + a_2(x-x_0)(x-x_1) + \cdots + a_n(x-x_0)(x-x_1)\cdots(x-x_{n-1})$.

Taking $x_i = x_0 + ih$, $i = 0, 1, 2, \ldots, n$, we obtain on putting $x = x_0$, $y_0 = a_0$. Again, using this result and putting $x = x_1$, we get $y_1 = y_0 + ha_1$

$$\Rightarrow a_1 = \frac{\Delta y_0}{h}.$$

While putting $x = x_2$ and making use of the values of a_0, a_1, one gets

$$y_2 = y_0 + 2h\frac{\Delta y_0}{h} + 2a_2h^2$$

$$\Rightarrow 2a_2h^2 = y_2 - y_0 - 2\Delta y_0$$

$$\Rightarrow 2a_2h^2 = y_2 - y_1 + y_1 - y_0 - 2\Delta y_0$$

$$= \Delta y_1 + \Delta y_0 - 2\Delta y_0 = \Delta y_1 - \Delta y_0 = \Delta^2 y_0$$

$$\Rightarrow a_2 = \frac{\Delta^2 y_0}{2!h^2}.$$

Similarly, at $x = x_3$, we get

$$y_3 = y_0 + 3h\frac{\Delta y_0}{h} + 6h^2\frac{\Delta^2 y_0}{2h^2} + a_3(3h)(2h)(h).$$

$$\Rightarrow y_3 = y_0 + 3(y_1 - y_0) + 3(y_2 - 2y_1 + y_0) + 6h^3a_3$$

$$\Rightarrow 6h^3a_3 = y_3 - 3(y_2 - 2y_1 + y_0) - 3(y_1 - y_0) - y_0$$

$$= y_3 - 3y_2 + 3y_1 - y_0 = \Delta^3 y_0$$

$$\Rightarrow a_3 = \frac{\Delta^3 y_0}{3!h^3}.$$

Proceeding in this way we get $a_4 = \dfrac{\Delta^4 y_0}{4!h^4}, \ldots, a_n = \dfrac{\Delta^n y_0}{n!h^n}$.

and putting the values of a_i, $0 \le i \le n$, we get

$$\varphi_n(x) = y_0 + ph\frac{\Delta y_0}{h} + ph(ph-h)\frac{\Delta^2 y_0}{2!h^2} + \cdots + ph(ph-h)(ph-2h)\cdots \{ph-(n-1)h\}\frac{\Delta^n y_0}{n!h^n}$$

$$= y_0 + p\Delta y_0 + p(p-1)\frac{\Delta^2 y_0}{2!} + p(p-1)(p-2)\frac{\Delta^3 y_0}{3!} + \cdots$$

$$+ p(p-1)(p-2)\cdots\{p-(n-1)\}\frac{\Delta^n y_0}{n!}. \quad (3.2)$$

This is the Newton's forward difference polynomial for equispaced $(n + 1)$ points
The error term in this formula is given by

$$f(x) - \varphi_n(x) = \Pi(x)\frac{f^{n+1}(\xi)}{(n+1)!}.$$

Now, $$f'(x) \approx \frac{f(x+h)-f(x)}{h} = \frac{\Delta f(x)}{h}.$$

So, $$\frac{d}{dx} = \frac{\Delta}{h} \rightarrow \frac{d^n}{dx^n} = \frac{\Delta^n}{h^n}.$$

Hence, for $x = x_0 + ph$, the error term becomes

$$\text{Error} = \frac{ph(ph-h)(ph-2h)\cdots(ph-nh)}{(n+1)!h^{n+1}}\Delta^{n+1} f(\xi)$$

$$= \frac{p(p-1)(p-2)\cdots(p-n)}{(n+1)!}\Delta^{n+1} f(\xi), \text{ where } x_0 < \xi < x_n.$$

3.5 NEWTON'S BACKWARD DIFFERENCE POLYNOMIAL FOR EQUISPACED POINTS

Let a set $n + 1$ points (x_i, y_i), $0 \le i \le n$ be given. We have to find an nth degree polynomial $\varphi_n(x)$ such that $\varphi_n(x_i) = y_i$ for $0 \le i \le n$.

Let us take $\varphi_n(x) = a_0 + a_1(x-x_n) + a_2(x-x_n)(x-x_{n-1}) + \cdots + a_n(x-x_n)(x-x_{n-1})\cdots(x-x_1)$.

Taking $x_i = x_0 + ih$, $i = 0, 1, 2, \ldots, n$, we obtain on putting $x = x_n$, $y_n = a_0$. Again, using this result and putting $x = x_{n-1}$, we get $y_{n-1} = y_n - ha_1 \Rightarrow a_1 = \dfrac{\nabla y_n}{h}$.

While putting $x = x_{n-2}$ and making use of the value of a_0, a_1, one gets

$$y_{n-2} = y_n - 2h\frac{\nabla y_n}{h} + 2a_2h^2 \Rightarrow 2a_2h^2 = y_{n-2} - y_n + 2\nabla y_n$$

$$\Rightarrow 2a_2h^2 = y_{n-2} - y_{n-1} + y_{n-1} - y_n + 2\nabla y_n$$

$$= -\nabla y_{n-1} - \nabla y_n + 2\nabla y_n = \nabla y_n - \nabla y_{n-1} = \nabla^2 y_n.$$

$$\Rightarrow a_2 = \frac{\nabla^2 y_n}{2!h^2}.$$

Similarly, at $x = x_{n-3}$, we get

$$y_{n-3} = y_n - 3h\frac{\nabla y_n}{h} + 6h^2\frac{\nabla^2 y_n}{2!h^2} - a_3(3h)(2h)(h)$$

$$\Rightarrow 6h^3 a_3 = y_n - 3\nabla y_n + 3\nabla^2 y_n - y_{n-3}$$

$$\Rightarrow 6h^3 a_3 = y_n + 3(y_n - 2y_{n-1} + y_{n-2}) - 3(y_n - y_{n-1}) - y_{n-3}$$

$$= y_n - 3y_{n-1} + 3y_{n-2} - y_{n-3}$$

$$= \nabla^3 y_n$$

$$\Rightarrow a_3 = \frac{\nabla^3 y_n}{3!h^3}.$$

Proceeding in this way, we get $a_4 = \dfrac{\nabla^4 y_n}{4!h^4}, \cdots, a_n = \dfrac{\nabla^n y_n}{n!h^n}$.

Next setting $x = x_n + ph$ and putting the values of a_i, $0 \le i \le n$, we get

$$\varphi_n(x) = y_n + ph\frac{\nabla y_n}{h} + ph(ph+h)\frac{\nabla^2 y_n}{2!h^2} + \cdots + ph(ph+h)(ph+2h)\cdots\{ph+(n-1)h\}\frac{\nabla^n y_n}{n!h^n}$$

$$= y_n + p\nabla y_n + p(p+1)\frac{\nabla^2 y_n}{2!} + p(p+1)(p+2)\frac{\nabla^3 y_n}{3!} + \cdots + p(p+1)(p+2)\cdots\{p+(n-1)\}\frac{\nabla^n y_n}{n!}. \qquad (3.3)$$

This is the Newton's backward difference polynomial for equi-spaced $(n + 1)$ points. Proceeding as before we get the error term in this case is

$$= \frac{p(p+1)(p+2)\ldots(p+n)}{(n+1)!}\Delta^{n+1} f(\xi), \text{ where } x_0 < \xi < x_n.$$

3.6 GAUSS'S FORWARD DIFFERENCE FORMULA

The Gauss's forward formula is based on the following difference table:

x_{-1}	y_{-1}						
		Δy_{-1}		$\Delta^3 y_{-2}$		$\Delta^5 y_{-3}$	
x_0	y_0		$\Delta^2 y_{-1}$		$\Delta^4 y_{-2}$		$\Delta^6 y_{-3}$
		Δy_0		$\Delta^3 y_{-1}$		$\Delta^5 y_{-2}$	
x_1	y_1		$\Delta^2 y_0$		$\Delta^4 y_{-1}$		$\Delta^6 y_{-2}$

The formula is assumed of the form

$$\varphi_n(x_p) = y_0 + G_1\Delta y_0 + G_2\Delta^2 y_{-1} + G_3\Delta^3 y_{-1} + G_4\Delta^4 y_{-2} + \cdots, \qquad (3.4)$$

where the constants G_1, G_2, G_3, $\cdots$ are to be determined.

Again, $\varphi_n(x_p) = E^p y_0 = (1+\Delta)^p y_0 = \sum_{i=0}^{p} {}^pC_i \Delta^i y_0$. (3.5)

$$\text{Now, } \Delta^2 y_{-1} = \Delta^2 E^{-1} y_0 = \Delta^2 (1+\Delta)^{-1} y_0 = \Delta^2 (1 - \Delta + \Delta^2 - \Delta^3 + \Delta^4 - \cdots) y_0$$
$$= \Delta^2 y_0 - \Delta^3 y_0 + \Delta^4 y_0 - \Delta^5 y_0 + \cdots$$
$$\Delta^3 y_{-1} = \Delta^3 E^{-1} y_0 = \Delta^3 (1+\Delta)^{-1} y_0 = \Delta^3 (1 - \Delta + \Delta^2 - \Delta^3 + \Delta^4 - \cdots) y_0$$
$$= \Delta^3 y_0 - \Delta^4 y_0 + \Delta^5 y_0 - \Delta^6 y_0 + \cdots$$
$$\Delta^4 y_{-2} = \Delta^4 E^{-2} y_0 = \Delta^4 (1+\Delta)^{-2} y_0 = \Delta^4 (1 - 2\Delta + 3\Delta^2 - 4\Delta^3 + \cdots) y_0$$
$$= \Delta^4 y_0 - 2\Delta^5 y_0 + 3\Delta^6 y_0 - 4\Delta^7 y_0 + \cdots.$$

Putting these values of $\Delta^2 y_{-1}, \Delta^3 y_{-1}, \Delta^4 y_{-2}, \cdots$ in (3.4) and comparing the result with (3.5), we get

$$G_1 = p,\ G_2 = \frac{p(p-1)}{2!},\ G_3 = \frac{(p+1)\,p(p-1)}{3!},\ G_4 = \frac{(p+1)\,p(p-1)(p-2)}{4!}, \text{ etc.}$$

Hence the result (3.4) becomes

$$\varphi_n(x_p) = y_0 + p\Delta y_0 + \frac{p(p-1)}{2!}\Delta^2 y_{-1} + \frac{(p+1)\,p(p-1)}{3!}\Delta^3 y_{-1} + \frac{(p+1)\,p(p-1)(p-2)}{4!}\Delta^4 y_{-2} + \cdots. \tag{3.6}$$

3.7 GAUSS'S BACKWARD DIFFERENCE FORMULA

The Gauss's backward formula is based on the following difference table

x_{-1}	y_{-1}						
		Δy_{-1}		$\Delta^3 y_{-2}$		$\Delta^5 y_{-3}$	
x_0	y_0		$\Delta^2 y_{-1}$		$\Delta^4 y_{-2}$		$\Delta^6 y_{-3}$
		Δy_0		$\Delta^3 y_{-1}$		$\Delta^5 y_{-2}$	
x_1	y_1		$\Delta^2 y_0$		$\Delta^4 y_{-1}$		$\Delta^6 y_{-2}$

The formula is assumed of the form

$$\varphi_n(x_p) = y_0 + G_1 \Delta y_{-1} + G_2 \Delta^2 y_{-1} + G_3 \Delta^3 y_{-2} + G_4 \Delta^4 y_{-2} + \cdots, \tag{3.7}$$

where the constants $G_1, G_2, G_3, \ldots$ are to be determined.

$$\text{Again, } \varphi_n(x_p) = E^p y_0 = (1+\Delta)^p y_0 = \sum_{i=0}^{p} {}^pC_i \Delta^i y_0. \tag{3.8}$$

$$\text{Now, } \Delta y_{-1} = \Delta E^{-1} y_0 = \Delta (1+\Delta)^{-1} y_0 = \Delta (1 - \Delta + \Delta^2 - \Delta^3 + \Delta^4 - \cdots) y_0$$
$$= \Delta y_0 - \Delta^2 y_0 + \Delta^3 y_0 - \Delta^4 y_0 + \cdots$$
$$\Delta^2 y_{-1} = \Delta^2 E^{-1} y_0 = \Delta^2 (1+\Delta)^{-1} y_0 = \Delta^2 (1 - \Delta + \Delta^2 - \Delta^3 + \Delta^4 - \cdots) y_0$$
$$= \Delta^2 y_0 - \Delta^3 y_0 + \Delta^4 y_0 - \Delta^5 y_0 + \cdots$$
$$\Delta^3 y_{-2} = \Delta^3 E^{-2} y_0 = \Delta^3 (1+\Delta)^{-2} y_0 = \Delta^3 (1 - 2\Delta + 3\Delta^2 - 4\Delta^3 + \cdots) y_0$$
$$= \Delta^3 y_0 - 2\Delta^4 y_0 + 3\Delta^5 y_0 - 4\Delta^6 y_0 + \cdots$$

$$\Delta^4 y_{-2} = \Delta^4 E^{-2} y_0 = \Delta^4 (1+\Delta)^{-2} y_0 = \Delta^4 (1 - 2\Delta + 3\Delta^2 - 4\Delta^3 + \cdots) y_0$$
$$= \Delta^4 y_0 - 2\Delta^5 y_0 + 3\Delta^6 y_0 - 4\Delta^7 y_0 + \cdots.$$

Putting these values of $\Delta^2 y_{-1}, \Delta^3 y_{-1}, \Delta^4 y_{-2}, \cdots$ in (3.7) and comparing the result with (3.8) we get

$$G_1 = p,\ G_2 = \frac{p(p+1)}{2!},\ G_3 = \frac{(p+1)p(p-1)}{3!},\ G_4 = \frac{(p+2)p(p+1)(p-1)}{4!}, \text{etc.}$$

Hence the result (3.7) becomes

$$\varphi_n(x_p) = y_0 + p\Delta y_{-1} + \frac{p(p+1)}{2!}\Delta^2 y_{-1} + \frac{(p+1)p(p-1)}{3!}\Delta^3 y_{-2} +$$
$$\frac{(p+2)(p+1)p(p-1)}{4!}\Delta^4 y_{-2} + \cdots. \qquad (3.9)$$

3.8 STIRLING INTERPOLATION FORMULA

Stirling Interpolation Formula is the average of Gauss's forward and backward formulae. Thus Stirling Interpolation Formula is

$$\varphi_n(x_p) = y_0 + \frac{p}{2}(\Delta y_{-1} + \Delta y_0) + \frac{p^2}{2}\Delta^2 y_{-1} + \frac{p(p^2-1)}{12}(\Delta^3 y_{-2} + \Delta^3 y_{-1}) +$$
$$\frac{p^2(p^2-1)}{24}\Delta^4 y_{-2} + \cdots. \qquad (3.10)$$

3.9 BESSEL'S INTERPOLATION FORMULA

The Bessel's interpolation formula is based on the following difference table:

x_{-1}	y_{-1}						
x_0	$\begin{pmatrix} y_0 \\ y_1 \end{pmatrix}$	Δy_0	$\begin{pmatrix} \Delta^2 y_{-1} \\ \Delta^2 y_0 \end{pmatrix}$	$\Delta^3 y_{-1}$	$\begin{pmatrix} \Delta^4 y_{-2} \\ \Delta^4 y_{-1} \end{pmatrix}$	$\Delta^5 y_{-2}$	$\begin{pmatrix} \Delta^6 y_{-3} \\ \Delta^6 y_{-2} \end{pmatrix}$
x_1							
x_2	y_2						

The formula is assumed of the form

$$\varphi_n(x_p) = \frac{1}{2}(y_0 + y_1) + G_1\Delta y_0 + \frac{1}{2}G_2(\Delta^2 y_{-1} + \Delta^2 y_0) +$$
$$G_3\Delta^3 y_{-1} + \frac{1}{2}G_4(\Delta^4 y_{-2} + \Delta^4 y_{-1}) \cdots. \qquad (3.11)$$

where the constants $G_1, G_2, G_3, \ldots$ are to be determined.

Again, $\varphi_n(x_p) = E^p y_0 = (1+\Delta)^p y_0 = \sum_{i=0}^{p} \binom{p}{i} \Delta^i y_0.$ (3.12)

Now, $\Delta^2 y_{-1} = \Delta^2 E^{-1} y_0 = \Delta^2(1+\Delta)^{-1} y_0 = \Delta^2(1-\Delta+\Delta^2-\Delta^3+\Delta^4-\cdots)\, y_0$

$$= \Delta^2 y_0 - \Delta^3 y_0 + \Delta^4 y_0 - \Delta^5 y_0 + \cdots$$

$$\Delta^3 y_{-1} = \Delta^3 E^{-1} y_0 = \Delta^3(1+\Delta)^{-1} y_0 = \Delta^3(1-\Delta+\Delta^2-\Delta^3+\cdots)\, y_0$$

$$= \Delta^3 y_0 - \Delta^4 y_0 + \Delta^5 y_0 - \Delta^6 y_0 + \cdots$$

$$\Delta^4 y_{-1} = \Delta^4 E^{-1} y_0 = \Delta^4(1+\Delta)^{-1} y_0 = \Delta^4(1-\Delta+\Delta^2-\Delta^3+\cdots)\, y_0$$

$$= \Delta^4 y_0 - \Delta^5 y_0 + \Delta^6 y_0 - \Delta^7 y_0 + \cdots$$

$$\Delta^4 y_{-2} = \Delta^4 E^{-2} y_0 = \Delta^4(1+\Delta)^{-2} y_0 = \Delta^4(1-2\Delta+3\Delta^2-4\Delta^3+\cdots)\, y_0$$

$$= \Delta^4 y_0 - 2\Delta^5 y_0 + 3\Delta^6 y_0 - 4\Delta^7 y_0 + \cdots.$$

Putting these values of $\Delta^2 y_{-1}, \Delta^3 y_{-1}, \Delta^4 y_{-1}, \Delta^4 y_{-2}, \cdots$ in (3.11) and comparing the result with (3.12), we get

$$G_1 = p - \frac{1}{2},\ G_2 = \frac{p(p-1)}{2!},\ G_3 = \frac{(p-1)\,p(p-1/2)}{3!},\ G_4 = \frac{(p+1)\,p(p-1)(p-2)}{4!}, \text{ etc.}$$

Hence the result (3.11) becomes

$$\varphi_n(x_p) = y_0 + p\Delta y_0 + \frac{p(p-1)}{4}\left(\Delta^2 y_{-1} + \Delta^2 y_0\right) + \frac{(p-1)\,p(2p-1)}{12}\Delta^3 y_{-1} + \frac{(p+1)\,p(p-1)(p-2)}{48}\left(\Delta^4 y_{-2} + \Delta^4 y_{-1}\right) + \cdots \quad (3.13)$$

3.10 LAGRANGE'S INTERPOLATION FORMULA

Let a set of $(n+1)$ points (x_i, y_i), $0 \le i \le n$ be given where x_i's are not necessarily equi spaced. The polynomial

$$P(x) = a_0 l_0(x) + a_1 l_1(x) + \cdots + a_n l_n(x) \quad (3.14)$$

with
$$l_k(x) = \prod_{\substack{i=0 \\ k \ne i}}^{n} \frac{x - x_i}{x_k - x_i}, \qquad k = 0, 1, 2, \cdots, n \quad (3.15)$$

is the Lagrange's polynomial for the points $x = x_i$, $i = 0, 1, \cdots, n$. The function $l_k(x)$ is the product of n linear factors, hence a polynomial of nth degree. Therefore, (3.14) does indeed describe a polynomial of degree n. Further, $l_k(x)$ vanishes at $x = x_i$ for all $i \ne k$ and takes the value 1 at $x = x_k$, i.e., $l_k(x_i) = 1$, if $i = k$, and $l_k(x_i) = 0$, if $i \ne k$. This shows that $P(x_i) = a_i$, $i = 0, 1, 2, \cdots, n$. i.e., the coefficients a_is are simply the values of the polynomial $P(x)$ at the points $x = x_i$, $i = 0, 1, 2, \cdots, n$. Consequently for an arbitrary function $y(x)$,

$$P(x) = \sum_{k=0}^{n} y_k l_k(x) \quad (3.16)$$

is a polynomial of degree n and interpolates the function $y(x)$ at $x = x_i$, $i = 0, 1, 2, \cdots, n$. The above form of polynomial can be written as

$$P(x) = w(x) \sum_{k=0}^{n} \frac{f(x_k)}{(x - x_k)\, w'(x_k)} = w(x) \sum_{k=0}^{n} \frac{y_k}{D_k},$$

where $w(x) = (x - x_0)(x - x_1)(x - x_2) \cdots (x - x_n)$
and $D_k = (x - x_k)(x_k - x_0) \ldots (x_k - x_{k-1})(x_k - x_{k+1}) \cdots (x_k - x_n)$

Lagrange's interpolation scheme

$$\begin{array}{cccccccc} & & & & & D & y & y/D \\ x - x_0 & x_0 - x_1 & x_0 - x_2 & \cdots & x_0 - x_n & D_0 & y_0 & y_0/D_0 \\ x_1 - x_0 & x - x_1 & x_1 - x_2 & \cdots & x_1 - x_n & D_1 & y_1 & y_1/D_1 \\ \cdots & \cdots & \cdots & \cdots & \cdots & \cdots & \cdots & \cdots \\ x_n - x_0 & x_n - x_1 & x_n - x_2 & \cdots & x - x_n & D_n & y_n & y_n/D_n \\ \hline & & & & & & & \Sigma y_n/D_n \end{array} \qquad (3.17)$$

Advantage and disadvantage of Lagrange's method

To determine an nth degree Lagrange's interpolation formula at a point x we need to perform at least $2(n+1)$ multiplications/divisions and $(2n + 1)$ additions/subtractions as compared to n multiplications and n additions necessary for evaluation of a polynomial of n degree in power form. Further a change of degree involves a completely new computation of all terms, which is done in other methods based on equally-spaced points just by adding a new term or by subtracting. On the other hand, Lagrange's polynomial does not use finite differences at all and has the advantage that they may be used on functions, which are not tabulated at equal intervals of the argument.

3.11 AITKEN'S INTERPOLATION FORMULA

Let a set of $(n + 1)$ points (x_i, y_i), $0 \le i \le n$ be given, where the values of x need not necessarily equi-spaced. To find the value of the function at point x, we proceed iteratively as follows:

(i) Obtain a first approximation of y taking first two points only; then obtain
(ii) the second approximation of y by adding next point and so on.

We denote

$$L_{01}(x) = \frac{1}{x_1 - x_0}\begin{vmatrix} y_0 & x_0 - x \\ y_1 & x_1 - x \end{vmatrix},$$

which is a polynomial of degree 1 such that $L_{01}(x_0) = y_0$ and $L_{01}(x_1) = y_1$, i.e., $L_{01}(x)$ is interpolation polynomial corresponding to the nodes x_0, x_1 only. Compute $L_{12}(x)$, $L_{23}(x)$, …, $L_{n-1,\,n}(x)$ in a similar manner.

Define again, $$L_{012}(x) = \frac{1}{x_2 - x_0}\begin{vmatrix} L_{01}(x) & x_0 - x \\ L_{12}(x) & x_2 - x \end{vmatrix},$$

which is a polynomial of degree 2 such that

$L_{012}(x_0) = L_{01}(x_0) = y_0$,
$L_{012}(x_1) = y_1$ and $L_{012}(x_2) = L_{12}(x_2) = y_2$

so that $L_{012}(x)$ is a polynomial corresponding to three nodes x_0, x_1, x_2.

Similarly, $L_{123}(x)$, $L_{234}(x)$, ..., $L_{n-2,\, n-1,\, n}(x)$ are calculated.

Proceeding in this way, we define finally

$$L_{0123\ldots n}(x) = \frac{1}{x_n - x_0}\begin{vmatrix} L_{012\ldots n-1}(x) & x_0 - x \\ L_{123\ldots n}(x) & x_n - x \end{vmatrix}$$

this coincides with the required interpolation polynomial.

Aitken's interpolation scheme

$$\begin{array}{llllllll}
x_0 & y_0 & x_0 - x & & & & & \\
x_1 & y_1 & x_1 - x & L_{01}(x) & & & & \\
x_2 & y_2 & x_2 - x & L_{12}(x) & L_{012}(x) & & & \\
x_3 & y_3 & x_3 - x & L_{23}(x) & L_{123}(x) & L_{0123}(x) & & \\
x_4 & y_4 & x_4 - x & L_{34}(x) & L_{234}(x) & L_{1234}(x) & L_{01234}(x) & \\
x_5 & y_5 & x_5 - x & L_{45}(x) & L_{345}(x) & L_{2345}(x) & L_{12345}(x) & L_{012345}(x)
\end{array} \quad (3.18)$$

Note: In the above discussions, we have derived several interpolation formulae. Out of those formulae Newton's forward and backward interpolation formulae, which are for equi-spaced arguments, are applicable for interpolation near the beginning and end of the tabulated values respectively. For interpolation near the middle of the set of equi-spaced arguments of a function, generally Stirling's and Bessel's formulae are used. Stirling's formula gives the better approximation if $p < 1/4$, while for $1/4 < p < 3/4$, Bessel's formula gives better approximation. Also Stirling's formula, in general, is more accurate than Bessel's formula, and if the number of given arguments is even and if the number of given arguments is odd the Bessel's is more accurate than Stirling formula. The above formulae are not applicable when the values of the independent variables are not equi-spaced. For such cases, we have derived Lagrange's interpolation formula and Aitken's method of interpolation by iteration.

Examples

1. From the following table of values of x and $f(x)$ calculate the value of $f(0.23)$:

x:	0.20	0.22	0.24	0.26	0.28	0.30
$f(x)$:	1.6596	1.6698	1.6804	1.6912	1.7024	1.7139.

Solution: Since the argument is closed to the beginning of the table, we use Newton's forward difference formula,

$$f(x) = y_0 + p\Delta y_0 + p(p-1)\frac{\Delta^2 y_0}{2!} + p(p-1)(p-2)\frac{\Delta^3 y_0}{3!} + \cdots$$

$$+ p(p-1)(p-2)\cdots\{(p-(n-1)\}\frac{\Delta^n y_0}{n!}.$$

Here $x = 0.23$, $x_0 = 0.20$, $h = 0.02$, $p = (0.23 - 0.20)/0.02 = 1.5$.

Now we construct the difference table.

x	$y = f(x)$	Δy	$\Delta^2 y$	$\Delta^3 y$	$\Delta^4 y$	$\Delta^5 y$
0.20	**1.65960**					
		0.01020				
0.22	1.66980		**0.00040**			
		0.01060		**–0.00020**		
0.24	1.68040		0.00020		**0.00040**	
		0.01080		0.00020		**–0.00070**
0.26	1.69120		0.00040		–0.00030	
		0.01120		–0.00010		
0.28	1.70240		0.00030			
		0.01150				
0.30	1.71390					

Coefficients	*Multipliers*	*+ ve terms*	*– ve terms*
1.00000	1.65960	1.65960	
1.50000	0.01020	0.01530	
0.37500	0.00040	0.00015	
–0.06250	–0.00020	0.0000125	
0.0234375	0.00040	0.000004375	
–0.1171875	–0.00070	0.000008203	

$$f(0.23) = 1.675080078 \approx 1.67508$$

2. From the following table of values of x and $f(x)$ calculate the value of $f(0.12)$:

x:	0.10	0.15	0.20	0.25	0.30
$f(x)$:	0.1003	0.1511	0.2027	0.2553	0.3093.

Solution: Since the argument lies close to the beginning of the table, we use Newton's forward difference formula,

$$f(x) = y_0 + p\Delta y_0 + p(p-1)\frac{\Delta^2 y_0}{2!} + p(p-1)(p-2)\frac{\Delta^3 y_0}{3!} + \cdots$$

$$+ p(p-1)(p-2)\cdots\{(p-(n-1)\}\frac{\Delta^n y_0}{n!}.$$

Here $x = 0.12$, $x_0 = 0.10$, $h = 0.05$, $p = (0.12 - 0.10)/0.05 = 0.4$.

Now we construct the difference table.

x	$y = f(x)$	Δy	$\Delta^2 y$	$\Delta^3 y$	$\Delta^4 y$
0.10	**0.10030**				
		0.05080			
0.15	0.15110		**0.00080**		
		0.05160		**0.00020**	
0.20	0.20270		0.00100		**0.00020**
		0.05060		0.00040	
0.25	0.25530		0.00140		
		0.05400			
0.30	0.30930				

Coefficients	*Multipliers*	*+ ve terms*	*− ve terms*
1.00000	0.10030	0.10030	
0.40000	0.05080	0.02032	
−0.12000	0.00080		0.000096
0.06400	0.00020	0.0000128	
−0.04160	0.00020		0.00000832
		0.1206328	0.00010432

$f(0.12) = 0.1206328 - 0.00010432 = 0.12052848 \approx 0.12053.$

3. From the following table of values of x and $f(x)$ calculate the value of $f(2.18)$:

x:	1.7	1.8	1.9	2.0	2.1	2.2
$f(x)$:	5.4739	6.0496	6.6859	7.3891	8.1662	9.0250.

Solution: Since the argument lies close to the beginning of the table, we use Newton's backward difference formula,

$$f(x) = y_n + p\nabla y_n + p(p+1)\frac{\nabla^2 y_n}{2!} + p(p+1)(p+2)\frac{\nabla^3 y_n}{3!} + \cdots$$

$$+ p(p+1)(p+2)\cdots\{(p+(n-1)\}\frac{\nabla^n y_n}{n!}$$

$$= y_n + p\Delta y_{n-1} + p(p+1)\frac{\Delta^2 y_{n-2}}{2!} + p(p+1)(p+2)\frac{\Delta^3 y_{n-3}}{3!} +$$

$$\cdots + p(p+1)(p+2)\cdots\{p+(n+1)\}\frac{\Delta^n y_0}{n!}$$

(using the result $\nabla^k y_n = \Delta^k y_{n-k}$)

Here $x = 2.18$, $x_n = 2.2$, $h = 0.1$, $p = (2.18 - 2.2)/0.1 = -0.2$.

Now we construct the difference table.

x	$y = f(x)$	Δy	$\Delta^2 y$	$\Delta^3 y$	$\Delta^4 y$	$\Delta^5 y$
1.70	5.47390					
		0.57570				
1.80	6.04960		0.06060			
		0.63630		0.00630		
1.90	6.68590		0.06690		0.00070	
		0.70320		0.00700		**0.00010**
2.00	7.38910		0.07390		**0.00080**	
		0.77710		**0.00780**		
2.10	8.16620		**0.08170**			
		0.85880				
2.20	**9.02500**					

Coefficients	*Multipliers*	*+ ve terms*	*– ve terms*
1.00000	9.02500	9.02500	
−0.20000	0.85880		0.171760
−0.08000	0.08170		0.006536
−0.04800	0.00780		0.0003744
−0.03360	0.00080		0.00002688
−0.025536	0.00010		0.000002553
		9.02500	0.178699833

$f(2.18) = 9.02500 - 0.178699833 = 8.846300166 \approx 8.84630.$

4. From the following table of values of x and $f(x)$ calculate the value of $f(0.29)$:

x:	0.20	0.22	0.24	0.26	0.28	0.30
$f(x)$:	1.6596	1.6698	1.6804	1.6912	1.7024	1.7139.

Solution: Since the argument is close to the end of the table, we use Newton's backward difference formula,

$$f(x) = y_n + p\Delta y_{n-1} + p(p+1)\frac{\Delta^2 y_{n-2}}{2!} + p(p+1)(p+2)\frac{\Delta^3 y_{n-3}}{3!} + \cdots$$
$$+ p(p+1)(p+2)\cdots\{(p+(n-1)\}\frac{\Delta^n y_0}{n!}.$$

Here $x = 0.29$, $x_n = 0.30$, $h = 0.02$, $p = (0.29 - 0.30)/0.02 = -0.5.$

Now we construct the difference table.

x	$y = f(x)$	Δy	$\Delta^2 y$	$\Delta^3 y$	$\Delta^4 y$	$\Delta^5 y$
0.20	1.65960					
		0.01020				
0.22	1.66980		0.00040			
		0.01060		−0.00020		
0.24	1.68040		0.00020		0.00040	
		0.01080		0.00020		**−0.00070**
0.26	1.69120		0.00040		**−0.00030**	
		0.01120		**−0.00010**		
0.28	1.70240		**0.00030**			
		0.01150				
0.30	**1.71390**					

Coefficients	*Multipliers*	*+ ve terms*	*– ve terms*
1.00000	1.71390	1.71390	
−0.50000	0.01150		0.005750
−0.12500	0.00030		0.0000375
−0.06250	−0.00010	0.00000625	
−0.0390625	−0.00030	0.0000117187	
−0.02734375	−0.00070		0.0000191406
		1.713937109	0.0057875

$f(0.29) = 1.713937109 - 0.0057875 = 1.708149609 \approx 1.7081.$

5. From the following table of values of x and $f(x)$ calculate the value of $f(37)$:

x:	25	30	35	40	45
$f(x)$:	1.39794	1.47712	1.54407	1.60206	1.65321.

Solution: Since the argument close to the middle of the table, we use Gauss's forward difference formula

$$f(x) = y_0 + p\Delta y_0 + \frac{p(p-1)}{2!}\Delta^2 y_{-1} \frac{(p+1)p(p-1)}{3!}\Delta^3 y_{-1}$$

$$+\frac{(p+1)p(p-1)(p-2)}{4!}\Delta^4 y_{-2} + \cdots$$

Let the origin be taken as 35. Here $h = 5$, $p = (37 - 35)/5 = 0.4$.

Now we construct the difference table

x	u	$10^5 y$	$10^5\Delta y$	$10^5\Delta^2 y$	$10^5\Delta^3 y$	$10^5\Delta^4 y$
25	−2	139794				
			7918			
30	−1	147712		−1223		
			6695		327	
35	0	**154407**		**−896**		−115
			5799		**212**	
40	1	160206		−684		
			5115			
45	2	165321				

Now $10^5 f(p) = 10^5 y_0 + 10^5 p\Delta y_0 + 10^5 \frac{p(p-1)}{2!}\Delta^2 y_{-1} + 10^5 \frac{(p+1)p(p-1)}{3!}\Delta^3 y_{-1} + \cdots$

$$10^5 f(0.4) = 154407 + (0.4)5799 + (0.4)(-0.6)(-896)/2 + (1.4)(0.4)(-0.6)(212)/6$$
$$= 156822.25.$$

$f(37) = 1.56822.$

6. From the following table of values of x and $f(x)$ calculate the value of $f(1.748)$:

x:	1.72	1.73	1.74	1.75	1.76	1.77
$f(x)$:	0.1790	0.1773	0.1755	0.1738	0.1720	0.1703.

Solution: Since the argument close to the middle of the table, we use Gauss's forward difference formula,

$$f(x) = y_0 + p\Delta y_0 + \frac{p(p-1)}{2!}\Delta^2 y_{-1} \frac{(p+1)p(p-1)}{3!}\Delta^3 y_{-1}$$

$$+\frac{(p+1)p(p-1)(p-2)}{4!}\Delta^4 y_{-2} + \cdots$$

Let the origin be taken as 1.74. Here $h = 0.01$, $p = (1.748 - 1.74)/0.01 = 0.8$.

Now we construct the difference table.

x	u	$10^4 y$	$10^4 \Delta y$	$10^4 \Delta^2 y$	$10^4 \Delta^3 y$	$10^4 \Delta^4 y$
1.72	−2	1790				
			−17			
1.73	−1	1773		−1		
			−18		2	
1.74	0	**1775**		**1**		**−4**
			−17		**−2**	
1.75	1	1738		−1		4
			−18		2	
1.76	2	1720		1		
			−17			
1.77	3	1703				

Now $10^4 f(p) = 10^4 y_0 + 10^4 p\Delta y_0 + 10^4 \dfrac{p(p-1)}{2!}\Delta^2 y_{-1} + 10^5 \dfrac{(p+1)p(p-1)}{3!}\Delta^3 y_{-1}$

$$+10^4 \frac{(p+1)p(p-1)(p-2)}{4!}\Delta^3 y_{-2} + \cdots$$

$$10^4 f(0.8) = 1755 + (0.8)(-17) + (0.8)(-17) + (0.8)(-0.2)(1)/2 + (1.8)(0.4)(-0.2)(-2)/6$$

$$+(1.8)(0.4)(-0.2)(-1.2)(-4)/24$$

$= 1741.3872.$

$f(1.748) = 0.17413872 \approx 0.1741.$

7. The population in a town is as follows:

Year	1931	1941	1951	1961	1971	1981
Population (in thousand)	15	20	27	39	52	70.

Find the population of the town in 1957.

Solution: Let the origin be taken as 1951. Here $h = 10$, so $p = (1957 - 1951)/10 = 0.6$

The difference table is as follows:

x	u	y	Δy	$\Delta^2 y$	$\Delta^3 y$	$\Delta^4 y$	$\Delta^5 y$
1931	−2	15					
			5				
1941	−1	20		2			
			7		**3**		
1951	0	**27**		**5**		**−7**	
			12		−4		15
1961	1	39		1		8	
			13		4		
1971	2	52		5			
			18				
1981	3	70					

Now using Gauss's backward formula, we get

$$f(p) = y_0 + p\Delta y_{-1} + \frac{p(p+1)}{2!}\Delta^2 y_{-1} + \frac{(p+1)\,p(p-1)}{3!}\Delta^3 y_{-2}$$

$$+\frac{(p+2)\,p(p+1)(p-1)}{4!}\Delta^4 y_{-2} + \cdots$$

$$= 27 + (0.6)(7) + (1.6)(0.6)(5)/2 + (1.6)(0.6)(-0.4)(3)/6$$

$$+ (2.6)(1.6)(0.6)(-0.4)(-7)/24 = 33.6992 \text{ (in thousand)}.$$

So, the population in 1957 is 33699.

8. The population in a town is as follows:

Year	1911	1921	1931	1941	1951	1961
Population (in thousand)	12	15	20	27	39	52.

Find the population of the town in 1946.

Solution: Let the origin be taken as 1931. Here $h = 10$, so $p = (1946 - 1931)/10 = 1.5$.

The difference table is as follows:

x	u	y	Δy	$\Delta^2 y$	$\Delta^3 y$	$\Delta^4 y$	$\Delta^5 y$
1911	−2	12					
			3				
1921	−1	15		2			
			5		**0**		
1931	0	**20**		**2**		**3**	
			7		3		−10
1941	1	27		5		−7	
			12		−4		
1951	2	39		1			
			13				
1961	3	52					

Now using Gauss's backward formula, we get

$$f(p) = y_0 + p\Delta y_{-1} + \frac{p(p+1)}{2!}\Delta^2 y_{-1} + \frac{(p+1)\,p(p-1)}{3!}\Delta^3 y_{-2}$$

$$+\frac{(p+2)\,p(p+1)(p-1)}{4!}\Delta^4 y_{-2} + \cdots$$

$$= 20 + (1.5)(5) + (2.5)(1.5)(2)/+(2.5)(1.5)(1.5)(0.5)(0)/6$$

$$+ (3.5)(2.5)(1.5)(0.5)(0.5)(3)/24 = 32.07 \text{ (in lakhs)}.$$

So, the population in 1946 is 3207000.

9. From the following table of values of x and $f(x)$ calculate the value of $f(1.91)$ by using Stirling's formula:

x:	1.7	1.8	1.9	2.0	2.1	2.2
$f(x)$:	5.4739	6.0496	6.6259	7.3891	8.1662	9.0250.

Solution: Stirling's formula is

$$f(x) = y_0 + \frac{p}{2}(\Delta y_{-1} + \Delta y_0) + \frac{p^2}{2}\Delta^2 y_{-1} + \frac{p(p^2-1)}{12}(\Delta^3 y_{-2} + \Delta^3 y_{-1}) +$$

$$\frac{p^2(p^2-1)}{24}\Delta^4 y_{-2} + \frac{p(p^2-1)(p^2-2)}{240}(\Delta^5 y_{-2} + \Delta^5 y_{-3}) + \cdots$$

Here $x = 1.91$, $x_0 = 1.90$, $h = 0.1$, $p = (1.91 - 1.90)/0.1 = 0.1$.

Now we construct the difference table.

x	n	y	Δy	$\Delta^2 y$	$\Delta^3 y$	$\Delta^4 y$	$\Delta^5 y$
1.7	–2	5.47390					
			0.57570				
1.8	–1	6.04960		0.06060			
			0.63630		**0.00630**		
1.9	0	**6.68590**		**0.06690**		**0.00070**	
			0.70320		**0.00700**		0.00010
2.0	1	7.38910		0.07390		0.00080	
			0.077710		0.00780		
2.1	2	8.16620		0.08170			
			0.85880				
2.2	3	9.02500					

$f(1.91) = \mathbf{6.68590} + (0.1)\,[\mathbf{0.63630} + \mathbf{0.70320}]/2 + (0.1)^2(\mathbf{0.06690})/2 +$

$(0.1)\,[(0.1)^2 - 1]\,[\mathbf{0.00630} + \mathbf{0.00700}]/2 + (0.1)^2[(0.1)^2 - 1]\,(\mathbf{0.00070})/24$

$= 6.753099486 \approx 6.75310$

10. From the following table of values of x and $f(x)$ calculate the value of $f(1.91)$ by using Stirling's formula:

x:	0.61	0.62	0.63	0.64	0.65	0.66
$f(x)$:	1.840431	1.858928	1.877610	1.896481	1.915541	1.934792

Solution: Stirling's formula is

$$f(x) = y_0 + \frac{p}{2}(\Delta y_{-1} + \Delta y_0) + \frac{p^2}{2}\Delta^2 y_{-1} + \frac{p(p^2-1)}{12}(\Delta^3 y_{-2} + \Delta^3 y_{-1}) +$$

$$\frac{p^2(p^2-1)}{24}\Delta^4 y_{-2} + \frac{p(p^2-1)(p^2-2)}{240}(\Delta^5 y_{-2} + \Delta^5 y_{-3}) + \cdots$$

Here $x = 0.638$, $x_0 = 0.630$, $h = 0.01$, $p = (0.638 - 0.630)/0.01 = 0.8$.

Now we construct the difference table.

x	n	y	Δy	$\Delta^2 y$	$\Delta^3 y$	$\Delta^4 y$	$\Delta^5 y$
0.61	−2	1.840431					
			0.018497				
0.62	−1	1.858928		0.000185			
			0.018682		**0.000004**		
0.63	0	**1.877610**		**0.000189**		**−0.000004**	
			0.018871		**0.000000**		0.000006
0.64	1	1.896481		0.000189		0.000002	
			0.019060		0.000002		
0.65	2	1.915541		0.000191			
			0.019251				
0.66	3	1.934792					

$f(0.638) = \mathbf{1.877610} + (0.8)\,[\mathbf{0.018682} + \mathbf{0.018871}]/2 + (0.8)^2(\mathbf{0.000189})/2 +$
$(0.8)[(0.8)^2 - 1]\,[\mathbf{0.000004} + 0]/2 + (0.8)^2 + [(0.8)^2 - 1]\,(\mathbf{0.000004})/24$
$= 1.892691622 \approx 1.892692.$

11. From the following table of values of x and $f(x)$ calculate the value of $f(9)$ by using Bessel's formula:

x:	0	5	10	15	20
$f(x)$:	0.196120	0.195293	0.192815	0.188690	0.182928.

Solution: Bessel's formula is

$$f(x) = y_0 + p\Delta y_0 + \frac{p(p-1)}{4}(\Delta^2 y_{-1} + \Delta^2 y_0) + \frac{p(p-1)(2p-1)}{12}\Delta^3 y_{-1}$$
$$+ \frac{(p+1)\,p(p-1)(p-2)}{48}(\Delta^4 y_{-2} + \Delta^4 y_{-1}) + \cdots$$

Here $x = 9$, $x_0 = 10$, $h = 5$, $p = (9 - 10)/5 = -0.2$.

Now we construct the difference table.

x	n	y	Δy	$\Delta^2 y$	$\Delta^3 y$	$\Delta^4 y$
0	−2	0.196120				
			−0.000827			
5.	−1	0.195293		−0.001651		
			−0.002478		0.000004	
10	0	**0.192815**		**−0.001647**		0.000006
			−0.004125		**0.000010**	
15	1	0.199690		**−0.001637**		
			−0.005762			
20	2	0.182928				

$$f(9) = \mathbf{0.192815} + (-0.2)(\mathbf{-0.004125} + (-0.2)(-1.2)(\mathbf{0.001637} - \mathbf{0.001647})/4$$
$$+ (-0.2)(-1.2)(-1.4)(\mathbf{0.000010})/12 = 0.19344268 \approx 0.193443.$$

12. From the following table of values of x and $f(x)$ calculate the value of $f(9)$ by using Bessel's formula:

x:	0	5	10	15	20
$f(x)$:	1.5708	1.5738	1.5828	1.5981	1.6200.

Solution: Bessel's formula is

$$f(x) = y_0 + p\Delta y_0 + \frac{p(p-1)}{4}(\Delta^2 y_{-1} + \Delta^2 y_0) + \frac{p(p-1)(2p-1)}{12}\Delta^3 y_{-1}$$
$$+ \frac{(p+1)p(p-1)(p-2)}{48}(\Delta^4 y_{-2} + \Delta^4 y_{-1}) + \cdots$$

Here $x = 9$, $x_0 = 10$, $h = 5$, $p = (9 - 10)/5 = -0.2$

Now we construct the difference table

x	n	y	Δy	$\Delta^2 y$	$\Delta^3 y$	$\Delta^4 y$
0	−2	1.5708				
			0.0030			
5	−1	1.5738		0.0060		
			0.0090		0.0003	
10	0	**1.5828**		**0.0063**		0.0000
			0.0153		**0.0003**	
15	1	1.5981		**0.0066**		
			0.0219			
20	2	1.6200				

$$f(9) = \mathbf{1.5828} + (-0.2)(\mathbf{-0.0153}) + (-0.2)(-1.2)(\mathbf{0.0063} + \mathbf{0.0066})/4$$
$$+ (-0.2)(-1.2)(-1.4)(\mathbf{0.0003})/12 = 1.5805056 \approx 1.5805.$$

13. From the following table of values of x and $f(x)$ calculate the value of $f(0.35)$ by using Lagrange's interpolation formula:

x:	0.3	0.5	0.6
$f(x)$:	0.6179	0.6915	0.7257.

Solution: Lagrange's interpolation formula is

$$L(x) = w(x)\sum_{k=0}^{n}\frac{f(x_k)}{(x - x_k)w'(x_k)} = w(x)\sum_{k=0}^{n}\frac{y_k}{D_k},$$

where $w(x) = (x - x_0)(x - x_1)(x - x_2) \cdots (x - x_n)$

and $D_k = (x - x_k)(x_k - x_0) \ldots (x_k - x_{k-1})(x_k - x_{k+1}) \cdots (x_k - x_n)$.

Now we construct the following table using (10.17):

			D	y	y/D
0.50	−0.20	−0.30	0.00300	0.61790	205.9666667
0.20	−0.15	−0.10	0.00300	0.69150	230.5000000
0.30	0.10	−0.25	−0.00750	0.72570	−96.7600000

Now

$w(0.35) = (0.05)(-0.15)(-0.25) = 0.001875.$

$y/D = 205.9666667 + 230.5000000 - 96.7600000 = 339.7066667.$

$f(0.35) = 0.001875 \times 339.7066667 = 0.63695.$

14. From the following table of values of x and $f(x)$ calculate the value of $f(1.5)$ by using Lagrange's interpolation formula:

x:	1.00	1.20	1.40	1.60	1.80	2.00
$f(x)$:	0.2420	0.1942	0.1497	0.1109	0.0790	0.0540.

Solution: Lagrange's interpolation formula is

$$L(x) = w(x)\sum_{k=0}^{n} \frac{f(x_k)}{(x - x_k)w'(x_k)} = w(x)\sum_{k=0}^{n} \frac{y_k}{D_k},$$

where $w(x) = (x - x_0)(x - x_1)(x - x_2) \cdots (x - x_n)$

and $D_k = (x - x_k)(x_k - x_0) \ldots (x_k - x_{k-1})(x_k - x_{k+1}) \cdots (x_k - x_n)$.

Now we construct the following table using (10.17):

						D	y	y/D
0.50	−0.20	−0.40	−0.60	−0.80	−1.00	−0.019200	0.2420	12.60416667
0.20	0.30	−0.20	−0.40	−0.60	−0.80	0.002304	0.19420	84.28819444
0.40	0.20	0.10	−0.20	−0.40	−0.60	−0.000384	0.14970	−389.84875000
0.60	0.40	0.20	−0.10	−0.20	−0.40	−0.000384	0.11090	−288.80208330
0.80	0.60	0.40	0.20	−0.30	−0.20	0.002304	0.07900	34.28819444
1.00	0.80	0.60	0.40	0.20	−0.50	−0.019200	0.05400	−2.81250000

Now

$w(1.5) = (0.5)(0.3)(0.1)(-0.1)(-0.3)(-0.5) = -0.000225$

$\Sigma y/D = 12.60416667 + 84.28819444 - 389.84875000 - 288.80208330$
$+ 34.28819444 - 2.8125000 = -575.486111$

$f(1.5) = -0.000225 \times (-575.486111) = 0.129484375 \approx 0.1295$

15. Given the following table, Calculate $f(0.35)$ by Aitken's method:

x:	0.3	0.4	0.5	0.6
$f(x)$:	0.955336	0.921061	0.877583	0.825336.

Solution: Using the Aitken's computational scheme given by (10.18), we get

x_i	y_i	L_{0i}	L_{01i}	L_{012i}	$x_i - x$
0.30	0.955336				−0.14
0.40	0.921061	0.9073510			−0.04
0.50	0.877583	0.9009089	0.90477416		0.06
0.60	0.825836	0.8946693	0.90481467	0.90474986	0.16

Hence, $f(0.44) = 0.90474986 \approx 0.904750$

EXERCISE

1. Definine interpolation. Why polynomials are used in interpolation?
2. Estimate the error in polynomial interpolation.
3. Show that there exists at most one polynomial which can interpolate a given function at specified points.
4. Derive Newtons' forward difference interpolation formula for equi-spaced arguments.
5. Derive Newtons' backward difference interpolation formula for equi-spaced arguments.
6. Derive Lagrange's interpolation formula.
7. What are the merits and demerits of Lagrange's interpolation formula?
8. Derive Aitken's interpolation formula.
9. In the following series estimate the value corresponding to 1982:

Year:	1981	1983	1985	1987
Value:	100	125	137	151.

10. The population in a town in the decennial census is given below. Estimate the population in 1925.

Year:	1891	1901	1911	1921	1931
Population: (In thousands)	46	66	81	93	101

11. The values of the annuities are given for the following ages. Find the value of the annuity at the age 28.5

Age:	25	26	27	28	29
Annuity:	16.2	15.9	15.6	15.3	15.

12. Apply Gauss' forward difference formula to obtain $f(32)$, given that

x:	25	30	35	40
$f(x)$:	0.2707	0.3027	0.3386	0.3794.

13. Find f(3.75) from the following table

x:	2.5	3.0	3.5	4.0	4.5	5.0
$f(x)$:	24.145	22.043	20.225	18.644	17.262	16.047.

14. Interpolate by using backward formula the population for the year 1936. Given the following data:

Year:	1901	1911	1921	1931	1941	1951
Population: (in thousands)	12	15	20	27	39	52.

15. Given the data:

θ^0:	0	5	10	15	20	25	30
$\tan\theta^0$:	0	0.0875	0.1763	0.2679	0.3640	0.4663	0.5774.

Find the value of $\tan 16^0$, with the help of Stirling's formula.

16. With the help of Stirling's formula, find the value of $f(1.22)$ from the following table:

x:	1.0	1.1	1.2	1.3	1.4
$f(x)$:	0.841417	0.89121	0.93204	0.96356	0.98545

17. Use Bessel's formula to find $f(25)$, given that

x:	20	24	28	32
$f(x)$:	2854	3162	3544	3992

18. Use Bessel's formula to find $f(1.2)$, given that

x:	0.2	0.6	1.0	1.4	1.8
$f(x)$:	0.39104	0.33322	0.24197	0.14973	0.07875

19. Apply the Lagrange's interpolation formula to determine $f(5)$. Given the data:

x:	1	2	3	4	7
$f(x)$:	2	4	8	16	128

20. Apply the Lagrange's interpolation formula to determine $f(3)$. Given the data:

x:	0	1	2	4
$f(x)$:	580	556	520	385

21. Apply the Aitken's interpolation formula to determine $f(6)$. Given the data:

x:	2	5	7	10	12
$f(x)$:	18	180	448	1210	2028

22. A certain function f, defined on (0,1) is such that $f(0) = 0$, $f(1/2) = -1$, $f(1) = 0$. Find the quadratic polynomial $p(x)$ which agrees with $f(x)$ for $x = 0, 1/2, 1$.

23. A certain function f, defined on (−1,5) is such that $f(-1) = 9$, $f(0) = 5$, $f(2) = 3$, $f(5) = 15$ .Find the quadratic polynomial $p(x)$, by Lagrange's formula.

ANSWERS

9. 110, **10.** 96.31 thousands, **11.** 15.15, **12.** 0.3165,
13. 19.407, **14.** 32070, **15.** 0.2867, **16.** 0.9391,
17. 3251, **18.** 0.19457, **19.** 32.93, **20.** 465.25,
21. 224. **22.** $4x(x-1)$, **23.** $x^2 - 3x + 5$

Chapter 4

DIVIDED DIFFERENCE

4.1 DEFINITION

The first divided difference at the points x_0, x_1 is denoted by $f[x_0, x_1]$ and given by

$$f[x_0, x_1] = \frac{f(x_0) - f(x_1)}{x_0 - x_1} = \frac{f(x_0)}{x_0 - x_1} + \frac{f(x_1)}{x_1 - x_0}.$$

The higher order differences are obtained in terms of lower order differences. For example, the second difference is at the points x_0, x_1, x_2 is given by

$$f[x_0, x_1, x_2] = \frac{f[x_0, x_1] - f[x_1, x_2]}{x_0 - x_2}.$$

Proceeding in this way we get the kth divided difference of $f(x)$ at $x_0, \cdots, x_k$ as

$$f[x_0, x_1, x_2, \cdots, x_k] = \frac{f[x_0, x_1, \cdots, x_{k-1}] - f[x_1, x_2, \cdots, x_k]}{x_0 - x_k} \tag{4.1}$$

A difference table is again a convenient way to display differences.

4.2 DIVIDED DIFFERENCE TABLE

x_0	$f(x_0)$				
		$f[x_0, x_1]$			
x_1	$f(x_1)$		$f[x_0, x_1, x_2]$		
		$f[x_1, x_2]$		$f[x_0, x_1, x_2, x_3]$	
x_2	$f(x_2)$		$f[x_1, x_2, x_3]$		$f[x_0, x_1, x_2, x_3, x_4]$
		$f[x_2, x_3]$		$f[x_1, x_2, x_3, x_4]$	
x_3	$f(x_3)$		$f[x_2, x_3, x_4]$		
		$f[x_3, x_4]$			
x_4	$f(x_4)$				

From the above definition of divided difference it is clear that $f[x_0, x_1] = f[x_1, x_0]$. Further we see that

$$f[x_0, x_1, x_2] = \frac{f[x_0, x_1] - f[x_1, x_2]}{x_0 - x_2} = \frac{1}{x_0 - x_2}\left[\frac{f(x_0) - f(x_1)}{x_0 - x_1} - \frac{f(x_1) - f(x_2)}{x_1 - x_2}\right]$$

$$= \frac{f(x_0)}{(x_0 - x_2)(x_0 - x_2)} + \frac{f(x_1)}{(x_1 - x_0)(x_1 - x_2)} + \frac{f(x_2)}{(x_2 - x_0)(x_2 - x_1)} \qquad (4.2)$$

from which we see that the expression remains unchanged if we interchange any two arguments, i.e., the second order divided difference is symmetric.

Proceeding in this way it can be shown that the nth order divided difference is also symmetric.

If we take $f(x) = x^2$, then $f[x_0, x_1] = \dfrac{x_0^2 - x_1^2}{x_0 - x_1} = x_0 + x_1$

$$f[x_0, x_1, x_2] = \frac{f[x_0, x_1] - f[x_1, x_2]}{x_0 - x_2} = \frac{x_0 + x_1 - x_1 - x_2}{x_0 - x_2} = 1$$

and $f[x_0, x_1, x_2, x_3] = \dfrac{f[x_0, x_1, x_2] - f[x_1, x_2, x_3]}{x_0 - x_3} = \dfrac{1 - 1}{x_0 - x_3} = 0,$ which shows that all higher order differences are zero.

Similarly, for $f(x) = x^3$, $f[x_0, x_1] = \dfrac{x_0^3 - x_1^3}{x_0 - x_1} = x_0^2 + x_1 x_0 + x_1^2,$

$$f[x_0, x_1, x_2] = \frac{f[x_0, x_1] - f[x_1, x_2]}{x_0 - x_2} = \frac{x_0^2 + x_0 x_1 + x_1^2 - x_1^2 - x_1 x_2 - x_2^2}{x_0 - x_2} = x_0 + x_1 + x_2$$

$$f[x_0, x_1, x_2, x_3] = \frac{f[x_0, x_1, x_2] - f[x_1, x_2, x_3]}{x_0 - x_3} = \frac{x_0 + x_1 + x_2 - x_1 - x_2 - x_3}{x_0 - x_3} = 1.$$

Again all the higher differences are zero. The above results shows that for a nth degree polynomial the nth divided difference is constant and all $(n + 1)$th or higher divided differences are zero which indicates that divided difference formula has a relation with the formula of derivatives.

4.3 RELATION BETWEEN DIVIDED DIFFERENCE AND DERIVATIVES

To illustrate it we start with the notion

$$f[x_0, x] = \frac{f(x_0) - f(x)}{x_0 - x} \Rightarrow f[x_0, x_0] = f'(x_0), \text{ if } f(x) \text{ is differentiable.}$$

Similarly, $f[x_0, x_0, x] = \dfrac{f[x_0, x_0] - f[x_0, x]}{x_0 - x} = \dfrac{1}{x_0 - x}\left[f'(x_0) - \dfrac{f(x_0) - f(x)}{x_0 - x}\right].$

Expanding $f(x)$ in the neighbourhood of x_0 and taking limit $x \to x_0$, we obtain

$$f[x_0, x_0, x_0] = f^2(x_0)/2!$$

Analogously, $$\underbrace{f[x_0, x_0, \cdots, x_0]}_{(k+1)\text{ times}} = f^k(x_0)/k! \qquad (4.3)$$

The expression $f(x_0, \cdots, x_0)$ is called *confluent divided difference*.

Again we see that

$$\frac{d}{dx} f(x, x_0) = \frac{d}{dx} \frac{f(x) - f(x_0)}{x - x_0} = \frac{f'(x) - f[x, x_0]}{x - x_0} = \frac{f[x, x] - f[x, x_0]}{x - x_0} = f[x, x, x_0]$$

$$\frac{d}{dx} f[x, x_0, x_1] = \frac{d}{dx} \frac{f[x, x_1] - f[x_1, x_0]}{x - x_0} = \frac{f'[x, x_1] - f[x, x_0, x_1]}{x - x_0} = \frac{f[x, x, x_1] - f[x_0, x, x_1]}{x - x_0}$$

$$= f[x, x, x_0, x_1].$$

Proceeding in this way, we get

$$\frac{d}{dx} f[x, x_0, x_1, \cdots, x_n] = f[x, x, x_0 x_1, \cdots, x_n]. \tag{4.4}$$

4.4 RELATION BETWEEN DIVIDED DIFFERENCE AND FINITE DIFFERENCE

For equally spaced points, i.e., when $x_1 - x_0 = x_2 - x_1 = \cdots = x_n - x_{n-1} = \cdots = h$, we see that

$$f[x_0, x_1] = \frac{f(x_0) - f(x_1)}{x_0 - x_1} = \frac{\Delta f(x_0)}{h} = \frac{\Delta f_0}{h},$$

$$f[x_0, x_1, x_2] = \frac{f(x_0, x_1) - f(x_1, x_2)}{x_0 - x_2} = \frac{\dfrac{\Delta f_0}{h} - \dfrac{\Delta f_1}{h}}{x_0 - x_2} = \frac{\Delta^2 f_0}{2! h^2}.$$

Let $f[x_0, x_1, \cdots, x_k] = \dfrac{\Delta^k f_0}{k! h^k}$, then

$$f[x_0, x_1, \cdots, x_k, x_{k+1}] = \frac{f[x_0, x_1, \cdots, x_k] - f[x_1, x_2, \cdots, x_{k+1}]}{x_0 - x_{k+1}} = \frac{\dfrac{\Delta^k f_0}{k! h^k} - \dfrac{\Delta^k f_1}{k! h^k}}{x_0 - x_{k+1}} = \frac{\Delta^{k+1} f_0}{(k+1)! h^{k+1}}.$$

So, in general, $f[x_0, x_1, \cdots, x_n] = \dfrac{\Delta^n f_0}{n! h^n}$. (4.5)

Further using the result

$$\frac{df(x)}{dx} = \frac{f(x+h) - f(x)}{h} = \frac{\Delta f(x)}{h} \text{ for small value of } h,$$

we see that

$$\frac{d}{dx} \approx \frac{\Delta}{h} \text{ we obtain } \frac{d^n}{dx^n} \approx \frac{\Delta^n}{h^n} \text{ and hence } \frac{\Delta^n f_0}{n! h^n} = \frac{f^n(x_0)}{n!}.$$

So from (4.4) and (4.5) we can write,

$$f[x, x_0, x_1, \cdots, x_n] = \frac{f^{n+1}(\xi)}{(n+1)!},$$

where $\min\{x, x_0, \cdots, x_n\} < \xi < \max\{x, x_0, \cdots, x_n\}$.

From this we can also write,

$$f[x_0, x_1, \cdots, x_n] = \frac{f^n(\xi)}{n!},$$

where $\min\{x_0, \cdots, x_n\} < \xi < \max\{x_0, \cdots, x_n\}$.

Now if we assume that $f^n(x)$ is continuous at x_0 and taking limit $x_i \to x_0$ for all i, we get

$$\underbrace{f[x_0, x_0, \cdots, x_0]}_{(n+1)\text{ times}} = \frac{f^n(x_0)}{n!}. \tag{4.6}$$

4.5 NEWTON'S INTERPOLATION FORMULA WITH DIVIDED DIFFERENCE

Now we shall write Newton's general interpolation polynomial using the notion of divided difference. From the definition of divided difference we have

$$f(x) = f(x_0) + (x - x_0)\, f[x, x_0]$$
$$f[x, x_0] = f[x_0, x_1] + (x - x_1)\, f[x, x_0, x_1]$$
$$f[x, x_0, x_1] = f[x_0, x_1, x_2] + (x - x_2)\, f[x, x_0, x_1, x_2]$$

....

....

$$f[x, x_0, \cdots, x_{n-1}] = f[x_0, x_1, \cdots, x_n] + (x - x_n)\, f[x, x_0, x_1, \cdots, x_n]$$

Next multiplying 2nd, 3rd, $\cdots$, $(n + 1)^{\text{th}}$ equations by $(x - x_0)$, $(x - x_0)(x - x_1)$, $\cdots$, $(x - x_0)(x - x_1) \cdots (x - x_{n-1})$ respectively and adding we get

$$f(x) = f(x_0) + (x - x_0)\, f[x_0, x_1] + (x - x_0)(x - x_1)\, f[x_0, x_1, x_2] + \cdots + (x - x_0)(x - x_1) \cdots (x - x_{n-1})\, f[x_0, x_1, \cdots, x_n] + (x - x_0)(x - x_1) \cdots (x - x_n)\, f[x, x_0, x_1, \cdots, x_n] \tag{4.7}$$

This is the Newton's interpolation formula with divided difference.

Again, let $h(x) = f[x, x_0, x_1, \cdots, x_n]$, then

$$f[x, x', x_0, x_1, \cdots, x_n] = f[x, x_0, x_1, \cdots, x_n, x'] = \frac{f[x, x_0, x_1, \cdots, x_n] - f[x_0, x_1, \cdots, x_n, x']}{x - x'}$$

$$= \frac{h(x) - h(x')}{x - x'} = h[x, x'].$$

Proceeding in this manner we get

$$h[x, t_1, t_2, \cdots, t_k] = f[x, t_1, t_2, \cdots, t_k, x_0, x_1, \cdots, x_n]$$

Taking limit $(t_1, t_2, \ldots, t_k) \to \underbrace{(x, x, \ldots\ldots, x)}_{k\text{ times}}$ we obtain

$$f[\underbrace{x, x, x, \cdots, x}_{(k+1)\text{ times}}, x_0, x_1, \cdots, x_n] = h[\underbrace{x, x, x, \cdots, x}_{(k+1)\text{ times}}] = \frac{h^k(x)}{k!} \qquad \text{[Using (4.6)]}$$

$$h^k(x) = k!\, f[\underbrace{x, x, x, \cdots, x}_{(k+1)\text{ times}}, x_0, x_1, \cdots, x_n]. \tag{4.8}$$

Examples

1. Define divided difference of a function $f(x)$ at the points $x_0, x_1, \cdots, x_k$.

Solution: The kth order divided difference of a function $f(x)$ at the points x_0, x_1, $\cdots$, x_k is given by

$$f[x_0, x_1, x_2, \cdots, x_k] = \frac{f[x_0, x_1, \cdots, x_{k-1}] - f[x_1, x_2, \cdots, x_k]}{x_0 - x_k}.$$

2. Show that if $f(x) = x^2$, then $f[x_0, x_1, x_2]$ is constant and all higher order differences are zero.

Solution: If $f(x) = x^2$, then $f[x_0, x_1] = \dfrac{x_0^2 - x_1^2}{x_0 - x_1} = x_0 + x_1$,

So, $f[x_0, x_1, x_2] = \dfrac{f[x_0, x_1] - f[x_1, x_2]}{x_0 - x_2} = \dfrac{x_0 + x_1 - x_1 - x_2}{x_0 - x_2} = 1$

and $f[x_0, x_1, x_2, x_3] = \dfrac{f[x_0, x_1, x_2] - f[x_1, x_2, x_3]}{x_0 - x_3} = \dfrac{1-1}{x_0 - x_3} = 0$, which shows that all higher order differences are zero.

3. Show that the nth order divided difference of x^n is constant and that of order $(n + 1)$ is zero.

Solution: Taking $f(x) = x^n$, we see that

$$f[x, x_0] = \frac{x^n - x_0^n}{x - x_0} = x^{n-1} + x^{n-2}x_0 + x^{n-3}x_0^2 + \cdots + x_0^{n-1} \tag{1}$$

$$f[x_0, x_1] = \frac{x_0^n - x_1^n}{x - x_1} = x_0^{n-1} + x_0^{n-2}x_1 + x_0^{n-3}x_1^2 + \cdots + x_1^{n-1}$$

$$\begin{aligned} f[x, x_0, x_1] &= \frac{f[x, x_0] - f[x_0, x_1]}{x - x_1} = \frac{x^{n-1} - x_1^{n-1}}{x - x_1} + x_0 \frac{x^{n-2} - x_1^{n-2}}{x - x_1} \\ &\quad + x_0^2 \frac{x^{n-3} - x_1^{n-3}}{x - x_1} + \ldots + x_0^{n-2} \\ &= x^{n-2} + x^{n-3}x_1 + x^{n-4}x_1^2 + \cdots + x_1^{n-2} \\ &\quad + x_0[x^{n-3} + x^{n-4}x_1 + x^{n-5}x_1^2 + \cdots + x_1^{n-3}] \\ &\quad + x_0^2[x^{n-4} + x^{n-5}x_1 + \cdots + x_1^{n-4}] \\ &\quad + \cdots + x_0^{n-2} \\ &= x^{n-2} + x^{n-3}(x_1 + x_0) + x^{n-4}(x_1^2 + x_0x_1 + x_0^2) + \cdots + x_0^{n-2}. \end{aligned} \tag{2}$$

From (1) and (2) we see that 1st and 2nd orders divided differences of x^n are of degree $(n–1)$ and $(n–2)$ respectively. Then by principle of induction we can say that the nth order divided difference operator is of degree $(n–n)$, i.e., a constant and the $(n+1)$th order difference is zero.

4. Show that the nth order divided difference of an nth degree polynomial is constant.

Solution: Since an nth degree polynomial contains x^n and terms having lower degree of x, the above result shows that the nth order divided difference of an nth degree polynomial is constant.

EXERCISE

1. Define nth order divided difference of $f(x)$ at x_k, $0 \le k \le n$.
2. Prove that divided difference of $f(x) + g(x)$ is sum of the divided differences of $f(x)$ and $g(x)$.
3. Prove that divided difference of $cf(x)$, c being a constant, is c times the divided differences of $f(x)$.
4. Any divided difference of a constant is zero.
5. Evaluate first few differences of x^2, x^3 and x^4.
6. Calculate the divided differences through third order for the following x_k, y_k.

x_k	0	1	4	6
y_k	1	–1	1	–1.

7. Calculate the divided differences through fourth order for the following x_k, y_k.

x_k	0	1	2	3	4
y_k	0	16	48	88	120.

8. If $f(x) = 1/x$ whose arguments are x_i, $i = 0, 1, 2, 3$, then calculate $f[x_0, x_1, x_2, x_3]$.
9. If $f(x) = 1/x^2$ whose arguments are x_i, $i = 0, 1, 2, 3$, then show that

$$f[x_0, x_1, x_2, x_3] = -\frac{\Sigma x_0 x_1 x_2}{x_0^2 x_1^2 x_2^2 x_3^2}$$

10. Show that nth order divided difference can be expressed as the quotient of two determinants each of order $(n + 1)$.
11. Prove that divided difference of any order is a symmetric function of its arguments.
12. Prove that

$$f[x_0, x_0, x_0, x_0] = -\frac{f'''(x_0)}{3!}.$$

13. Show that for equally spaced arguments

$$f[x_0, x_1, \cdots, x_n] = \frac{\Delta^k f_0}{n!h^n}.$$

14. Prove that

$$\frac{d}{dx} f[x, x_0, x_1, \cdots, x_n] = f[x, x, x_0 x_1, \cdots, x_n]$$

15. Express Newton's Interpolation formula with divided difference.
16. Show that the Lagrange's interpolation formula and Newton's interpolation formula with divided difference are equal.

Chapter 5

NON-LINEAR EQUATION

5.1 INTRODUCTION

The first non-linear equation encountered in algebra courses is well known quadratic equation $ax^2 + bx + c = 0$ ($a \neq 0$) and all the students are familiar with the solution of this equation. The formula for the roots of general cubic is somewhat more complicated. Accordingly, except in special cases, we prefer in practice to use a numerical method to solve polynomial equation of degree higher than two. Another class of non-linear equations are transcendental equations, which consist of functions like e^x, $\log x$, $\tan x$, $\sin x$. Useful analytical solutions of such equations are rare, so we are usually compelled to adopt numerical methods.

Procedure: To find the roots of the equation $f(x) = 0$ is one of the most frequently occurring problem in engineering and scientific work, where $f(x)$ may be algebraic or transcendental and for which algebraic method are not applicable. In rare cases, one can obtain only approximate solutions which mean that a point $x = x^*$ for which $|f(x^*)| < \varepsilon$, ε being a small number. An approximate solution has always an error due to round off and though the exact solution of an equation is unique, there may be several approximate solutions, which are valid.

In the next section we will discuss various iterative methods for finding the approximate roots of an equation.

5.2 ITERATIVE METHODS

Iterative methods start with an initial approximation and which, by applying a suitably chosen algorithm, lead to successively better approximations. Solving an equation by iterative method we can only hope an approximate solution (if the method converges). Different iterative methods have different order of convergence. The simplicity and uniformity of the operations to be performed is the advantage of iterative method. Let $\{x_n\}$ be a sequence converging to the root x^* of the equation $f(x) = 0$ produced by a given iterative method, the error ε_n at nth step of the calculation is defined by $\varepsilon_n = x^* - x_n$. Then the iterative method converges if $\varepsilon_n \to 0$ as $n \to \infty$, otherwise the method diverges. If an iterative method converges and two constants $p \geq 1$ and $C > 0$ exist such that

$$|\varepsilon_{n+1}| = C\,|\varepsilon_n^p| \text{ as } n \to \infty,$$

then p is called the order of convergence. The convergence of an iterative method depends on initial approximation of the root. An iterative method diverges if the initial estimate is not good, in such cases we have to start with another fresh approximation.

5.3 METHOD OF TABULATION

This method is used for finding a rough location of roots and is based on the fact that if $f(x)$ is continuous in $[a, b]$ such that $f(a)\,f(b) < 0$, then there exists at least one root of the equation $f(x) = 0$ in the interval $[a, b]$. So, we choose an interval $[a, b]$ containing a root of the equation $f(x) = 0$ for which $f'(x)$ must have same sign in $[a, b]$. The values of $f(x)$ are then tabulated in this interval with a suitable size of the of the step length, say h. It is to be mentioned that as we are interested in signs of the function value apart from the values of function at the end points of the subinterval, say $[a_1, b_1]$, in which sign of the function value changes, its values in the table need not be calculated accurately. Now consider the subinterval $[a_1, b_1]$ only, which is of length h and calculate the values of the function using a smaller value of the step length, say $h/10$. By noting a change in the values of $f(x)$ we get an interval of length $h/10$ containing the root of $f(x) = 0$. As the length of the intervals containing the root is being decreased step by step, repeated application of this method gives the approximate value of the root. As the method is laborious and needs more computational work, we can apply this method two or three times to get a rough approximation of the root which can serve the purpose of selecting an initial approximation of any iterative method.

Scheme of computation: Let us choose a, b for the continuous function $f(x)$ such that $f(a)\,f(b) < 0$ and a suitable step length h and construct the following table.

TABLE 1

x:	a	$a + h$	$a + 2h$	$\cdots$	$a + (r-1)\,h$	$a + rh$	$\cdots$	$a + nh = b$
$f(x)$:	–ve	–ve	–ve		–ve	+ve		+ve

TABLE 2

x:	a_1	$a_1 + h/10$	$a_1 + 2h/10$	$\cdots$	$a_1 + (r-1)h/10$	$a_1 + rh/10$	$\cdots$	$a_1 + nh = b_1$
$f(x)$:	–ve	–ve	–ve		–ve	+ve		+ve

Next choose the subinterval $[a_1 + (r-1)h/10,\, a_1 + rh/10] = [a_2, b_2]$, say, and repeat the process till the root is obtained to the desired accuracy.

5.4 BISECTION METHOD

The procedure of locating a solution of the equation $f(x) = 0$ in a sequence of intervals of decreasing size obtained by halving it, is known as bisection method. The basic idea behind this method is a property of a continuous function which states that if a function $f(x)$ is continuous in $[a, b]$ and if $f(a)\, f(b) < 0$, then there exists a root of $f(x) = 0$ in $[a, b]$*. Let us take $f(a) < 0$ and $f(b) > 0$. If $f(x_0) = 0$, where $x_0 = (a + b)/2$, we say that x_0 is root of $f(x) = 0$. Otherwise, the root lies either in (a, x_0) or in (x_0, b) according as $f(x_0)$ is positive or negative. Then as before, we bisect the interval and repeat the process until the root is known to be of desired accuracy.

Algorithm of bisection method

Given a function $f(x)$ continuous on $[a_0, b_0]$ such that $f(a_0)f(b_0) < 0$. For $n = 0, 1, 2, \ldots$ until satisfied, calculate

$$m = \frac{a_n + b_n}{2}.$$

If $f(a_n) \,.\, f(m) < 0$, set $a_{n+1} = a_n$, $b_n = m$, otherwise $a_{n+1} = m$, $b_{n+1} = b_n$. Then $f(x) = 0$ has root in the interval $[a_{n+1}, b_{n+1}]$.

* The condition is necessary only as there are functions which have roots in $[a, b]$ but which do not satisfy the condition $f(a)\, f(b) < 0$, e.g., $f(x) = x^2 = 0$ has a root in $[-1, 1]$ but $f(-1)\, f(1) > 0$.

Fig.5.1 Bisection method

Advantage and disadvantage of bisection method

The bisection method is suitable for automatic computation and is almost certain to give a root, provided the conditions of the theorem mentioned therein hold. It can only fail if the accumulated error in the calculation of $f(x)$ at a bisection point gives a small negative value when actually it should have a small positive value (or vice versa); the interval subsequently chosen would therefore be wrong. This can be avoided by working to sufficient accuracy, and this almost assured convergence is not true for many other methods of finding a root. Other drawback of this method is that it can be applied to determine only simple root. Finally, we note that this method is too slow. After nth step the interval containing the root is of length $|b-a|/2^n$ showing that a large number of calculations are required to get a root of desired accuracy.

Scheme of computation:

n	$a_n[f(a_n)<0]$	$b_n[f(b_n)>0]$	$x_{n+1}=(a_n+b_n)/2$	$f(x_{n+1})$
0	a_0	b_0	$x_1=(a_0+b_0)/2$	$f(x_1)<0$
1	$a_1=x_1$	$b_1=b_0$	$x_2=(a_1+b_1)/2$	$f(x_2)>0$
2	$a_2=a_1$	$b_2=x_2$	$x_3=(a_2+b_2)/2$	$f(x_3)>0$
3	$a_3=a_2$	$b_3=x_3$	$x_4=(a_3+b_3)/2$	$f(x_4)<0$
4	$a_4=x_4$	$b_4=b_3$	$x_5=(a_4+b_4)/2$	

5.5 REGULA FALSI METHOD

In this method, we choose two points a_0, b_0 such that $f(a_0)$ is negative and $f(b_0)$ is positive. This means that the continuous curve $y=f(x)$ crosses the X-axis between these two points, so a root must be in between these points. Now the equation of the chord joining $[a_0, f(a_0)]$ and $[b_0, f(b_0)]$ is

$$\frac{y-f(a_0)}{f(b_0)-f(a_0)}=\frac{x-a_0}{b_0-a_0}.$$

It cuts the X-axis at $(x_0, 0)$ where

$$x_0=\frac{a_0 f(b_0)-b_0 f(a_0)}{f(b_0)-f(a_0)}.$$

As in the bisection method, there are three possibilities:

(1) $f(x_0)=0$, in this case x_0 is a root of the equation $f(x)=0$.

(2) $f(x_0)<0$, in this case the root of the equation $f(x)=0$ lies in the interval $[x_0, b_0]$.

(3) $f(x_0)>0$, in this case the root of the equation $f(x)=0$ lies in the interval $[a_0, x_0]$.

If the case (1) occurs, the process terminates and for other cases the process is to be repeated until the root is obtained to the desired accuracy.

Algorithm of regula falsi method

Given a function continuous in $[a_0, b_0]$ such that $f(b_0)\ f(a_0)<0$. For $n=0, 1, 2, \cdots$ until satisfied calculate

$$x^*=\frac{a_n f(b_n)-b_n f(a_n)}{f(b_n)-f(a_n)}.$$

If $f(a_n)f(x^*)<0$, set $a_{n+1}=a_n$, $b_{n+1}=x^*$ otherwise $a_{n+1}=x^*$, $b_{n+1}=b_n$. Then $f(x)=0$ has root in the interval $[a_{n+1}, b_{n+1}]$.

Geometrical significance of regula falsi method

If we join two points on the curve, say, $\{x_0, f(x_0)\}$ and $\{x_1, f(x_1)\}$, such that $f(x_0)\,f(x_1) < 0$, then the next iteration will be the point where the chord joining those points cuts the X-axis.

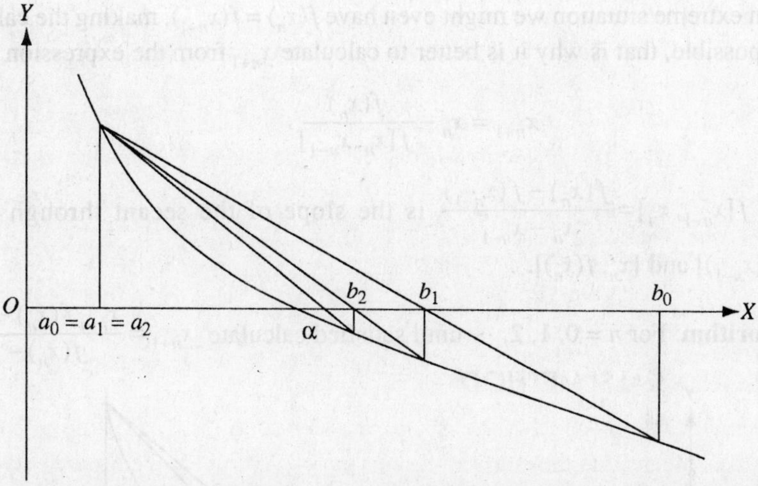

Fig.5.2 Regula falsi method

Advantage and disadvantage of regula falsi method

Like bisection method, the regula falsi method has almost assured convergence and it usually converges to a root faster. However, it may happen that most or all of the calculated values x^* are on the same side of the root, in which case convergence may be slow. This is avoided in the next method, which is a modification of regula falsi method and in this method no attempt is made to ensure that the root is enclosed.

Scheme of computation

n	a	b	$f(a)\,(>0)$	$f(b)\,(<0)$	$x = \dfrac{af(b) - bf(a)}{f(b) - f(a)}$	$f(x)$
0	a_0	b_0	$f(a_0)\,(>0)$	$f(b_0)\,(<0)$	$x = \dfrac{a_0 f(b_0) - b_0 f(a_0)}{f(b_0) - f(a_0)}$	$f(x) > 0$
1	$a_1 = x$	$b_1 = b_0$	$f(a_1)\,(>0)$	$f(b_1)\,(<0)$	$x = \dfrac{a_1 f(b_1) - b_1 f(a_1)}{f(b_1) - f(a_1)}$	$f(x) < 0$
2	$a_2 = a_1$	$b_2 = x$	$f(a_2)\,(>0)$	$f(b_2)\,(<0)$	$x = \dfrac{a_2 f(b_2) - b_2 f(a_2)}{f(b_2) - f(a_2)}$	$f(x) < 0$
3	$a_3 = a_2$	$b_3 = x$	$f(a_3)\,(>0)$	$f(b_3)\,(<0)$	$x = \dfrac{a_3 f(b_3) - b_3 f(a_3)}{f(b_3) - f(a_3)}$	$f(x) > 0$

5.6 SECANT METHOD

In this method we start with two points x_{-1} and x_0 where $f(x_{-1})$ and $f(x_0)$ need not be of different sign. We calculate x_{n+1}, $n = 0, 1, 2, \cdots$ using the formula

$$x_{n+1} = \frac{x_{n-1}f(x_n) - x_n f(x_{n-1})}{f(x_n) - f(x_{n-1})}.$$

In an extreme situation we might even have $f(x_n) = f(x_{n-1})$, making the calculation of x_{n+1} impossible, that is why it is better to calculate x_{n+1} from the expression

$$x_{n+1} = x_n - \frac{f(x_n)}{f[x_n, x_{n-1}]},$$

where $f[x_{n-1}, x_n] = \dfrac{f(x_n) - f(x_{n-1})}{x_n - x_{n-1}}$ is the slope of the secant through the points $[x_{n-1}, f(x_{n-1})]$ and $[x_n, f(x_n)]$.

Algorithm: For $n = 0, 1, 2, \cdots$ until satisfied calculate $x_{n+1} = \dfrac{x_{n-1}f(x_n) - x_n f(x_{n-1})}{f(x_n) - f(x_{n-1})}$.

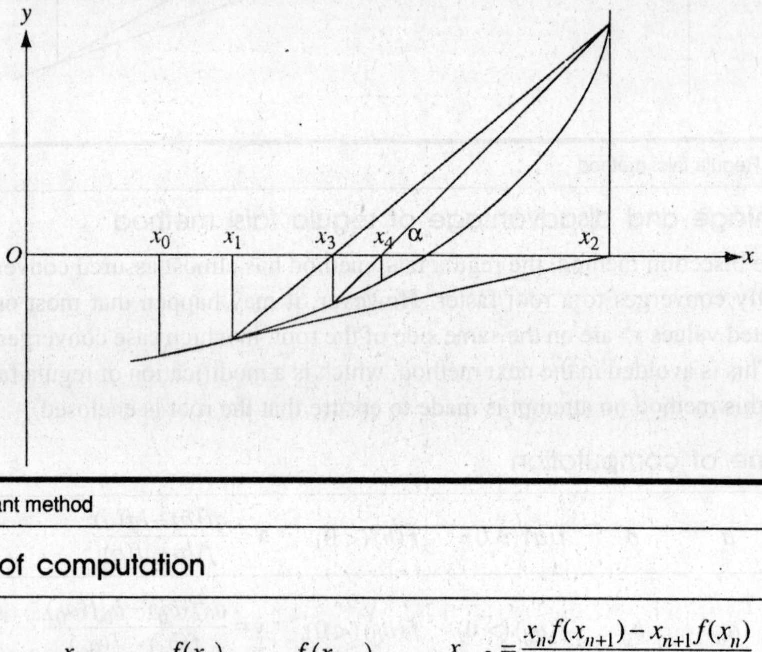

Fig.5.3 Secant method

Scheme of computation

n	x_n	x_{n+1}	$f(x_n)$	$f(x_{n+1})$	$x_{n+2} = \dfrac{x_n f(x_{n+1}) - x_{n+1} f(x_n)}{f(x_{n+1}) - f(x_n)}$
0	x_0	x_1	$f(x_0)$	$f(x_1)$	$x_2 = \dfrac{x_0 f(x_1) - x_1 f(x_0)}{f(x_1) - f(x_0)}$
1	x_1	x_2	$f(x_1)$	$f(x_2)$	$x_3 = \dfrac{x_1 f(x_2) - x_2 f(x_1)}{f(x_2) - f(x_1)}$
2	x_2	x_3	$f(x_2)$	$f(x_3)$	$x_4 = \dfrac{x_2 f(x_3) - x_3 f(x_2)}{f(x_3) - f(x_2)}$

Advantage and disadvantage of secant method

Secant method needs one evaluation per step and depends on the value of the function at different points which are easy to calculate. This method converges (order of convergence is 1.617) more rapidly than the usual fixed point iteration (order of convergence is 1) but less rapidly than Newton's method (order of convergence is 2).

5.7 METHOD OF ITERATION

In this method we rewrite the equation $f(x) = 0$ in the form $x = g(x)$, where the function $g(x)$ is called iteration function, suitable for the construction of a sequence of approximations (under some conditions) to some root in repetitive fashion. In some way we obtain a rough approximation x_0 of the desired root, which then may be substituted into the right hand side to give a new approximation, $x_1 = g(x_0)$. This new approximation when again substituted into the right hand side yield $x_2 = g(x_1)$, and so on until a sufficiently accuurate approximation is obtained. This repetitive process, based on $x_{n+1} = g(x_n)$, $n = 0, 1, 2, \cdots$ what is called method of iteration, provided that $|x_{n+1} - x_n|$ decreases as n increases. Finally we get, $\alpha = g(\alpha)$, where α is called fixed point and denotes the root.

Algorithm: Given an iteration function $g(x)$ and an initial approximation x_0. For $n = 0, 1, 2, \cdots$ until satisfied, calculate

$$x_{n+1} = g(x_n).$$

To make the above algorithm clear we must consider the following points:

(a) Whether the sequence $\{x_n\}$ convergent for any initial approximation x_0 or not.

(b) If the sequence convergent to some α, will it be a root?

(c) How $g(x)$ should be chosen so that the sequence $\{x_n\}$ converges to the root?

Answer of the first question follows from the example. If we go to solve the equation $x = e^x$ starting with $x_0 = 1$, we shall get $x_1 = 2.7183$, $x_2 = 15.1543 \cdots$ Showing that the sequence $\{x_n\}$ is divergent. Answer of the second question is positive, since if $\lim x_n = \alpha$, then

$$\begin{aligned} \lim x_{n+1} &= \lim g(x_n) \\ \Rightarrow \lim x_{n+1} &= g(\lim x_n) \\ \Rightarrow \alpha &= g(\alpha). \end{aligned}$$

The following important theorem will give the answer of the third question.

Theorem: If the iteration function $g(x)$ maps the interval $[a, b]$ into itselt and is differentiable there and further, if there exists a non-negative constant $k < 1$ such that for all $x \in [a, b]$, $|g'(x)| \le k$, then $g(x)$ has exactly one fixed point α in $[a, b]$ and the sequence $\{x_n\}$ converges to α.

Proof: Since α is a root of the equation $x = g(x)$, we have $\alpha = g(\alpha)$. Also from the algorithm of method of iteration $x_{n+1} = g(x_n)$, if we denote the error in the nth step by e_n, then we have

$$x_{n+1} - \alpha = g(x_n) - g(\alpha) \Rightarrow x_{n+1} - \alpha = g'(\eta)(x_n - \alpha),$$

by first mean value theorem, where $\eta \in (x_n, \alpha)$, which implies $$e_{n+1} = g'(\eta) e_n \Rightarrow |e_{n+1}| \le k |e_n|.$$

Proceeding in this way, we get $|e_{n+1}| \le k^{n+1} |e_0|$.

As $k < 1$, we have $\lim k^{n+1} = 0$, therefore, $\lim |e_{n+1}| = \lim k^{n+1} |e_0| = 0$, whatever the initial error may be. Which shows that the sequence $\{x_n\}$ converges to the root. Further if possible let the sequence $\{x_n\}$ has two limits, say α and β.

i.e., $\alpha = g(\alpha)$ and $\beta = g(\beta)$,

which give that $$|\alpha - \beta| = |g(\alpha) - g(\beta)|$$
$$\Rightarrow |\alpha - \beta| = |g'(\eta)| \, |\alpha - \beta|,$$

as $|g'(\eta)| < 1$, only possibility is that $\alpha = \beta$.

Scheme of computation

n	x_n	$x_{n+1} = g(x_n)$	$\lvert x_{n+1} - x_n \rvert$
0	x_0	$x_1 = g(x_0)$	$\lvert x_1 - x_0 \rvert$
1	x_1	$x_2 = g(x_1)$	$\lvert x_2 - x_1 \rvert$
2	x_2	$x_3 = g(x_2)$	$\lvert x_3 - x_2 \rvert$

From the result $e_{n+1} = g'(\eta)\, e_n$, it is clear that the methods of iteration is linearly convergent.

Advantage and disadvantage of method of iteration

If the iteration function $g(x)$ satisfies, $|g'(x)| \le k < 1$ for all $x \in [a, b]$, then we know that $g(x)$ has exactly one fixed point in $[a, b]$ and the sequence $\{x_n\}$ obtained by using $x_{n+1} = g(x_n)$, converges to α which shows that the method converges surely but the result $e_{n+1} = g'(\eta)\, e_n$, shows that the method is linearly convergent, i.e., a large number of iterations are to be performed to get desired accuracy.

This slow rate of convergence can be accelerated by using Aitken's Δ^2-method which is described below.

5.8 AITKEN'S Δ^2-METHOD

Let x_{i-1}, x_i, x_{i+1} be three successive approximations to the desired root $x = \alpha$ of the equation $x = g(x)$, then we have

$$x_{i+1} - \alpha = g'(\eta)(x_i - \alpha)$$

and $$x_i - \alpha = g'(\eta)(x_{i-1} - \alpha).$$

Dividing, we get $$\frac{x_{i+1} - \alpha}{x_i - \alpha} = \frac{x_i - \alpha}{x_{i-1} - \alpha}.$$

On simplification we get $$\alpha = x_{i+1} - \frac{(x_i - x_{i-1})^2}{x_{i+1} - 2x_i + x_{i-1}}.$$

Using the result,

$$\Delta x_i = x_{i+1} - x_i \text{ and } \Delta^2 x_{i-1} = \Delta(x_i - x_{i-1}) = \Delta x_i - \Delta x_{i-1} = x_{i+1} - 2x_i + x_{i-1},$$

$$\alpha = x_{i+1} - \frac{(\Delta x_i)^2}{\Delta^2 x_{i-1}},$$

which is called Aitken's Δ^2-method. This is the process, which can be applied to any linearly convergent sequence to get a faster converging sequence.

Algorithm: Given a sequence $\{x_n\}$ converges to α. Calculate the sequence $\{x_n^*\}$ using

$$x_n^* = x_{n+1} - \frac{(\Delta x_n)^2}{\Delta^2 x_{n-1}}.$$

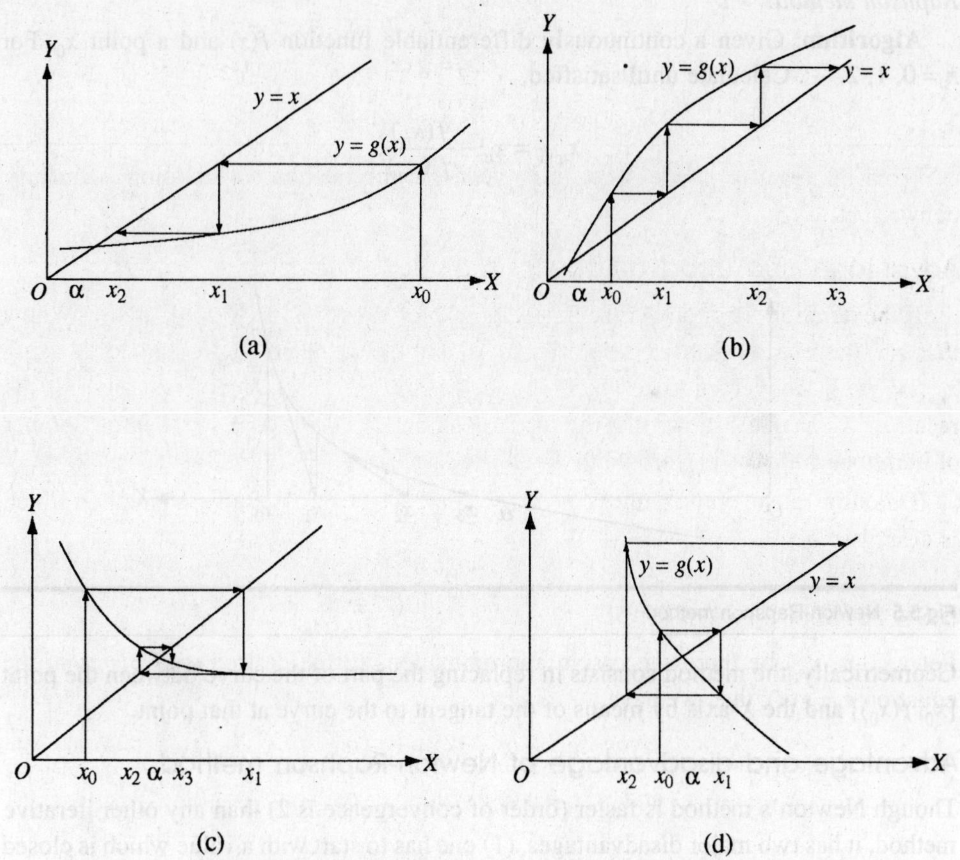

Convergent : (a) $0 < g'(x) < 1$; (c) $-1 < g'(x) < 0$

Divergent : (b) $g'(x) > 1$; (d) $g'(x) < -1$

Fig.5.4 The method of iteration

5.9 NEWTON-RAPHSON METHOD

This method is generally used to improve the result obtained by one of the previous methods and can be viewed as a limiting case of the secant method or a special case of method of iteration. Let α, x_n be the exact and approximate values of a root of the equation $f(x) = 0$, where $\alpha = x_n + h$ (say). Then

$$0 = f(\alpha) = f(x_n + h) = f(x_n) + hf'(x_n) + \frac{1}{2}h^2 f''(x_n) + \cdots$$

If all the terms of 2nd and higher degree in h are neglected, we get

$$0 = f(x_n) + hf'(x_n) \rightarrow h = -\frac{f(x_n)}{f'(x_n)}.$$

So, $x_{n+1} = x_n - \dfrac{f(x_n)}{f'(x_n)}$ should be a better estimate of the root. Even better approximation may be obtained by repetition of the above process, which is known as *Newton-Raphson method.*

Algorithm: Given a continuously differentiable function $f(x)$ and a point x_0. For $n = 0, 1, 2, \cdots$. Calculate until satisfied,

$$x_{n+1} = x_n - \frac{f(x_n)}{f'(x_n)}.$$

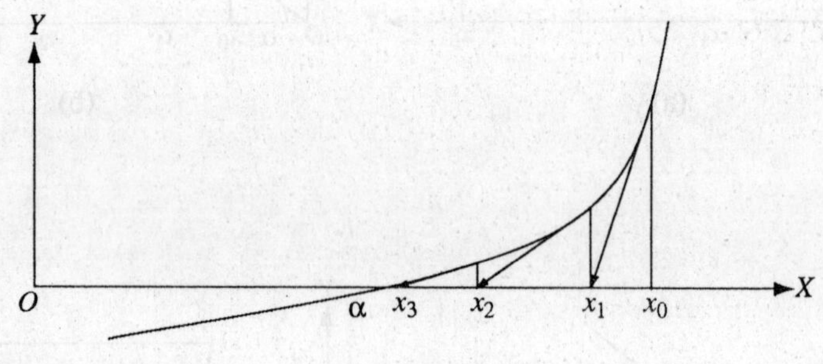

Fig.5.5 Newton-Rapshon method

Geometrically, the method consists in replacing the part of the curve between the point $[x_0, f(x_0)]$ and the X-axis by means of the tangent to the curve at that point.

Advantage and disadvantage of Newton-Raphson method

Though Newton's method is faster (order of convergence is 2) than any other iterative method, it has two major disadvantages, (1) one has to start with a value which is closed to the root of $f(x) = 0$. Since one usually does not know this value, this method might be difficult to do in practice, unless one has already obtained a good estimate of the root by some other method, (2) in Newton's method it is necessary to evaluate two values per step, out of which one is $f'(x)$ and which may not be available explicitly and in some cases it needs considerable calculation.

Scheme of computation

n	x_n	$f(x_n)$	$f'(x_n)$	$h_n = -\dfrac{f(x_n)}{f'(x_n)}$	$x_{n+1} = x_n + h_n$
0	x_0	$f(x_0)$	$f'(x_0)$	$h_0 = -\dfrac{f(x_0)}{f'(x_0)}$	$x_1 = x_0 + h_0$
1	x_1	$f(x_1)$	$f'(x_1)$	$h_1 = -\dfrac{f(x_1)}{f'(x_1)}$	$x_2 = x_1 + h_1$
2	x_2	$f(x_2)$	$f'(x_2)$	$h_2 = -\dfrac{f(x_2)}{f'(x_2)}$	$x_3 = x_2 + h_2$

5.10 CONDITION OF CONVERGENCE OF NEWTON-RAPHSON METHOD

If we write $g(x) = x - \dfrac{f(x)}{f'(x)}$, then Newton-Raphson iteration formula can be written as $x_{n+1} = g(x_n)$ and this method converges if $|g'(x)| < 1$ near the root. In this case,

$$g'(x) = 1 - \frac{[f'(x)]^2 - f(x)\,f''(x)}{[f'(x)]^2} = \frac{f(x)\,f''(x)}{[f'(x)]^2}.$$

So the condition of convergence is $f(x)f''(x) < [f'(x)]^2$.

5.11 ORDER OF CONVERGENCE OF SECANT METHOD AND NEWTON-RAPHSON METHOD

Secant algorithm for determining the root of a non-linear equation $f(x) = 0$ is: given a function $f(x) = 0$ with two approximations x_0 and x_1. Calculate, for $n = 0, 1, 2, \cdots$

$$x_{n+1} = \frac{x_{n-1}f(x_n) - x_n f(x_{n-1})}{f(x_n) - f(x_{n-1})}.$$

If e_n denotes the error in the nth step, then

$$e_{n+1} = x - x_{n+1} = x - \frac{x_{n-1}f(x_n) - x_n f(x_{n-1})}{f(x_n) - f(x_{n-1})}$$

$$= \frac{(x - x_{n-1})f(x_n) - (x - x_n)f(x_{n-1})}{f(x_n) - f(x_{n-1})}$$

$$= \frac{e_{n-1}f(x - e_n) - e_n f(x - e_{n-1})}{f(x - e_n) - f(x - e_{n-1})}$$

$$= \frac{e_{n-1}\left[f(x) - e_n f'(x) + \frac{e_n^2}{2} f''(x) - \cdots\right] - e_n\left[f(x) - e_{n-1}f'(x) + \frac{e_{n-1}^2}{2} f''(x) - \cdots\right]}{\left[f(x) - e_n f'(x) + \frac{e_n^2}{2} f''(x) - \cdots\right] - \left[f(x) - e_{n-1}f'(x) + \frac{e_{n-1}^2}{2} f''(x) - \cdots\right]}$$

$$= \frac{(e_{n-1} - e_n)\, f(x) + \frac{1}{2}(e_{n-1}e_n^2 - e_n e_{n-1}^2)\, f''(x)}{(e_{n-1} - e_n)\, f'(x)} \quad \text{[when } e_n \text{ is sufficiently small]}$$

$$\approx -e_{n-1}\, e_n \frac{f''(x)}{2f'(x)} \approx e_{n-1}e_n.$$

If the order of convergence be k, then $e_n = e^{\,k}_{n-1}$ and $e_{n+1} = e_n^k$. So $e_{n+1} \approx e_{n-1}\, e_n$ gives $k^2 = k + 1 \Rightarrow k = (1 + \sqrt{5})/2 = 1.618$, which shows that secant method is faster than linear but slower than quadratic.

Newton-Raphson algorithm gives $x_{n+1} = x_n - \dfrac{f(x_n)}{f'(x_n)}$. If e_n denotes the error in the nth step, then

$$e_{n+1} = x - x_{n+1} = x - x_n + \frac{f(x_n)}{f'(x_n)} = e_n + \frac{f(x - e_n)}{f'(x - e_n)}$$

$$= e_n + \frac{f(x) - e_n f'(x) + \frac{e_n^2}{2} f''(x) - \cdots}{f'(x) - e_n f''(x) + \frac{e_n^2}{2} f'''(x) - \cdots}$$

$$= e_n - e_n \frac{f'(x)\left[1 - \dfrac{e_n f''(x)}{2f'(x)}\right]}{f'(x)\left[1 - \dfrac{e_n f''(x)}{f'(x)}\right]} \quad \text{[when } e_n \text{ is sufficiently small]}$$

$$= e_n - e_n \left[1 - \frac{e_n f''(x)}{2f'(x)}\right]\left[1 - \frac{e_n f''(x)}{f'(x)}\right]^{-1}$$

$$= e_n - e_n \left[1 - \frac{e_n f''(x)}{2f'(x)}\right]\left[1 + \frac{e_n f''(x)}{f'(x)}\right] = e_n - e_n \left[1 - e_n\left(\frac{f''(x)}{2f'(x)} - \frac{f''(x)}{f'(x)}\right)\right]$$

$$= e_n - e_n - e_n^2 \frac{f''(x)}{2f'(x)} = -e_n^2 \frac{f''(x)}{2f'(x)}.$$

This result shows that Newton-Raphson method is quadrically convergent.

Note: Different iterative methods of solving non-linear equations have been discussed in the above sections out of which the bisection and regula falsi method are known as root bracketing methods as in these methods we always work with a bracket containing root. The regula falsi method is modified to obtain secant method which may give up the bracketing of the root and which is more rapidly convergent than the aforesaid methods. If $f(x)$ is differentiable, the slope of the secant used in secant method can be replaced by the slope of the tangent to get a new faster method, namely Newton's method. The methods of iteration and Aitken's Δ^2-method have also been presented in the above sections. The Aitken's Δ^2-method is used to improve the result obtained by the method of iteration. Though Newton's method is faster than any other method presented in the above sections, it has two major disadvantages, (1) one has to start with a value which is closed to the root of $f(x) = 0$. Since one usually does not know this value, this method might be difficult to do in practice, unless one has already obtained a good estimate of the root by some other method, (2) in Newton's method it is necessary to evaluate two values per step, out of which one is $f'(x)$ and which may not be available explicitly and in some cases it needs considerable calculation. As compared to Newton's method, secant method needs one evaluation per step and this method converges (order of convergence is 1.617) more rapidly than the usual fixed point iteration (order of convergence is 1) but less rapidly than Newton's method (order of convergence is 2). In fact, no method can be said to be superior to another method. The bisection and regula falsi methods are slow in convergence but these are certain to converge, while Newton's may diverge unless the initial approximation is good enough.

Examples

1. Compute the positive root of the equation $x^3–x–1 = 0$ by the method of tabulation correct to two decimal places.

Solution: Let $f(x) = x^3 - x - 1$, then $f(1) = -1$ and $f(2) = 5$. Further $f'(x) = 3x^2 - 1 > 0$ for all x in [1, 2]. Hence $f(x) = 0$ has only one positive root in the interval [1, 2]. Now we tabulate the values of $f(x)$ for consecutive values of x from 1 to 2 with step length 0.1.

TABLE 1

x:	1.0	1.1	1.2	1.3	1.4	1.5	1.6	1.7	1.8	1.9	2.0
$f(x)$:	–1.0	–0.77	–0.47	–0.10	0.34	0.88	1.50	2.20	3.03	3.96	5.00

From this table we see that $f(1.3)\, f(1.4) < 0$. So we consider the subinterval [1.3,1.4] and tabulate the values of $f(x)$ for consecutive values of x from 1.3 to 1.4 with step length 0.01.

TABLE 2

x:	1.30	1.31	1.32	1.33	1.34	1.35	1.36	1.37	1.38	1.39	1.40
$f(x)$:	–1.0	–0.06	–0.02	0.02	0.07	0.11	0.16	0.20	0.25	0.30	0.34

The above table shows that $f(1.32)\, f(1.33) < 0$. So we consider the subinterval [1.32, 1.33] and tabulate the values of $f(x)$ for consecutive values of x from 1.32 to 1.33 with step length 0.001.

TABLE 3

x:	1.320	1.321	1.322	1.323	1.324	1.325	1.326	1.327	1.328	1.329	1.330
$f(x)$:	–0.02	–0.016	–0.012	–0.007	–0.003	0.001	0.005	0.010	0.014	0.018	0.020

From this table we see that the positive root of $f(x) = 0$ lies in the interval (1.324, 1.325). So $x = 1.32$ is a positive root of the given equation correct to two decimal places.

2. Find a real root of the equation $x^3 - 4x - 9 = 0$ correct up to four significant digits using bisection method.

Solution: Let $f(x) = x^3 - 4x - 9$, then we see that $f(x)$ has only one variation in sign which indicates that $f(x) = 0$ has exactly one +ve root. Now $f(2) = -9 < 0$ and $f(3) = 6 > 0$, so in the interval [2, 3], $f(x) = 0$ has one root.

Scheme of computation

n	$a_n[f(a_n) < 0]$	$b_n[f(b_n) > 0]$	$x_{n+1} = (a_n + b_n)/2$	$f(x_{n+1})$
1	2.000000	3.000000	2.500000	–3.375000
2	2.500000	3.000000	2.750000	0.796875
3	2.500000	2.750000	2.625000	–1.412109
4	2.625000	2.750000	2.687500	–0.339111
5	2.687500	2.750000	2.718750	0.220917
6	2.687500	2.718750	2.703125	–0.061077
7	2.703125	2.718750	2.710938	0.079423
8	2.703125	2.710938	2.707031	0.009049
9	2.703125	2.707031	2.705078	–0.026045
10	2.705078	2.707031	2.706055	–0.008505
11	2.706055	2.707031	2.706543	0.000270
12	2.706055	2.706543	2.706299	–0.004118

Here we see that $|x_{12} - x_{11}| < 0.0005$, so the required root corrected to four significant figures is 2.706.

3. Find a real root of the equation $x = \ln 2(x + 1)$ correct up to four significant digits using bisection method.

Solution: Let $f(x) = x - \ln 2(x + 1)$, then $f(1) = -0.39 < 0$ and $f(2) = 0.21 > 0$, so in the interval [1, 2], $f(x) = 0$ has one root.

Scheme of computation

n	$a_n[f(a_n) < 0]$	$b_n[f(b_n) > 0]$	$x_{n+1} = (a_n + b_n)/2$	$f(x_{n+1})$
1	1.000000	2.000000	1.500000	–0.1094379
2	1.500000	2.000000	1.750000	0.0452519
3	1.500000	1.750000	1.625000	–0.0332281
4	1.625000	1.750000	1.687500	0.0057414
5	1.625000	1.687500	1.656250	–0.0138125
6	1.656250	1.687500	1.671875	–0.0040527
7	1.671875	1.687500	1.679688	0.0008401
8	1.671875	1.679688	1.675781	–0.0016073
9	1.675781	1.679699	1.677734	–0.0003839
10	1.677734	1.679688	1.678711	0.0002281
11	1.6777734	1.678711	1.678223	–0.0000779
12	1.678223	1.678711	1.678467	0.0000751

Here we see that $|x_{12} - x_{11}| < 0.0005$, so the required root corrected to four significant figures is 1.678.

4. Find a real root of the equation $x^3 - x - 4 = 0$ correct up to four significant digits using bisection method.

Solution: Let $f(x) = x^3 - x - 4$, then $f(1) = -4 < 0$ and $f(2) = 2 > 0$, so in the interval [1, 2], $f(x) = 0$ has one root.

Scheme of computation

n	$a_n[f(a_n) < 0]$	$b_n[f(b_n) > 0]$	$x_{n+1} = (a_n + b_n)/2$	$f(x_{n+1})$
1	1.000000	2.000000	1.500000	–2.125000
2	1.500000	2.000000	1.750000	–0.390625
3	1.750000	2.000000	1.875000	0.716797
4	1.750000	1.875000	1.812500	0.141846
5	1.750000	1.812500	1.781250	–0.129608
6	1.781250	1.812500	1.796875	0.004803
7	1.781250	1.796875	1.789063	–0.062730
8	1.789063	1.796875	1.792969	–0.029046
9	1.792969	1.796875	1.794922	–0.012142
10	1.794922	1.796875	1.795898	–0.003675
11	1.795898	1.796875	1.796387	0.000563
12	1.795898	1.796387	1.796143	–0.001556

Here we see that $|x_{12} - x_{11}| < 0.0005$, so the required root corrected to four significant figures is 1.796.

5. Find a real root of the equation $x^3 - 2x - 5 = 0$ by regula falsi method, corrected to 4D.

Solution: Let $f(x) = x^3 - 2x - 5$, then we observe that $f(2) = -1 < 0$ and $f(3) = 16 > 0$ and hence $f(x) = 0$ has a root in the interval [2, 3]

Scheme of computation

n	a	b	$f(a)$	$f(b)$	$x = \dfrac{af(b) - bf(a)}{f(b) - f(a)}$	$f(x)$
1	2.000000	3.000000	–1.000000	16.000000	2.058824	–0.390799
2	2.058824	3.000000	–0.390800	16.000000	2.081264	–0.147203
3	2.081264	3.000000	–0.147203	16.000000	2.089639	–0.054677
4	2.089639	3.000000	–0.054677	16.000000	2.092740	–0.020203
5	2.092740	3.000000	–0.020203	16.000000	2.093884	–0.007450
6	2.093884	3.000000	–0.007450	16.000000	2.094306	–0.002745
7	2.094306	3.000000	–0.002745	16.000000	2.094461	–0.001010
8	2.094461	3.000000	–0.001010	16.000000	2.094518	–0.000372
9	2.094518	3.000000	–0.000372	16.000000	2.094539	–0.000137

Hence the required root corrected to four decimal places is 2.0945.

6. Find a real root of the equation $x^2 + \ln x = 0$ by regula falsi method, corrected to 4D.

Solution: Let $f(x) = x^2 + \ln x$, then we observe that $f(0.6) = -0.151 < 0$ and $f(1) = 1 > 0$ and hence $f(x) = 0$ has a root in the interval [0.6, 1]

Scheme of computation

n	a	b	$f(a)$	$f(b)$	$x=\dfrac{af(b)-bf(a)}{f(b)-f(a)}$	$f(x)$
1	0.600000	1.000000	–0.150826	1.000000	0.652423	–0.001405
2	0.652423	1.000000	–0.001405	1.000000	0.652911	–0.000021
3	0.652911	1.000000	–0.000021	1.000000	0.652919	0.000000

Hence the required root corrected to four decimal places is 0.6529.

7. Find a real root of the equation $xe^x - 2 = 0$ by regula falsi method, corrected to 4D.

Solution: Let $f(x) = xe^x - 2$, then we observe that $f(0) = -2 < 0$ and $f(1) = 0.7183 > 0$ and hence $f(x) = 0$ has a root in the interval [0, 1]

Scheme of computation

n	a	b	$f(a)$	$f(b)$	$x=\dfrac{af(b)-bf(a)}{f(b)-f(a)}$	$f(x)$
1	0.000000	1.000000	–2.000000	0.718282	0.735759	–0.464423
2	0.735759	1.000000	–0.464423	0.718282	0.839521	–0.056293
3	0.839521	1.000000	–0.056293	0.718282	0.851184	–0.006171
4	0.851184	1.000000	–0.006171	0.718282	0.852452	–0.000669
5	0.852452	1.000000	–0.000669	0.718282	0.852589	–0.000072
6	0.852589	1.000000	–0.000072	0.718282	0.852604	–0.000008
7	0.852604	1.000000	–0.000008	0.718282	0.852605	–0.000001

Hence the required root corrected to four decimal places is 0.8526.

8. Find a real root of the equation $x - \sin(x) - 0.5 = 0$ correct to 5D by secant method.

Solution: Let $f(x) = x - \sin(x) - 0.5$, then we observe that $f(1) = -0.341471$ and $f(2) = 0.590703$.

Scheme of computation

n	x_n	x_{n+1}	$f(x_n)$	$f(x_{n+1})$	$x_{n+2}=\dfrac{x_{n-1}f(x_n)-x_nf(x_{n-1})}{f(x_n)-f(x_{n-1})}$
1	1.000000	2.000000	–0.341471	0.590703	1.366317
2	2.000000	1.366317	0.590703	–0.112850	1.467960
3	1.366317	1.467960	–0.112850	–0.026757	1.499550
4	1.467960	1.499550	–0.026757	0.002087	1.497264
5	1.499550	1.497264	0.002087	–0.000033	1.497300
6	1.497264	1.497300	–0.000033	0.000000	1.497300

The required is 1.49730 (corrected to 5D).

9. Find a real root of the equation $10^x + x - 4 = 0$, correct to 5D by secant method.

Solution: Let $f(x) = 10^x + x - 4$, then we observe that $f(0.5) = -0.3377$ and $f(0.6) = 0.5811$.

Scheme of computation

n	x_n	x_{n+1}	$f(x_n)$	$f(x_{n+1})$	$x_{n+2} = \dfrac{x_{n-1}f(x_n) - x_n f(x_{n-1})}{f(x_n) - f(x_{n-1})}$
1	0.500000	0.600000	–0.337722	0.581072	0.536757
2	0.600000	0.536757	0.581072	–0.021668	0.539031
3	0.536757	0.539031	–0.021668	–0.001331	0.539182
4	0.539031	0.539182	–0.001331	0.000029	0.539188
5	0.539031	0.539182	–0.001331	0.000029	0.539188

The required is 0.53919 (corrected to 5D).

10. Use method of iteration to find the root of the equation $\cos x - 3x + 1 = 0$, correct to 4D.

Solution: We rewrite the given equation as

$$x = \frac{1 + \cos x}{3} = \varphi(x) \text{ (say)}$$

so that

$$|\varphi'(x)| = \frac{\sin(x)}{3} < 1 \text{ for } x = \frac{\pi}{2}.$$

Scheme of computation

n	x_n	$x_{n+1} = \varphi(x_n)$	$\lvert x_{n+1} - x_n \rvert$
0	1.57079633	0.33333333	1.237463
1	0.33333333	0.64831898	0.314986
2	0.64831898	059903333	0.049286
3	0.59903333	0.60862702	0.009594
4	0.60862702	0.60681126	0.001816
5	0.60681126	0.60715686	0.003456
6	0.60715686	0.60709115	0.000066
7	0.60709115	0.60710364	0.000013
8	0.60710364	0.60710126	0.000002

The required root is 0.6071 (correct to 4D).

11. Use method of iteration to find the root of the equation $10^x + x - 4 = 0$, correct to 5D.

Solution: We rewrite the given equation as

$$x = \log_{10}(4 - x) = \varphi(x) \text{ (say)}$$

so that

$$|\varphi'(x)| = \frac{0.4343}{4 - x} < 1 \text{ for } x = 0.5.$$

Scheme of computation

n	x_n	$x_{n+1} = \varphi(x_n)$	$\lvert x_{n+1} - x_n \rvert$
0	0.500000	0.544068	0.044068
1	0.544068	0.538565	0.005503
2	0.538565	0.539256	0.000691
3	0.539256	0.539168	0.000088
4	0.539168	0.539181	0.000013
5	0.539181	0.539179	0.000002

Hence the required root is 0.53918 (correct to 5D).

12. Use method of iteration to find the root of the equation $xe^x = 1$, correct to 3D.

Solution: We rewrite the given equation as

$$x = e^{-x} = \varphi(x) \text{ (say)}$$

so that

$$|\varphi'(x)| = e^{-x} < 1 \text{ for } x < 1.$$

We start with $x_0 = 1$

Scheme of computation

n	x_n	$x_{n+1} = \varphi(x_n)$	$\lvert x_{n+1} - x_n \rvert$
1	1.0000000	0.3678795	0.6321205
2	0.3678795	0.6922006	0.3243212
3	0.6922006	0.5004735	0.1917271
4	0.5004735	0.6062436	0.1057701
5	0.6062436	0.5453958	0.0608478
6	0.5453958	0.5796123	0.0342165
7	0.5796123	0.5601155	0.0194969
8	0.5601155	0.5711431	0.0110276
9	0.5711431	0.5648794	0.0062637
10	0.5648794	0.5684287	0.0035493
11	0.5684287	0.5664148	0.0020139
12	0.5664148	0.5675566	0.0011418
13	0.5675566	0.5669089	0.0006477
14	0.5669089	0.5672762	0.0003673
15	0.5672762	0.5670679	0.0002083

Hence the required root is 0.567 (correct to 3D).

13. Apply the Aitken's acceleration method to the earlier approximations obtained by method of iteration to find the root of the equation $xe^x = 1$, correct to 4D.

Solution: Here we start with three consecutive approximations, viz., x_0, x_1, x_2 taken from the above table and construct the following difference table where x_n, for $n = 3$, 6, 9, 12, 15 are obtained using the result

$$\alpha = x_{i+1} - \frac{(\Delta x_i)^2}{\Delta^2 x_{i-1}}$$

and the next two iterations are obtained using iteration method.

n	x_n	Δx_n	$\Delta^2 x_n$
0	1		
1	0.36788	–0.63212	
2	**0.69220**	**0.32432**	**0.95644**
3	0.58223		
4	0.55865	–0.02358	
5	**0.57198**	**0.01333**	**0.03691**
6	0.55717		
7	0.56713	–0.00004	
8	**0.56715**	**0.00002**	**0.00006**
9	0.56714		
10	0.56714	0.00000	

Hence the required root is 0.5671 (correct to 4D).

14. Apply the Aitken's acceleration method to the earlier approximations obtained by method of iteration to find the root of the equation $10^x + x - 4 = 0$, correct to 5D.

Solution: Here we start with three consecutive approximations viz, x_0, x_1, x_2 taken from the above table and construct the following difference table where x_n, for $n = 3$, is obtained using the result

$$\alpha = x_{i+1} - \frac{(\Delta x_i)^2}{\Delta^2 x_{i-1}}$$

and the next two iterations are obtained using iteration method.

n	x_n	Δx_n	$\Delta^2 x_n$
0	0.5		
1	0.544068	0.044068	
2	**0.538565**	**–0.005503**	**–0.049571**
3	0.539176		
4	0.539180	0.000004	
5	0.539180	0.000000	

Hence the required root is 0.53918 (correct to 5D).

15. Find the root of $2x - \log_{10}x - 7 = 0$ using Newton's method, correct to 4D.

Solution: Let $f(x) = 2x - \log_{10}x - 7$, then $f(3) = -1.47712 < 0$ and $f(4) = 0.39794$.

So $f'(x) = 2 - \frac{1}{x}\log_{10} e$.

Scheme of computation

n	x_n	$f(x_n)$	$f'(x_n)$	$h_n = -\dfrac{f(x_n)}{f'(x_n)}$	$x_{n+1} = x_n + h_n$
0	3	–1.477121	1.855235	0.7961916	3.796192
1	3.796192	0.013035	1.885596	–0.0069131	3.789279
2	3.789279	0.000001	1.883890	–0.0000002	3.789278

Hence the required root is 3.78928 (correct to 5D).

16. Find the root of $e^x - 3x = 0$ using Newton's method, correct to 5D.

Solution: Let $f(x) = e^x - 3x$, then $f(0) = 1 > 0$ and $f(1) = -0.2817182 < 0$.

Again, $f'(x) = e^x - 3$.

Scheme of computation

n	x_n	$f(x_n)$	$f'(x_n)$	$h_n = -\dfrac{f(x_n)}{f'(x_n)}$	$x_{n+1} = x_n + h_n$
0	0.000000	1.000000	–2.000000	0.500000	0.500000
1	0.500000	0.148721	–1.351279	0.110060	0.610060
2	0.610060	0.010362	–1.159459	0.008937	0.618997
3	0.618997	–0.000074	–1.142936	0.000064	0.619063
4	0.619063	0.000000	–0.142816	0.000000	0.619063

Hence the required root is 0.61906 (correct to 5D).

17. Find the root of $x + \ln x = 2$ using Newton's method, correct to 5D.

Solution: Let $f(x) = x + \ln x - 2$, then $f(1) = -1 < 0$ and $f(2) = 0.69315 > 0$.

Again, $f'(x) = 1 + 1/x$, we start with $x_0 = 1.5$.

Scheme of computation

n	x_n	$f(x_n)$	$f'(x_n)$	$h_n = -\dfrac{f(x_n)}{f'(x_n)}$	$x_{n+1} = x_n + h_n$
0	1.500000	–0.094535	1.666667	0.056721	1.556721
1	1.556721	–0.000697	1.642376	0.000425	1.557146
2	1.557146	0.000000	1.642201	0.000000	1.557146

Hence the required root is 1.55715 (correct to 5D).

EXERCISE

1. Define iterative method and order of convergence of an iterative method.
2. Describe following methods for finding the real root of the function $f(x) = 0$ with their merits and demerits.
   i. Bisection method,
   ii. Regula falsi method,
   iii. Secant method,
   iv. Method of iteration,
   v. Newton-Raphson method.
3. Show that if the iteration function of the equation $f(x) = 0$ is such that $|g'(x)| \leq k < 1$ for all x in [a, b], then the sequence $\{x_n\}$ generated by $x_n = g(x_{n-1})$; $n = 1, 2, 3, \cdots$ converges to the real root of $f(x) = 0$ uniquely for any choice of x_0 in [a, b].
4. What is the condition of convergence of Newton-Rapshon method? Show that this method converges quadrically.
5. Show that the order of convergence of secant method is approximately 1.618.
6. Apply bisection method to solve the following equations, correct to 3D.
   (a) $x^3 - 3x - 5 = 0$, (b) $x^3 - 4x - 9 = 0$, (c) $x^3 - x - 1 = 0$, (d) $x^3 - 9x + 1 = 0$, (e) $3x = 1 + \cos x$, (f) $x + \ln x = 2$, (g) $xe^x = 1$, (h) $e^x = 2$.
7. Apply regula falsi method to solve the following equations, correct to 3D.
   (a) $x^2 + 2x - 2 = 0$, (b) $x + \ln x - 2 = 0$, (c) $\tan x + x = 0$, (d) $x^3 - 8x + 4 = 0$, (e) $3x = 1 + \cos x$, (f) $x = \ln 2(x + 1)$.
8. Apply secant method to solve the following equations, correct to 3D.
   (a) $x^2 + 4\sin x = 0$, (b) $2x + 3\sin x - 5 = 0$, (c) $\tan x + x = 0$, (d) $x^3 - 8x - 4 = 0$, (e) $x = 1/2 + \sin x$, (f) $x = \ln 2(x + 1)$.
9. Apply method of iteration to solve the following equations, correct to 3D.
   (a) $\cos x = \log x$, (b) $e^x = 1 + 2x$, (c) $e^{-x} = 10x$, (d) $x^3 - 8x + 4 = 0$, (e) $10(x - 1) = \sin x$, (f) $\cos x = \ln x$, (g) $\log x = 1/x$, (h) $\tan x = 2e^x$.
10. Apply Newton's method to solve the following equations correct to 3D.
    (a) $x\sin x = 1$, (b) $e^x = 1 + 2x$, (c) $e^{-x} = 10x$, (d) $x = 1/(x + 1)^2$, (e) $\sin x = 1 - x$, (f) $\cos x = x$, (g) $x + \log x = 2$, (h) $e^x \tan x = 1$.
11. Apply the Aitken's acceleration method to the earlier approximations obtained by method of iteration to find the root of the equation $e^x = 1 + 2x$, correct to 5D.
12. Apply the Aitken's acceleration method to the earlier approximations obtained by method of iteration to find the root of the equation $\tan x = 2e^x$ correct to 5D.

ANSWERS

6. (a) 2.279, (b) 2.706, (c) 1.325, (d) 2.943, (e) 0.607, (f) 1.557, (g) 0.567, (h) 0.693,
7. (a) 0.732, (b) 1.557, (c) 2.029, (d) 3.051, (e) 0.607, (f) 1.679.
8. (a) 0, (b) 1.138, (c) 2.029, (d) 2.534, (e) 1.497, (f) 1.679.
9. (a) 1.418, (b) 1.256, (c) 0.091, (d) 3.051, (e) 1.088, (f) 1.303, (g) 2.506, (h) 1.454
10. (a) –0.932, (b) 1.256, (c) 0.091, (d) 0.466, (e) 0.511, (f) 0.739, (g) 1.756, (h) 0.531
11. 1.25643.
12. 1.45457.

Chapter 6

NUMERICAL DIFFERENTIATION

6.1 INTRODUCTION

In analysis, the methods of elementary calculus are employed to obtain the derivatives of a function $f(x)$. If the function is complicated or only the values of a function at a set points are given, it may be necessary to use numerical differentiation which is the method to find the derivatives of a function at a point when a set of functional values at a set of points are given. Formulae for numerical differentiation may easily be obtained by differentiating the interpolation polynomial. The derivatives $f'(x), f''(x), \cdots$ of a function are obtained by differentiating the collocation polynomial $\varphi_n(x)$ one time, two times, $\cdots$ respectively.

The choice of an interpolating polynomial for computing the numerical value of the derivative of the function is the same as in the case of interpolation. For example, for equi-spaced arguments we use (i) Newton's forward difference formula. (ii) Newton's backward difference formula, (iii) Stirling's formula for the values of x closed to beginning, end or middle of the table. On the other hand, we use (i) Newton's divided difference formula, (ii) Lagrange's formula for unequi-spaced arguments as discussed below. So the basic idea of numerical differentiation is: Choose an appropriate interpolating polynomial for the given set of data and also for the point at which the value of the derivative is calculated, and differentiate it as many times as you need.

$$\frac{dy}{dx} = f'(x) \approx \phi'_n(x) = \frac{1}{h}\frac{d\phi_n}{dp} = \frac{1}{h}\left(\Delta y_0 + \left(p - \frac{1}{2}\right)\Delta^2 y_0 + \frac{3p^2 - 6p + 2}{6}\Delta^3 y_0 + \frac{2p^3 - 9p^2 + 11p - 3}{12}\Delta^4 y_0 + \ldots\right)$$

$$\frac{d^2y}{dx^2} \approx \frac{1}{h^2}\frac{d^2\phi_n}{dp^2} = \frac{1}{h^2}\left(\Delta^2 y_0 + (p-1)\,\Delta^3 y_0 + \frac{6p^2-18p+11}{12}\Delta^4 y_0 + \cdots\right)$$

and $$\frac{d^3y}{dx^3} \approx \frac{1}{h^3}\frac{d^3\phi_n}{dp^3} = \frac{1}{h^3}\left(\Delta^3 y_0 + \left(p-\frac{3}{2}\right)\Delta^4 y_0 + \cdots\right)$$ and so on.

Similarly, Newton's backward difference formula (3.3) gives

$$\frac{dy}{dx} = f'(x) \approx \phi_n'(x) = \frac{1}{h}\frac{d\phi_n}{dp} = \frac{1}{h}\left(\nabla y_n + \left(p+\frac{1}{2}\right)\nabla^2 y_n + \frac{3p^2+6p+2}{6}\nabla^3 y_n + \frac{2p^3+9p^2+11p+3}{12}\nabla^4 y_n + \cdots\right)$$

$$\frac{d^2y}{dx^2} \approx \frac{1}{h^2}\frac{d^2\phi_n}{dp^2} = \frac{1}{h^2}\left(\nabla^2 y_n + (p+1)\,\nabla^3 y_n + \frac{6p^2+18p+11}{12}\nabla^4 y_n + \cdots\right)$$

and $$\frac{d^3y}{dx^3} \approx \frac{1}{h^3}\frac{d^3\phi_n}{dp^3} = \frac{1}{h^3}\left(\nabla^3 y_n + \left(p+\frac{3}{2}\right)\nabla^4 y_n + \cdots\right)$$ and so on.

Similarly Stirling's formula given by (3.10) can be differentiated to obtain

$$\frac{dy}{dx} \approx \frac{1}{h}\frac{d\phi_n}{dp} = \frac{1}{h}\left(\frac{1}{2}(\Delta y_{-1} + \Delta y_0) + p\Delta^2 y_{-1} + \frac{3p^2-1}{12}(\Delta^3 y_{-1} + \Delta^3 y_{-2}) + \frac{p(2p^2-1)}{12}\Delta^4 y_{-2} + \cdots\right)$$

$$\frac{d^2y}{dx^2} \approx \frac{1}{h^2}\frac{d^2\phi_n}{dp^2} = \frac{1}{h^2}\left(\Delta^2 y_{-1} + \frac{p}{2}\left(\Delta^3 y_{-1} + \Delta^3 y_{-2}\right) + \frac{6p^2-1}{12}\Delta^4 y_{-2} + \cdots\right)$$

The above formulae of numerical differentiation are based on differentiation of interpolation polynomial of equi-spaced arguments. In case of non equi-spaced arguments we have to use Lagrange's polynomial and Newton's formula with divided difference. From the result (3.16) it is easy to see that

$$\frac{dy}{dx} \approx \left(\frac{1}{x-x_1} + \frac{1}{x-x_2} + \cdots + \frac{1}{x-x_n}\right) l_0(x) y_0$$

$$+\left(\frac{1}{x-x_0} + \frac{1}{x-x_2} + \cdots + \frac{1}{x-x_n}\right) l_1(x) y_1 +$$

....

$$+\left(\frac{1}{x-x_0} + \frac{1}{x-x_1} + \ldots + \frac{1}{x-x_{n-1}}\right) l_n(x) y_n$$

Again, on differentiating the result (3.6), we get

$$\frac{dy}{dx} \approx f[x_0, x_1] + (2x - x_0 - x_1)\, f[x_0, x_1, x_2] + \{3x^2 - 2\,(x_0 + x_1 + x_2)\,x + x_0x_1 + x_1x_2 + x_0x_2\}\, f[x_0, x_1, x_2, x_3] + \cdots$$

Differentiating successively the above results, the higher order derivatives can be obtained.

6.2 ERROR IN NUMERICAL DIFFERENTIATION

As errors occur in calculation in every step, interpolation of a function using collocation polynomial always gives an approximate value of the function. The same is true in cases of derivatives as the basic difficulty is that $f(x) - \varphi(x)$ may be small while $f'(x) - \varphi_n'(x)$, $f''(x) - \varphi_n''(x)$, etc. may be very large. Geometrically, we may have two close curves which may differ considerably in slope, variation in slope, etc. It should also be noted that the formulae of numerical differentiation involve the terms h^{-k}, $k = 1,2,3, \cdots$ which is caused by replacing the tabulated function by means of interpolating polynomial, and is estimated by (3.1), without having any information of $f^{n+1}(\xi)$. The result (3.1) also shows that for small values of h, the truncation error will be tolerable. So form this discussion, it is advisable to check the errors, which occur in numerical differentiation based on interpolation polynomial.

Examples

1. Find $\dfrac{dy}{dx}$ at $x = 7$ and $\dfrac{d^2y}{dx^2}$ at $x = 0$ using the following table:

x	0	2	4	6	8
$f(x)$	15	20	30	40	60

Solution: Here, to find the value of dy/dx at $x = 7$, we shall use Newton's backward formula,

$$\frac{dy}{dx} \approx \frac{1}{h}\frac{d\phi_n}{dp} = \frac{1}{h}\left(\nabla y_n + \left(p+\frac{1}{2}\right)\nabla^2 y_n + \frac{3p^2+6p+2}{6}\nabla^3 y_n + \frac{2p^3+9p^2+11p+3}{12}\nabla^4 y_n + \cdots\right)$$

$$= \frac{1}{h}\left(\Delta y_{n-1} + \left(p+\frac{1}{2}\right)\Delta^2 y_{n-2} + \frac{3p^2+6p+2}{6}\Delta^3 y_{n-3} + \frac{2p^3+9p^2+11p+3}{12}\Delta^4 y_{n-4} + \cdots\right)$$

as the points are equi spaced and $x = 7$. Again, given that $h = 2$, $p = (7 - 8)/2 = -1/2$.

Now we construct the difference table

x	$f(x)$	$\Delta f(x)$	$\Delta f^2(x)$	$\Delta f^3(x)$	$\Delta f^4(x)$
0	15				
		5			
2	20		5		
		10		–5	
4	30		0		**15**
		10		10	
6	40		**10**		
		20			
8	60				

So, by Newton's backward formula at $x = 7$

$$\frac{dy}{dx} = \frac{1}{2}[20 + \{2(-0.5)+1\}\frac{10}{2} + \{3(-0.5)^2 + 6(-0.5)+2\}\frac{10}{6} +$$

$$\{2(-0.5)^3 + 9(0.5)^2 + 11(-0.5) + 3\}\frac{15}{12}]$$

$$= \frac{1}{2}[20 - \frac{5}{12} - \frac{5}{8}] = 9.479167$$

To find d^2y/dx^2 at $x = 0$, we use Newton's forward formula

$$\frac{d^2y}{dx^2} \approx \frac{1}{h^2}\frac{d^2\phi_n}{dp^2} = \frac{1}{h^2}\left(\Delta^2 y_0 + (p-1)\,\Delta^3 y_0 + \frac{6p^2 - 18p + 11}{12}\,\Delta^4 y_0 + \cdots\right),$$

where $p = (0 - 0)/2 = 0$.

$$\frac{d^2y}{dx^2} = \frac{1}{4}[5 + (-1)(-5) + 11\frac{15}{12}] = 5.94.$$

2. The following table gives the values of a function at equal intervals:

x	0.0	0.5	1.0	1.5	2.0
$f(x)$	0.3989	0.3521	0.2420	0.1295	0.0540

Find the values of dy/dx at $x = 1.5$.

Solution: Here as we want to evaluate dy/dx near the end point, so we make use of Newton's backward formula.

Given that $h = 0.5$, taking $x_p = 1.5$, we see $p = 0$.

Now we construct the difference table

x	$f(x)$	$\nabla f(x)$	$\nabla f^2(x)$	$\nabla f^3(x)$	$\nabla f^4(x)$
0.0	0.3989				
		–0.0468			
0.5	0.3521		–0.0633		
		–0.1101		**0.0609**	
1.0	0.2420		**–0.0024**		–0.0215
		–0.1125		0.0394	
1.5	0.1295		0.0370		
		–0.0755			
2.0	0.0540				

So, by Newton's backward formula at $x = 1.5$,

$$\frac{dy}{dx} = 2[-0.1125 - 0.0024/2 + 0.0609/3] = -0.1868.$$

3. Find $\frac{dy}{dx}$ at $x = 0.5$ using the following table:

x	0.1	0.3	0.5	0.7	0.9
$f(x)$	1.5	1.8	2.2	2.6	3.0

Solution: Here, as the points are equi-spaced, to find the value of $\frac{dy}{dx}$ at $x - 0.5$, we shall use Stirling's formula given by

$$\frac{dy}{dx} \approx \frac{1}{h}\frac{d\phi_n}{dp} = \frac{1}{h}\left(\frac{1}{2}(\Delta y_{-1} + \Delta y_0) + p\Delta^2 y_{-1} + \frac{3p^2 - 1}{12}\left(\Delta^3 y_{-1} + \Delta^3 y_{-2}\right) + \frac{p(2p^2 - 1)}{12}\Delta^4 y_{-2} + \cdots\right).$$

Again given that $h = 0.2$, $p = (0.5 - 0.5)/0.2 = 0$.

Now we construct the difference table

x	$f(x)$	$\Delta f(x)$	$\Delta f^2(x)$	$\Delta f^3(x)$	$\Delta f^4(x)$
0.1	1.5				
		0.3			
0.3	1.8		0.1		
		0.4		**−0.1**	
0.5	2.2		0.0		0.1
		0.4		**0.0**	
0.7	2.6		0.0		
		0.4			
0.9	3.0				

Hence using the above formulae, we get, at $x = 0.5$,

$$\frac{dy}{dx} = \left[(0.4 + 0.4)\frac{1}{2} - (-0.1)\frac{1}{12}\right]\frac{1}{0.2} = 2.04167 \approx 2.04.$$

4. Find $\frac{dy}{dx}$ and $\frac{d^2y}{dx^2}$ at $x = 2$ using the following table:

x	0	1	2	4
$f(x)$	1	1	2	5

Solution: Here the points are not equi-spaced, so we have to use Lagrange's formula. The Lagrange's polynomial corresponding to the given data is

$$p(x) = \frac{(x-1)(x-2)(x-4)}{(0-1)(0-2)(0-4)} \times 1 + \frac{(x-0)(x-2)(x-4)}{(1-0)(1-2)(1-4)} \times 1 + \frac{(x-0)(x-1)(x-4)}{(2-1)(2-1)(2-4)} \times 2 + \frac{(x-0)(x-1)(x-2)}{(4-0)(4-1)(4-2)} \times 5$$

$$= \frac{1}{12}(-x^3 + 9x^2 - 8x + 12).$$

$p'(x) = (-3x^2 + 18x - 8)/12$ and $p''(x) = (-6x + 18)/12$.

So, $p'(2) = 1.3333333$ and $p''(2) = 0.5$.

5. Prove that

(i) $y' = \frac{1}{h}\left[\Delta y - \frac{1}{2}\Delta^2 y + \frac{1}{3}\Delta^3 y - \frac{1}{4}\Delta^4 y + \cdots\right]$

(ii) $y'' = \frac{1}{h^2}\left[\nabla^2 y + \nabla^3 y + \frac{11}{12}\nabla^4 y + \cdots\right]$.

Solution:

(i) We know that

$$E = 1 + \Delta = e^{hD} \text{ or, } hD = \log(1 + \Delta).$$

$$y' = Dy = \frac{1}{h}\left[\Delta - \frac{\Delta^2}{2} + \frac{\Delta^3}{3} - \frac{\Delta^4}{4} + \cdots\right] y$$

$$= \frac{1}{h}\left[\Delta y - \frac{\Delta^2 y}{2} + \frac{\Delta^3 y}{3} - \frac{\Delta^4 y}{4} + \cdots\right].$$

(ii) Again, from the result $(1 + \Delta)(1 - \nabla) = 1$, we get $1 - \nabla = e^{-hD}$.

So $D = -\frac{1}{h}\log(1 - \nabla) \Rightarrow D^2 = \frac{1}{h^2}[\log(1 - \nabla)]^2$ and hence

$$y'' = \frac{1}{h^2}\left[\nabla + \frac{\nabla^2}{2} + \frac{\nabla^3}{3} - \frac{\nabla^4}{4} + \cdots\right]^2 y$$

$$= \frac{1}{h^2}\left[\nabla^2 + \nabla^3 + \frac{2\nabla^4}{3} + \frac{\nabla^4}{4} + \cdots\right] y$$

$$= \frac{1}{h^2}\left[\nabla^2 y + \nabla^3 y + \frac{11}{12}\nabla^4 y + \cdots\right].$$

EXERCISE

1. What is the basic principle of numerical differentiation.

2. Find $\frac{dy}{dx}, \frac{d^2y}{dx^2}$ and $\frac{d^3y}{dx^3}$ at $x = 1$ using the following table:

x	1.00	1.05	1.10	1.15	1.20	1.25	1.30
$f(x)$	1.0000	1.0247	1.0488	1.0724	1.0954	1.1180	1.1402

3. Find $\frac{dy}{dx}$ and $\frac{d^2y}{dx^2}$ at $x = 1.1$ using the following table:

x	1.0	1.2	1.4	1.6	1.8	2.0
$f(x)$	0.000	0.128	0.544	1.296	2.432	4.000

4. Find $\frac{dy}{dx}$ at $x = 0.4$ using the following table:

x	0.1	0.2	0.3	0.4
$f(x)$	1.10517	1.22140	1.34986	1.49182

5. Find $\frac{dy}{dx}$ at $x = 6$ using the following table:

x	4.5	5.0	5.5	6.0	6.5	7.0	7.5
$f(x)$	9.69	12.90	16.71	21.18	26.37	32.34	39.15

6. Find $\frac{dy}{dx}$ and $\frac{d^2y}{dx^2}$ at $x = 1.2$ using the following table:

x	1.0	1.2	1.4	1.6	1.8	2.0
$f(x)$	2.7183	3.3201	4.0552	4.9530	6.0496	7.3891

7. Find $\frac{dy}{dx}$ at $x = 1$, 3 and 6 using the following table:

x	0	1	2	3	4	5	6
$f(x)$	6.9897	7.4036	7.7815	8.1291	8.4510	8.7506	9.0309

8. Find $\frac{dy}{dx}$ and $\frac{d^2y}{dx^2}$ at $x = 2$ using the following table:

x	1	3	6
$f(x)$	5	10	20

9. Find $\frac{dy}{dx}$ and $\frac{d^2y}{dx^2}$ at $x = 5$ using the following table:

x	2	4	9	13
$f(x)$	57	1345	66340	402052

10. Prove that

(i) $y' = \frac{1}{h}\left[\nabla y + \frac{1}{2}\nabla^2 y + \frac{1}{3}\nabla^3 y + \frac{1}{4}\nabla^4 y + \cdots\right]$

(ii) $y'' = \frac{1}{h^2}\left[\Delta^2 y - \Delta^3 y + \frac{11}{12}\Delta^4 y + \cdots\right]$.

ANSWERS

2. 0.50024, –0.256, 0.4 **3.** 0.63, 6.65 **4.** 1.49133

5. 9.66 **6.** 3.3203, 3,3192 **7.** 0.3950, 0.3341, 0.2719

8. 2.5, 0.33333 **9.** 476; 3530.

Chapter 7

NUMERICAL INTEGRATION

The analytical method of evaluating a definite integral $\int_a^b f(x)\,dx$ is to find a primitive $g(x)$ of $f(x)$ such that $g'(x) = f(x)$ and then calculate $g(b) - g(a)$. But in many cases it is found that the primitive of $f(x)$ cannot be obtained in terms of known function, i.e., the integration cannot be carried out exactly. In some cases, only the values of function at a set of points are given. In these cases we have to use numerical methods of integration.

As a definite integral is interpreted as the area under a curve $y = f(x)$ for $a \le x \le b$, and may be evaluated by taking subdivision of the interval $[a, b]$ and by taking summation of the components area, the additive property of the definite integral allows evaluation in a piecewise sense.

The basic idea of numerical integration is to divide the interval $[a, b]$ into sub-intervals by inclusion of points $x_0, x_1, \cdots, x_n$ and then calculate the values of the function $f(x)$ at these points, approximate $f(x)$ by interpolation polynomial $\varphi_n(x)$, then the given integral will be approximately $\int_a^b \varphi_n(x)\,dx$, which can be directly calculated.

From this it is clear that using different interpolation formulae it is possible to find different quadrature formulae having varying degree of accuracy and varying features of usefulness.

7.1 NEWTON COTE'S FORMULA (CLOSED TYPE)

Let the nodes $a = x_0, x_1, \cdots, x_n = b$ be equi-spaced, we set $x_p = x_0 + ph$, $p = 0, 1, 2, \cdots$ where $h = (b - a)/n$ and we take

$$\phi_n(x) = y_0 l_0(x) + y_1 l_1(x) + \cdots + y_n l_n(x),$$

where $l_k(x) = \prod_{i=0}^{n} \frac{x - x_i}{x_k - x_i}$, $k = 0, 1, 2, \cdots, n$ and $k \neq i$.

Further we put $x = x_0 + sh$, $dx = hds$, then

$$l_k(s) = \frac{(-1)^{n-k}}{k!(n-k)!} \frac{s(s-1)\cdots(s-n)}{(s-k)}$$

and we get $$\int_a^b f(x)\,dx \approx \int_{x_0}^{x_n} \phi_n(x)\,dx = h\sum_{k=0}^{n} y_k \int_0^n l_k(s)\,ds = (b-a)\sum_{k=0}^{n} C_k^n y_k, \tag{7.1}$$

where $$nC_k^n = \int_0^n l_k(s)\,ds \tag{7.2}$$

are called Cote's coefficients and they satisfy the following properties:

P1. $$\sum_{k=0}^{n} C_k^n = 1 \tag{7.3}$$

- Putting $f(x) = 1$, i.e., $y_k = 1$ for all k in the result (7.1), we obtain

$$(x_n - x_0) = (b-a) = (b-a)\sum_{k=0}^{n} C_k^n, \text{ i.e., } \sum_{k=0}^{n} C_k^n = 1.$$

P2. $C_k^n = C_{n-k}^n$ for all k satisfying $0 \le k \le n$. (7.4)

- Using the result (7.2), we get

$$nC_k^n = \int_0^n l_k(s)\,ds$$

$$= \frac{(-1)^{n-k}}{k!(n-k)!}\int_0^n \frac{s(s-1)(s-2)\cdots(s-n)}{s-k}\,ds$$

$$= \frac{(-1)^{n-k}}{k!(n-k)!}\int_n^0 \frac{(n-u)(n-u-1)\cdots(-u)}{n-u-k}(-du) \text{ [putting } s = n-u.]$$

$$= \frac{(-1)^{2n-k}}{k!(n-k)!}\int_0^n \frac{u(u-1)(u-2)\cdots(u-n)}{u-(n-k)}\,du$$

$$= \frac{(-1)^{n-(n-k)}}{k!(n-k)!} \int_0^n \frac{u(u-1)(u-2)\dots(u-n)}{u-(n-k)} du$$

$$= nC_{n-k}^n .$$

The above results help us to determine the values of Cote's coefficients as follows:

For $n = 1$, we get

$$C_0^1 = C_1^1 \text{ and } C_0^1 + C_1^1 = 1 \text{ and thus } C_0^1 = C_1^1 = \frac{1}{2}.$$

For $n = 2$, we get

$$C_0^2 = C_2^2 \text{ and } C_0^2 + C_1^2 + C_2^2 = 1.$$

Now $$2C_1^2 = \frac{-1}{1!1!} \int_0^2 \frac{u(u-1)(u-2)}{u-1} du = -\int_0^2 u(u-2)\,du = \frac{4}{3}.$$

So $$C_1^2 = \frac{2}{3} \text{ and } C_0^2 = C_2^2 = \frac{1}{6}.$$

For $n = 3$, we get

$$C_0^3 = C_3^3;\ C_1^3 = C_2^3 \text{ and } C_0^3 + C_1^3 + C_2^3 + C_3^3 = 1.$$

Now $$3C_1^3 = \frac{1}{1!2!} \int_0^3 \frac{u(u-1)(u-2)(u-3)}{u-1} du = \frac{1}{2} \int_0^3 u(u-2)(u-3)\,du = \frac{9}{8}.$$

So $$C_1^3 = C_2^3 = \frac{3}{8} \text{ and } C_0^3 = C_3^3 = \frac{1}{8}$$

Hence $$C_1^3 = \frac{3}{8} = C_2^3 \text{ and } C_0^3 = C_3^3 = \frac{1}{8}.$$

In this way we can determine the values of the Cote's coefficients for different values of n. The following table shows a few Cote's coefficients.

TABLE 7.1: NEWTON-COTE'S COEFFICIENTS (Closed Type)

n	C_0^n	C_1^n	C_2^n	C_3^n	C_4^n	C_5^n	C_6^n	C_7^n	C_8^n
1	1/2	1/2							
2	1/6	4/6	1/6						
3	1/8	3/8	3/8	1/8					
4	7/90	32/90	12/90	32/90	7/90				
5	19/288	75/288	50/288	50/288	75/288	19/288			
6	41/840	216/840	27/840	272/840	27/840	216/840	41/840		
7	$\frac{989}{28350}$	$\frac{5888}{28350}$	$\frac{-928}{28350}$	$\frac{10496}{28350}$	$\frac{-4540}{28350}$	$\frac{10496}{28350}$	$\frac{-928}{28350}$	$\frac{5888}{28350}$	$\frac{989}{28350}$

It is observed that the coefficients for $n \geq 8$ are of mixed sign and round off erros in Newton-Cotes formula of order 8 or more becomes large and unsuitable for use.

The above formula is called Newton-Cotes formula of closed type as the end points a, b are the extreme abscissae in the formula. The error term in Newton-Cote's integration formula is obtained by integrating both sides of equation (7.1) with respect to x taking limit x_0 to x_n. Thus the error committed in this formula is given by

$$\varepsilon_{n+1} = \int_{x_0}^{x_n} (x - x_0)(x - x_1) \cdots (x - x_n) \frac{f^{n+1}(\xi)}{(n+1)!} dx$$

$$= \frac{h^{n+2}}{(n+1)!} \int_0^n u(u-1)(u-2) \cdots (u-n) f^{n+1}(\xi) du,$$

where $x = x_0 + uh$. If $f^{n+1}(x)$ vary slowly over (x_0, x_n), then $\Delta^{n+1} y_0 = h^{n+1} f^{n+1}(\xi)$, where $\xi \in (x_0, x_n)$, then

$$\varepsilon_{n+1} = \frac{h}{(n+1)!} \int_0^n u(u-1)(u-2) \cdots (u-n) \Delta^{n+1} y_0 \, du.$$

When n is even then the above result can be modified as

$$\varepsilon_{n+1} = \frac{h^{n+3}}{(n+2)!} \int_0^n u^2 (u-1)(u-2) \cdots (u-n) f^{n+2}(\xi) du.$$

Next we derive some standard techniques for numerical integration from Newton-Cote's formula.

(a) *Trapezoidal rule*:

This is a particular case of the above formula (n = 1) and known as two points closed type Newton-Cote's formula. For $n = 1, h = b - a$

$$I_T = \int_a^b f(x) dx \approx (b-a) \sum_{k=0}^{1} y_k C_k^1 = (b-a)[y_0 C_0^1 + y_1 C_1^1] = \frac{b-a}{2}(y_0 + y_1) = \frac{h}{2}(y_0 + y_1),$$

which is known as trapezoidal rule for numerical integration and the error term of this formula is

$$\varepsilon_T = \frac{h^3}{2} f''(\xi) \int_0^1 u(u-1) du = -\frac{h^3}{12} f''(\xi), (a < \xi < b)$$

(b) *Simpson's one-third rule*:

This is a particular case of Newton-Cote's formula (n = 2) and known as three points closed type Newton-Cote's formula. For $n = 2, h = \dfrac{b-a}{2}$

$$I_S = \int_a^b f(x) dx \approx (b-a) \sum_{k=0}^{2} y_k C_k^2 = (b-a)\left[y_0 C_0^2 + y_1 C_1^2 + y_2 C_2^2\right]$$

$$= \frac{b-a}{6}(y_0 + 4y_1 + y_2) = \frac{h}{3}(y_0 + 4y_1 + y_2),$$

which is known as Simpson's one-third rule for numerical integration and the error term is

$$\varepsilon_S = \frac{h^5}{4!} f^{iv}(\xi) \int_0^2 u^2(u-1)(u-2)\,du = -\frac{h^5}{90} f^{iv}(\xi), (a < \xi < b).$$

(c) *Simpson's three-eighth rule*:

This is a particular case of Newton-Cote's formula ($n = 3$) and known as four points closed type Newton-Cote's formula. For $n = 3, h = \dfrac{b-a}{3}$

$$I = \int_a^b f(x)\,dx \approx (b-a) \sum_{k=0}^{3} y_k C_k^3 = (b-a)\left[y_0 C_0^3 + y_1 C_1^3 + y_2 C_2^3 + y_3 C_3^3\right]$$

$$= \frac{b-a}{8}(y_0 + 3y_1 + 3y_2 + y_3) = \frac{3h}{8}(y_0 + 3y_1 + 3y_2 + y_3),$$

which is known as Simpson's three-eighth rule for numerical integration and the error term of this formula is

$$\varepsilon = \frac{h^5}{4!} f^{iv}(\xi) \int_0^3 u(u-1)(u-2)(u-3)\,du = -\frac{3h^5}{80} f^{iv}(\xi), (a < \xi < b)$$

7.2 NEWTON COTE'S FORMULA (OPEN TYPE)

On the other hand, if we take $a = x_{-1}, x_0, \cdots, x_n, x_{n+1} = b$ and $h = \dfrac{b-a}{n+2}$, we get the following results:

$$\int_{x_0}^{x_n} f(x)\,dx \approx (b-a) \sum_{k=0}^{n} C_k^n y_k, \tag{7.5}$$

where
$$(n+2)\, C_k^n = \int_{-1}^{n+1} l_k(s)\,ds, \tag{7.6}$$

and these coefficients also satisfy (7.3) and (7.4).

After calculating C_k^n for different values of n and k, we get.

TABLE 7.2: NEWTON-COTE'S COEFFICIENTS (Open Type)

n	C_0^n	C_1^n	C_2^n	C_3^n	C_4^n
1	1/2	1/2			
2	2/3	–1/3	2/3		
3	11/24	1/24	1/24	11/24	
4	11/20	–24/20	26/20	–24/20	11/20

From the above table it is clear that higher order Newton-Cotes formula (open type) is not suitable for numerical integration.

7.3 A GENERAL FORMULA FOR NUMERICAL INTEGRATION

In order to evaluate $\int_a^b f(x)\,dx$, we divide the interval $[a, b]$ into n equal sub-intervals such that $a = x_0, \cdots, x_n = b$. Clearly $x_p = x_0 + ph, p = 0, 1, 2, \cdots, n$. Hence we replace $f(x)$ by the collocation polynomial of degree n, i.e., by

$$\varphi_n(x) = y_0 + p\Delta y_0 + p(p-1)\frac{\Delta^2 y_0}{2!} + p(p-1)(p-2)\frac{\Delta^3 y_0}{3!} + \cdots + p(p-1)(p-2)\cdots\{p-(n-1)\}\frac{\Delta^n y_0}{n!}$$

and we get

$$\int_a^b f(x)\,dx \approx h\int_0^n [y_0 + p\Delta y_0 + p(p-1)\frac{\Delta^2 y_0}{2!} + p(p-1)(p-2)\frac{\Delta^3 y_0}{3!} + \cdots + p(p-1)(p-2)\cdots\{p-(n-1)\}\frac{\Delta^n y_0}{n!}]\,dp$$

$$= nh\left[y_0 + \frac{n}{2}\Delta y_0 + \frac{n(2n-3)}{12}\Delta^2 y_0 + \frac{n(n-2)^2}{24}\Delta^3 y_0 + \ldots\right] \tag{7.7}$$

Taking $n = 1$, we see that

$$\int_{x_0}^{x_1} f(x)\,dx \approx h\left[y_0 + \frac{\Delta y_0}{2}\right] = \frac{h}{2}(y_0 + y_1). \tag{7.8}$$

For $n = 2$, we get

$$\int_{x_0}^{x_2} f(x)\,dx \approx 2h\left[y_0 + \Delta y_0 + \frac{\Delta^2 y_0}{6}\right] = 2h\left(y_1 + \frac{\Delta(y_1 - y_0)}{6}\right) = \frac{h}{3}(y_0 + 4y_1 + y_2). \tag{7.9}$$

Similarly, for $n = 3$,

$$\int_{x_0}^{x_3} f(x)\,dx \approx 3h\left[y_0 + \frac{3}{2}\Delta y_0 + \frac{3}{4}\Delta^2 y_0 + \frac{1}{8}\Delta^3 y_0\right]$$

$$= \frac{3h}{8}(8y_0 + 12(y_1 - y_0) + 6(y_2 - 2y_1 + y_0) + y_3 - 3y_2 + 3y_1 - y_0) \tag{7.10}$$

$$= \frac{3h}{8}(y_0 + 3y_1 + 3y_2 + y_3).$$

Proceeding in this way we get the result for nth degree polynomial as

$$\int_a^b f(x)\,dx \approx Ch\sum_{k=0}^{n} c_k y_k. \tag{7.11}$$

The values of C and c_k, $k = 0, 1, 2, \cdots$ for the first few values of n are found.

TABLE 7.3: NEWTON-COTES COEFFICIENTS (Closed Type)

n	C	c_0	c_1	c_2	c_3	c_4	c_5	c_6
1	1/2	1	1					
2	1/3	1	4	1				
3	3/8	1	3	3	1			
4	2/45	7	32	12	32	7		
6	1/140	41	216	27	272	27	216	41

This is same as Table 7.1.

7.4 TRAPEZOIDAL RULE

From (7.8) we see that

$$\int_{x_0}^{x_1} f(x)\,dx \approx h\left[y_0 + \frac{\Delta y_0}{2}\right] = \frac{h}{2}(y_0 + y_1).$$

Proceeding in this way we obtain

$$\int_{x_1}^{x_2} f(x)\,dx \approx \frac{h}{2}(y_1 + y_2)$$

$$\int_{x_2}^{x_3} f(x)\,dx \approx \frac{h}{2}(y_2 + y_3)$$

$$\cdots\cdots\cdots\cdots\cdots\cdots$$

$$\int_{x_{n-1}}^{x_n} f(x)\,dx \approx \frac{h}{2}(y_{n-1} + y_n)$$

Adding all we get $$\int_a^b f(x)\,dx \approx \frac{h}{2}[y_0 + y_n + 2(y_1 + y_2 + \cdots + y_{n-1})],$$

which is known as Trapezoidal formula for numerical integration.

Error estimate

Expanding $f(x)$ in Taylor's series around the point x_0, we get

$$\int_{x_0}^{x_1} f(x)\,dx = \int_{x_0}^{x_1} [y_0 + (x - x_0)\,y_0' + \frac{1}{2}(x - x_0)^2\,y_0'' + \cdots]\,dx$$

$$= \left. xy_0 + \frac{1}{2}(x - x_0)^2 y_0' + \frac{1}{6}(x - x_0)^3 y_0'' + \cdots \right)_{x_0}^{x_1}$$

$$= (x_1 - x_0)\, y_0 + \frac{1}{2}(x_1 - x_0)^2 y_0' + \frac{1}{6}(x_1 - x_0)^3 y_0'' + .$$

$$= hy_0 + \frac{1}{2}h^2 y_0' + \frac{1}{6}h^3 y_0'' + .$$

Again, $\frac{h}{2}(y_0 + y_1) = \frac{h}{2}[y_0 + y_0 + hy_0' + \frac{1}{2}h^2 y_0'' + \cdots]$.

Hence the truncation error in the interval $[x_0, x_1] = \int_{x_0}^{x_1} f(x)\,dx - \frac{h}{2}(y_0 + y_1) = -\frac{h^3 y_0''}{12}$.

Proceeding in this way we can calculate the errors in the intervals $[x_1, x_2]$, $[x_2, x_3]$, ..., $[x_{n-1}, x_n]$.

So the total truncation error $E = -\frac{h^3}{12}[y_0'' + y_1'' + \cdots + y_{n-1}'']$.

$$= -\frac{nh^3}{12} y''(\xi), \text{ where } y''(\xi) = \max\{y_0'', y_1'', \ldots, y_{n-1}''\}$$

$$= -\frac{(b-a)h^2}{12} y''(\xi).$$

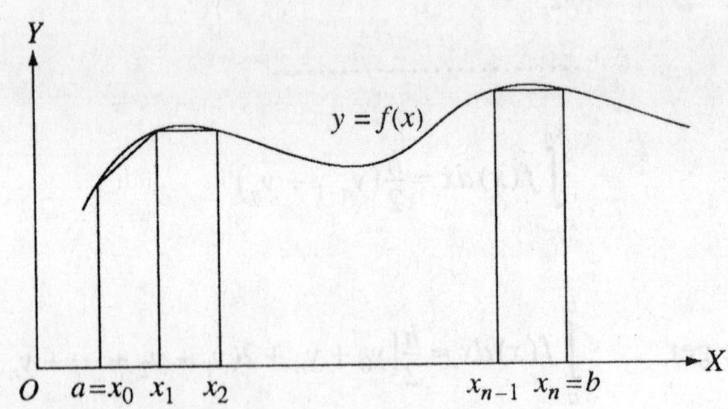

Fig.7.1 Geometrical representation of trapezoidal rule

Geometrical significance of trapezoidal rule

The geometrical significance of trapezoidal rule is that the curve $y = f(x)$ is replaced by n straight lines joining the points (x_0, y_0) and (x_1, y_1); (x_1, y_1) and (x_2, y_2), ⋯; (x_{n-1}, y_{n-1}) and (x_n, y_n). The area bounded by the curve $y = f(x)$, the ordinates $x = x_0$ and $x = x_n$ and the X-axis is then approximately equivalent to the sum of the areas of the n trapeziums thus obtained.

7.5 SIMPSON'S 1/3 RULE

From (7.9) we see that

$$\int_{x_0}^{x_2} f(x)\,dx \approx \frac{h}{3}(y_0 + 4y_1 + y_2).$$

Proceeding in this way we obtain

$$\int_{x_2}^{x_4} f(x)\,dx \approx \frac{h}{3}(y_2 + 4y_3 + y_4)$$

$$\int_{x_4}^{x_6} f(x)\,dx \approx \frac{h}{3}(y_4 + 4y_5 + y_6)$$

....

....

$$\int_{x_{n-2}}^{x_n} f(x)\,dx \approx \frac{h}{3}(y_{n-2} + 4y_{n-1} + y_n)$$

Adding all we get

$$\int_a^b f(x)\,dx \approx \frac{h}{3}[y_0 + y_n + 4(y_1 + y_3 + \ldots + y_{n-1}) + 2(y_2 + y_4 + \ldots + y_{n-2})],$$

which is known as Simpson's 1/3 rule for numerical integration.

Error estimate

Expanding $f(x)$ in Taylor's series around the point x_0 and integrating taking limits from x_0 to x_2, we get

$$\int_{x_0}^{x_2} f(x)\,dx = \int [y_0 + (x - x_0)\,y_0' + \frac{1}{2}(x - x_0)^2 y_0'' + \frac{1}{6}(x - x_0)^3 y_0''' + \frac{1}{24}(x - x_0)^4 y_0^{IV} \cdots]dx$$

$$= xy_0 + \frac{1}{2}(x - x_0)^2 y_0' + \frac{1}{6}(x - x_0)^3 y_0'' + \frac{1}{24}(x - x_0)^4 y_0''' + \frac{1}{120}(x - x_0)^5 y_0^{IV} + \cdots \Big)_{x_0}^{x_2}$$

$$= 2hy_0 + 2h^2 y_0' + \frac{4}{3}h^3 y_0'' + \frac{2}{3}h^4 y_0''' + \frac{4}{15}h^5 y_0^{IV} + \cdots.$$

Again,

$$\frac{h}{3}(y_0 + 4y_1 + y_2) = \frac{h}{3}[y_0 + 4\{y_0 + hy_0' + \frac{1}{2}h^2 y_0'' + \frac{1}{6}h^3 y_0''' + \frac{1}{24}h^4 y_0^{IV} + \cdots\}$$

$$+ \{y_0 + 2hy_0' + 2h^2 y_0'' + \frac{4}{3}h^3 y_0''' + \frac{2}{3}h^4 y_0^{IV} + \cdots\}]$$

$$= \frac{h}{3}[6y_0 + 6hy_0' + 4h^2 y_0'' + 2h^3 y_0''' + \frac{5}{6}h^4 y_0^{IV} + \cdots]$$

$$= 2hy_0 + 2h^2 y_0' + \frac{4}{3}h^3 y_0'' + \frac{2}{3}h^4 y_0''' + \frac{5}{18}h^5 y_0^{IV} + \cdots.$$

Hence the truncation error in the interval $[x_0, x_2]$

$$= \int_{x_0}^{x_2} f(x)\,dx - \frac{h}{3}(y_0 + 4y_1 + y_2) = -\frac{h^5 y_0^{\text{IV}}}{90}.$$

Proceeding in this way we can calculate the errors in the intervals $[x_2, x_4]$, $[x_4, x_6]$, …, $[x_{n-2}, x_n]$.

So the total truncation error $E = -\frac{h^5}{90}[y_0^{\text{IV}} + y_2^{\text{IV}} + \cdots + y_{n-2}^{\text{IV}}]$.

If $y^{\text{IV}}(\xi)$ is the largest value of all 4th order derivatives of $y = f(x)$ in (x_0, x_n), then the total truncation error $E = -\frac{nh^5}{180} y^{\text{IV}}(\xi) = -\frac{(b-a)h^4}{180} y^{\text{IV}}(\xi)$.

Geometrical significance of Simpson's 1/3rd rule

The geometrical significance of Simpson's 1/3rd rule is that the curve $y = f(x)$ is replaced by n parabolic chains joining three consecutive points i.e., (x_0, y_0), (x_1, y_1), (x_2, y_2); (x_2, y_2) (x_3, y_3) and (x_4, y_4); …; (x_{2n-2}, y_{2n-2}), (x_{2n-1}, y_{2n-1}) and (x_{2n}, y_{2n}). The sum of these n areas thus obtained is then approximately equivalent to the area bounded by the curve $y = f(x)$, the ordinates $x = x_0$ and $x = x_{2n}$ and X-axis.

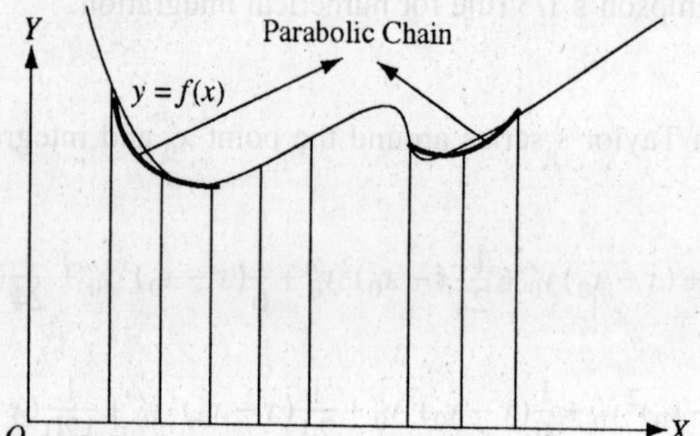

Fig.7.2 Geometrical representation of Simpson's 1/3rd rule

7.6 SIMPSON'S 3/8 RULE

From (7.10) we see that

$$\int_{x_0}^{x_3} f(x)\,dx \approx \frac{3h}{8}(y_0 + 3y_1 + 3y_2 + y_3).$$

Proceeding in this way we obtain

$$\int_{x_3}^{x_6} f(x)\,dx \approx \frac{3h}{8}(y_3 + 3y_4 + 3y_5 + y_6)$$

$$\int_{x_6}^{x_9} f(x)\,dx \approx \frac{3h}{8}(y_6 + 3y_7 + 3y_8 + y_9)$$

....

....

$$\int_{x_{n-3}}^{x_n} f(x)\,dx \approx \frac{3h}{8}(y_{n-3} + +3y_{n-2} + 3y_{n-1} + y_n).$$

Adding all, we get

$$\int_a^b f(x)dx \approx \frac{3h}{8}[y_0 + y_n + 3(y_1 + y_2 + y_4 + y_5 + \cdots + y_{n-1} + y_{n-2}) + 2(y_3 + y_6 + \ldots + y_{n-3})]$$

which is known as Simpson's 3/8 rule for numerical integration and is not useful as Simpson's 1/3 rule.

Error estimate

Expanding $f(x)$ in Taylor's series around the point x_0 and integrating taking limits from x_0 to x_3, we get

$$\int_{x_0}^{x_3} f(x)dx = \int_{x_0}^{x_3}[y_0 + (x-x_0)y_0' + \frac{1}{2}(x-x_0)^2 y_0'' + \frac{1}{6}(x-x_0)^3 y_0''' + \frac{1}{24}(x-x_0)^4 y_0^{IV} \cdots]dx$$

$$= xy_0 + \frac{1}{2}(x-x_0)^2 y_0' + \frac{1}{6}(x-x_0)^3 y_0'' + \frac{1}{24}(x-x_0)^4 y_0''' + \frac{1}{120}(x-x_0)^5 y_0^{IV} + \cdots\Big)_{x_0}^{x_3}$$

$$= 3hy_0 + \frac{9}{2}h^2 y_0' + \frac{9}{2}h^3 y_0'' + \frac{27}{8}h^4 y_0''' + \frac{81}{40}h^5 y_0^{IV} + \cdots.$$

Again,

$$\frac{3h}{8}(y_0 + 3y_1 + 3y_2 + y_3) = \frac{3h}{8}[y_0 + 3\{y_0 + hy_0' + \frac{1}{2}h^2 y_0'' + \frac{1}{6}h^3 y_0''' + \frac{1}{24}h^4 y_0^{IV} + \cdots\}$$

$$+ 3\{y_0 + 2hy_0' + 2h^2 y_0'' + \frac{4}{3}h^3 y_0''' + \frac{2}{3}h^4 y_0^{IV} + \cdots\}$$

$$+ y_0 + 3hy_0' + + \frac{9}{2}h^2 y_0'' + \frac{9}{2}h^3 y_0''' + \frac{27}{8}h^4 y_0^{IV} + \cdots]$$

$$= \frac{3h}{8}[8y_0 + 12hy_0' + 12h^2 y_0'' + 9h^3 y_0''' + \frac{11}{2}h^4 y_0^{IV} + \cdots]$$

$$= 3hy_0 + \frac{9}{2}h^2 y_0' + \frac{9}{2}h^3 y_0'' + \frac{27}{8}h^4 y_0''' + \frac{33}{16}h^5 y_0^{IV} + \cdots.$$

Hence, the truncation error in the interval $[x_0, x_3]$

$$= \int_{x_0}^{x_3} f(x)\,dx - \frac{3h}{8}(y_0 + 3y_1 + 3y_2 + 3y_3) = -\frac{3h^5 y_0^{IV}}{80}.$$

Proceeding in this way we can calculate the errors in the intervals $[x_3, x_6], [x_6, x_9], \cdots, [x_{n-3}, x_n]$.

So the total truncation error $E = -\frac{3h^5}{80}[y_0^{IV} + y_3^{IV} + \cdots + y_{n-3}^{IV}]$.

If $y^{IV}(\xi)$ is the largest value of all 4th order derivatives of $y = f(x)$ in (x_0, x_n), then the total truncation error $E = -\frac{nh^5}{80} y^{IV}(\xi) = -\frac{(b-a)h^4}{80} y^{IV}(\xi)$.

7.7 WEDDLE'S RULE

From [7.7] we see that, for $n = 6$

$$\int_{x_0}^{x_6} f(x)\,dx \approx h[6y_0 + 18\Delta y_0 + 27\Delta^2 y_0 + 24\Delta^3 y_0 + \frac{123}{10}\Delta^4 y_0 + \frac{33}{10}\Delta^5 y_0 + \frac{41}{140}\Delta^6 y_0]$$

$$= h[6y_0 + 18\Delta y_0 + 27\Delta^2 y_0 + 24\Delta^3 y_0 + \frac{123}{10}\Delta^4 y_0 + \frac{33}{10}\Delta^5 y_0 + \frac{3}{10}\Delta^6 y_0] - \frac{h}{140}\Delta^6 y_0$$

$$\approx h[6y_0 + 18(y_1 - y_0) + 27(y_2 - 2y_1 + y_0) + 24(y_3 - 3y_2 + 3y_1 - y_0)$$
$$+ \frac{123}{10}(y_4 - 4y_3 + 6y_2 - 4y_1 - y_0) + \frac{33}{10}(y_5 - 5y_4 + 10y_3 - 10y_2 + 5y_1 - y_0)$$
$$+ \frac{3}{10}(y_6 - 6y_5 + 15y_4 - 20y_3 + 15y_2 - 6y_1 + y_0)]$$

$$= \frac{3h}{10}[y_0 + 5y_1 + y_2 + 6y_3 + y_4 + 5y_5 + y_6].$$

Proceeding in this way we obtain

$$\int_{x_6}^{x_{12}} f(x)\,dx \approx \frac{3h}{10}[y_6 + 5y_7 + y_8 + 6y_9 + y_{10} + 5y_{11} + y_{12}]$$

$$\int_{x_{12}}^{x_{18}} f(x)\,dx \approx \frac{3h}{10}[y_{12} + 5y_{13} + y_{14} + 6y_{15} + y_{16} + 5y_{17} + y_{18}]$$

....

....

$$\int_{x_{n-6}}^{x_n} f(x)\,dx \approx \frac{3h}{10}[y_{n-6} + 5y_{n-5} + y_{n-4} + 6y_{n-3} + y_{n-2} + 5y_{n-1} + y_n]$$

Adding all, we get

$$\int_a^b f(x)\,dx \approx \frac{3h}{10}[y_0 + y_n + (y_2 + y_4 + y_8 + y_{10} + \cdots + y_{n-4} + y_{n-2})$$

$$+2(y_6 + y_{12} + y_{18} + \cdots + y_{n-6}) + 5(y_1 + y_5 + y_7 + y_{11} + \cdots + y_{n-5} + y_{n-1})$$

$$+6(y_3 + y_9 + y_{15} + \cdots + y_{n-3})].$$

This is known as *Weddle's rule* for numerical integration.

Proceeding as before it can be shown that if $y^{IV}(\zeta)$ is the largest value of all 6th order derivatives of $y = f(x)$ in (x_0, x_n), then the total truncation error

$$E = -nh^7 y^{IV}(\zeta)/140 = -(b-a)h^6 y^{IV}(\zeta)/140.$$

7.8 COMPUTATIONAL TABLE FOR TRAPEZOIDAL AND SIMPSON'S 1/3 RULE

$x_r\ (0 \le r \le n)$	$y_r\ (r = 0 \text{ and } n)$	$y_r\ (r = 1, 3, 5, \ldots, n-1)$	$y_r\ (r = 2, 4, \ldots, n-2)$
x_0	y_0		
x_1		y_1	
x_2			y_2
x_3		y_3	
x_4			y_4
....			
x_{n-2}			y_{n-2}
x_{n-1}		y_{n-1}	
x_n	y_n		
	$\Sigma y_r\ (= S)$	$\Sigma y_r\ (= S')$	$\Sigma y_r\ (= S'')$

$$I_T = \frac{h}{2}[S + 2(S' + S'')] \text{ and } I_S = \frac{h}{3}[S + 4S' + 2S'']$$

Note: In Simpson's 1/3 rule we have established the formula by taking two consecutive intervals at a time, so the total number of intervals in this case, i.e., *n* must be an even number but in Trapezoidal formula there is no such restriction.

7.9 COMPUTATIONAL TABLE FOR WEDDLE'S RULE

$x_r\ (0 \le r \le n)$	$y_r\ (r = 0, n)$	$y_{6r \pm 1}$	$y_{6r \pm 2}$	y_{6r-3}	y_{6r}
x_0	y_0				
x_1		y_1			
x_2			y_2		
x_3				y_3	

Contd. to next page

$x_r\ (0 \le r \le n)$	$y_r\ (r = 0, n)$	$y_{6r \pm 1}$	$y_{6r \pm 2}$	y_{6r-3}	y_{6r}
x_4			y_4		
x_5		y_5			
x_6					y_6
x_7		y_7			
x_8			y_8		
x_9				y_9	
x_{10}			y_{10}		
x_{11}		y_{11}			
x_{12}					y_{12}
x_{13}		y_{13}			
x_{14}			y_{14}		
x_{15}				y_{15}	
...	...	...	...	...	...
x_{n-2}			y_{n-2}		
x_{n-1}		y_{n-1}			
x_n	y_n				
	$\Sigma y_r\ (= S_1)$	$\Sigma y_r\ (= S_2)$	$\Sigma y_r\ (= S_3)$	$\Sigma y_r\ (= S_4)$	$\Sigma y_r\ (= S_5)$

$$I_W = 3h[S_1 + 5S_2 + S_3 + 6S_4 + 2S_5]/10$$

Note: In Weddle's rule we have established the formula by taking six consecutive intervals at a time, so the total number of intervals in this case, i.e., n must be multiple of six.

Note: Newton-Cote's, Trapezoidal rule and Simpson's 1/3 rule do not give satisfactory approximation to $\int_a^b f(x)\,dx$. Increasing the degree of the polynomial used, we can get better result but it is difficult to use a single interpolating polynomial to evaluate the above integral. That is why it is better to use piecewise polynomial of lower degree, which interpolates $f(x)$. However, above all these methods Simpson's 1/3 rule is preferred whenever the function is sufficiently smooth*.

7.10 GAUSSIAN INTEGRATION

In the preceding integration formulae we have used the ordinates in equidistant points and the weights were determined in such a way that the formula obtained as high accuracy as possible. Gauss derived a formula, which uses same number of function values and gives better accuracy. The main idea behind the Gaussian integration is that in selection of a formula

$$\int_{-1}^{1} f(x)\,dx \approx \sum_{k=0}^{n} A_k f(x_k). \tag{7.13}$$

* A function is said to be smooth at a point if higher order derivatives exist at that point. If it is smooth at all points of an interval $[a, b]$, then it is said to be smooth in $[a, b]$.

If the limits of integration be a to b, then it is customary to normalize the interval (a, b) to (–1, 1) by putting

$$x = \frac{b+a}{2} + \frac{b-a}{2}t$$

$$\text{i.e.,} \quad \int_a^b f(x)\,dx \approx \frac{b-a}{2}\sum_{k=0}^{n} A_k f\left(\frac{b+a}{2} + \frac{b-a}{2}t_k\right), \tag{7.14}$$

where x_k and t_k are the roots of nth degree equation $P_n(x) = 0$ and $P_n(x)$ is given by

$$P_n(x) = \frac{1}{2^n n!}\frac{d^n}{dx^n}(x^2-1)^n,$$

which is called nth degree *Legendre polynomial* and the weights A_i are given by

$$A_i = \int_{-1}^{1} l_i(x)\,dx, \text{ where } l_i(x) \text{ is given by (3.15).}$$

An advantage of this formula is that the abscissa and weights are symmetrical about the middle point of the interval. We list below the abscissas and weights for different values of n.

n	t_k	A_k	n	b_k	A_k
2	±0.57735027	1.00000000	4	±0.86113631	0.34785485
6	±0.93246951	0.17132449		±0.33998104	0.65214515
	±0.66120939	0.36076157	8	±0.96028986	0.10122854
	±0.23861919	0.46791393		±0.79666648	0.22238103
10	±0.97390653	0.06667134		±0.52553241	0.31370665
	±0.86506337	0.14945135		±0.18343464	0.36268378
	±0.67940957	0.21908636	12	±0.98156063	0.04717534
	±0.43339539	0.26926672		±0.90411725	0.10693933
	±0.14887434	0.29552422		±0.76990267	0.16007833
14	±0.98628381	0.03511946		±0.58731795	0.20316743
	±0.92843488	0.08015809		±0.36783150	0.23349254
	±0.82720132	0.12151857		±0.12533341	0.24914705
	±0.68729290	0.15720317			
	±0.51524864	0.18553840			
	±0.31911237	0.20519846			
	±0.10805495	0.21526385			

Examples

1. Calculate $\int_1^{1.5} e^x dx$ by (i) Trapezoidal rule and (ii) Simpson's 1/3 rule, taking $n = 10$.

Solution: Here $h = (1.5 - 1)/10 = 0.05$. Now we construct the following table:

x_i $(i = 0, 1, \ldots, 10)$	y_i $(i = 0$ and $10)$	y_i $(i = 1, 3, 5, 7, 9)$	y_i $(i = 2, 4, 6, 8)$
1.00	2.71828		
1.05		2.85765	
1.10			3.00417
1.15		3.15819	
1.20			3.32012
1.25		3.49034	
1.30			3.66930
1.35		3.85743	
1.40			4.05520
1.45		4.26311	
1.50	4.48169		
	7.19997	17.62672	14.04879

Hence, using Trapezoidal formula, we get

$$\int_1^{1.5} e^x dx \approx \frac{h}{2}[y_0 + y_{10} + 2(y_1 + y_2 + \cdots + y_9)]$$

$$= \frac{0.05}{2}[7.1997 + 2(17.62672 + 14.04879)]$$

$$= 1.76377 \approx 1.7638 \text{ (correct to 4 decimal places)}$$

and using Simpson's 1/3 formula, we get

$$\int_1^{1.5} e^x dx \approx \frac{h}{3}[y_0 + y_{10} + 4(y_1 + y_3 + y_5 + y_7 + y_9) + 2(y_2 + y_4 + y_6 + y_8)]$$

$$= \frac{0.05}{3}[7.19997 + 4(17.62672 + 4(14.04879)]$$

$$= 1.7634 \text{ (correct to 4 decimal places).}$$

2. Calculate $\int_0^1 \frac{dx}{1+x^2}$ by (i) Trapezoidal and (ii) Simpson's 1/3 rule, taking $n = 8$.

Solution: Here $h = (1 - 0)/8 = 0.125$. Now we construct the following table

x_i $(i = 0, 1, \ldots, 8)$	y_i $(i = 0$ and $8)$	y_i $(i = 1, 3, 5, 7)$	y_i $(i = 2, 4, 6)$
0.00	1.000000000		
0.125		0.984615384	

Contd. to next page

x_i ($i = 0, 1, \ldots, 8$)	y_i ($i = 0$ and 8)	y_i ($i = 1, 3, 5, 7$)	y_i ($i = 2, 4, 6$)
0.250			0.941176470
0.375		0.876712328	
0.500			0.800000000
0.625		0.719101123	
0.750			0.640000000
0.875		0.566371681	
1.000	0.500000000		
	1.500000000	3.146800516	2.381176470

Hence, using Trapezoidal formula, we get

$$\int_0^1 \frac{dx}{1+x^2} \approx \frac{h}{2}[y_0 + y_8 + 2(y_1 + y_2 + \ldots + y_7)]$$

$$= \frac{0.125}{2}[1.5000000 + 2(3.146800516 + 2.381176470)]$$

$$= 0.784747123 \approx 0.78475 \text{ (correct to 5 decimal places)}$$

and using Simpson's 1/3 formula, we get

$$\int_0^1 \frac{dx}{1+x^2} \approx \frac{h}{3}[y_0 + y_8 + 4(y_1 + y_3 + y_5 + y_7) + 2(y_2 + y_4 + y_6)]$$

$$= \frac{0.125}{2}[1.5000000 + 4(3.146800516) + 2(2.381176470)]$$

$$= 0.785406639 \approx 0.78541 \text{ (correct to 5 decimal places),}$$

where the correct result is 0.785398 (correct to 6 decimal places).

3. If $f(x)$ is a polynomial of degree 2, prove that

$$\int_0^1 f(x)\,dx = \frac{1}{12}[5f(0) + 8f(1) - f(2)].$$

Solution: Since $f(x)$ is a polynomial of degree 2, $\Delta^n f = 0$ for $n \geq 3$.

Using the result,

$$f(x) = f(0 + x.1) = E^x f(0)$$

$$= (1 + \Delta)^x f(0)$$

$$= [1 + x\Delta + x(x-1)\Delta^2/2 + \cdots] f(0)$$

$$= f(0) + x\Delta f(0) + x(x-1)\Delta^2 f(0)/2 + \cdots$$

we get $$\int_0^1 f(x)\,dx = \int_0^1 \left[f(0) + x\Delta f(0) + \frac{x(x-1)}{2}\Delta^2 f(0) \right] dx$$

$$= f(0) + \frac{1}{2}[\Delta f(1) - f(0)] - \frac{1}{12}[f(2) - 2f(1) + f(0)].$$

$$= \frac{1}{12}[5f(0) + 8f(1) - f(2)].$$

4. Apply Romberg's method to the integral $I = \int_0^{\pi/2} \sin x dx$.

Solution: Taking $\frac{2h}{\pi} = 0.25, 0.125, 0.0625, 0.03125$ successively and using Trapezoidal rule, we get the following table:

$\frac{2h}{\pi}$	0.25	0.125	0.0625	0.03125
Value of the integral	0.987116	0.996785	0.999196	0.999799
		1.000008	1.000000	1.000000
			1.000000	1.000000
				1.000000

Hence $I = 1.00000$.

5. Prove that the formula $\int_{-1}^{1} F(t)\,dt = F\left(-\frac{1}{\sqrt{3}}\right) + F\left(\frac{1}{\sqrt{3}}\right)$ is exact for any polynomial of degree 3 or less.

Solution: Using Gaussian formula for numerical integration, we get

$$\int_{-1}^{1} F(t)\,dt = A_0 F(x_0) + A_1 F(x_1) + \cdots + A_n F(x_n),$$

where x_i's are the roots of $(n + 1)$th degree Legendre polynomial.

Now the 2nd order Legendre polynomial is given by $P_2(x) = \frac{1}{8}(12x^2 - 4)$, hence roots of $P_2(x) = 0$ are

$$x = \pm\frac{1}{\sqrt{3}} \text{ (say } x_0, x_1).$$

Then $$A_1 = \int_{-1}^{1} \frac{x - x_0}{x_1 - x_0} dx = \int_{-1}^{1} \frac{x - 1/\sqrt{3}}{-2/\sqrt{3}} dx = 1$$

and $$A_0 = \int_{-1}^{1} \frac{x - x_1}{x_0 - x_1} dx = \int_{-1}^{1} \frac{x + 1/\sqrt{3}}{2/\sqrt{3}} dx = 1.$$

Hence, $\int_{-1}^{1} F(t)\,dt = F\left(-\frac{1}{\sqrt{3}}\right) + F\left(\frac{1}{\sqrt{3}}\right).$

As the given result is satisfied identically for an odd function, the given result is exact for any polynomial of degree three or less.

6. Construct a rule of the form:

$$\int_{-1}^{1} F(t)\,dt = A_0 F\left(-\frac{1}{2}\right) + A_1 F(0) + A_2 F\left(\frac{1}{2}\right).$$

Solution: Taking $x_0 = -1/2$, $x_1 = 0$, $x_2 = 1/2$, we get

$$A_0 = \int_{-1}^{1} \frac{(x-0)(x-1/2)}{(-1/2-0)(-1/2-1/2)}\,dx = \int_{-1}^{1} \frac{x(x-1/2)}{1/2}\,dx = \frac{4}{3}$$

$$A_1 = \int_{-1}^{1} \frac{(x+1/2)(x-1/2)}{(0-1/2)(0+1/2)}\,dx = \int_{-1}^{1} \frac{x^2-1/4}{-1/4}\,dx = -\frac{2}{3}$$

$$A_2 = \int_{-1}^{1} \frac{(x-0)(x+1/2)}{(1/2-0)(1/2+1/2)}\,dx = \int_{-1}^{1} \frac{x(x+1/2)}{1/2}\,dx = \frac{4}{3}.$$

Hence $\int_{-1}^{1} F(t)\,dt = \frac{4}{3} F\left(-\frac{1}{2}\right) + \frac{2}{3} F(0) + \frac{4}{3} F\left(\frac{1}{2}\right).$

7. Apply two-point Gaussian integration formula to evaluate $\int_{0}^{\pi/2} \sin x dx$.

Solution: To change the limits of integrations from $0 \to \pi/2$ to $-1 \to 1$, we put $x = \pi(t + 1)/4$, then the given integral becomes

$$\frac{\pi}{4}\int_{-1}^{1} \sin\left(\frac{\pi(t+1)}{4}\right)dt = \frac{\pi}{4}[1 \cdot f(0.57735027) + 1 \cdot f(-0.57735027)]$$

$$= \frac{\pi}{4}[0.32589 + 0.9451]$$

$= 0.99848$, which is correct to almost three decimal places.

8. Apply four-point Gaussian integration formula to evaluate $\int_{0}^{1} x dx$.

Solution: To change the limits of integrations from $0 \to 1$ to $-1 \to 1$, we put $x = (t + 1)/2$, then the given integral becomes

$$\frac{1}{4}\int_{-1}^{1}(t+1)\,dt = \frac{1}{4}[0.34785(-0.86114+1)+0.65214(-0.33998+1)$$

$$+0.34785(0.86114+1)+0.65214(0.33998+1)$$

$$= 0.499999999.$$

9. Compute the value of the integral $\int_1^2 \frac{dx}{\sqrt{2+x^2}}$ by the Weddle's rule.

Solution: Dividing [1, 2] into 12 sub-intervals we get $h = 1/12$, $a = 1$, $b = 2$, $x_i = 1 + i/12$ and $f(x_i) = 12/\sqrt{[288+(i+12)^2]}$

$x_r\ (0 \le r \le n)$	$y_r\ (r = 0, n)$	$y_{6r\pm1}$	$y_{6r\pm2}$	y_{6r-3}	y_{6r}
x_0	0.57735024				
x_1		0.56133636			
x_2			0.54545460		
x_3				0.52981296	
x_4			0.51449580		
x_5		0.49956648			
x_6					0.48507120
x_7		0.47104116			
x_8			0.45749568		
x_9				0.44444448	
x_{10}			0.43188949		
x_{11}		0.41982684			
x_{12}	0.40824828				
	0.98559852	1.95177084	1.94933556	0.97425744	0.48507120

$$I_W = 3\,[0.98559852 + 5(1.95177084) + 1.94933556$$

$$+ 6(0.97425744) + 2(0.48507120)]/120$$

$$= 0.487736883 \approx 0.48774 \text{ (correct to 5 decimal places).}$$

EXERCISE

1. Describe the following methods for numerical integration for evaluating the integral

$$\int_a^b f(x)\,dx.$$

(a) Composite trapezoidal rule,

(b) Composite Simpson's one-third rule,

(c) Simpson's three-eighth rule,

(d) Gaussian method.

2. Estimate the error in trapezoidal rule for numerical integration.
3. Estimate the error in Simpson's one-third rule for numerical integration.
4. Compute $\int_0^1 \frac{dx}{1-0.25\sin^2 x}$ by Simpson's 1/3 rule with the step size $h = 0.0625$ correct to 5D.
5. Show that for a polynomial $f(x)$ of degree 2

$$\int_{-1}^{1} f(x)\,dx = \frac{1}{3}[f(1)+4f(0)+f(-1)].$$

6. If $f(x)$ is a polynomial of degree 2, prove that

$$\int_0^1 f(x)\,dx = \frac{1}{12}[5f(1)+8f(0)+f(-1)].$$

7. If $f(x)$ is a polynomial of degree 2, prove that

$$\int_0^3 f(x)\,dx = \frac{1}{12}[f(0)+22f(2)+f(4)].$$

8. Evaluate $\int_0^1 \frac{x\,dx}{1+x^2}$ using Simpson's 1/3 rule correct up to 5 decimal places taking $n = 6$.
9. Evaluate $\int_0^1 \frac{x\,dx}{1+x}$ using Simpson's 1/3 rule correct up to 3 decimal places taking $n = 6$.
10. Evaluate $\int_1^2 f(x)\,dx$ by Trapezoidal rule and Simpson's 1/3 rule using the following table:

x	1.0	1.1	1.2	1.3	1.4	1.5	1.6	1.7	1.8	1.9	2.0
$f(x)$	1.0000	1.0488	1.0954	1.1402	1.1832	1.2247	1.2649	1.3038	1.3416	1.3784	1.4142

11. Evaluate $\int_0^1 \frac{\sin^2 x}{x}\,dx$ using Simpson's 1/3 rule correct up to 5 decimal places taking $n = 16$.
12. Evaluate $\int_{-1}^{1} \frac{dx}{1+x^2}$ using Gaussian integration correct up to 5 decimal places.

13. Evaluate $\int_0^1 \frac{dx}{1+x^2}$ using Simpson's 1/3 rule and obtain the approximate value of π.

14. Evaluate $\int_0^{\pi/2} \sqrt{\sin x}\, dx$ using Simpson's 1/3 rule correct up to 4 decimal places.

15. Use Simpson's 3/8 rule to obtain an approximate value of $\int_0^1 \sqrt{1-x^3}\, dx$.

16. Evaluate $\int_0^{\pi/2} \cos^2 x dx$, correct to 6D, by Simpson's 3/8 rule.

ANSWERS

4. 1.07711 **8.** 0.34661 **9.** 0.307 **10.** 1.21881, 1.21895
11. 0.42369 **12.** 1.57079 **13.** 0.78541 **14.** 1.1873
15. 0.8316 **16.** 0.785398

Chapter 8

SYSTEM OF LINEAR EQUATIONS

8.1 INTRODUCTION

In this chapter, we will discuss the methods of solving a system of n linear equations in n unknowns of the form

$$\begin{aligned} a_{11}x_1 + a_{12}x_2 + \dots + a_{1n}x_n &= b_1 \\ a_{21}x_1 + a_{22}x_2 + \dots + a_{2n}x_n &= b_2 \\ \dots\dots\dots\dots\dots\dots\dots\dots \\ \dots\dots\dots\dots\dots\dots\dots\dots \\ a_{n1}x_1 + a_{n2}x_2 + \dots + a_{nn}x_n &= b_n \end{aligned} \tag{8.1}$$

where the coefficients a_{ij}, $i, j = 1,2, \cdots, n$ and the constants b_i, $i = 1, 2, \cdots, n$ are given numbers. It is known that the above system of equations will have a unique solution for every right-hand side $\boldsymbol{b}$ if the coefficients determinant does not vanish, i.e., if the coefficients matrix is invertible. We will, therefore, assume that all linear systems, of equations under discussion have an invertible coefficients matrix.

Numerical methods for solving linear systems are divided into two types, direct methods and iterative methods. Direct methods are those, which in absence of round-off errors, will yield the exact solution in a finite number of elementary arithmetic operations i.e., addition, subtraction, multiplication and division. For example, Cramer's rule, Gauss's elimination method, etc. In practice, as a computing machine like calculator, computer works with a finite word length, direct method does not give exact solution when n is large. Indeed, errors arising from round-off and loss of significance may give poor and even useless results. On the other hand, the methods which start with an initial approximation and which, by applying a chosen algorithm, lead to successively better approximations are known as iterative methods. Example, Gauss-Jacobi method, Gauss-

Seidel method, etc. Applying such methods we can only hope for an approximate solution provided the method converges. As in some cases the method diverges or converges so slowly that they become useless for computational purpose. The main advantage of iterative methods are that these are very simple and easy to use.

8.2 GAUSSIAN ELIMINATION METHOD

Assuming that the system of equations (8.1) has a unique solution we rewrite the equations (8.1) as

$$\begin{aligned} a_{11}^1 x_1 + a_{12}^1 x_2 + a_{13}^1 x_3 \cdots + a_{1n}^1 x_n &= b_1^1 \\ a_{21}^1 x_1 + a_{22}^1 x_2 + a_{23}^1 x_3 \cdots + a_{2n}^1 x_n &= b_2^1 \\ \cdots\cdots\cdots\cdots\cdots\cdots\cdots\cdots\cdots & \\ \cdots\cdots\cdots\cdots\cdots\cdots\cdots\cdots\cdots & \\ a_{n1}^1 x_1 + a_{n2}^1 x_2 + a_{n3}^1 x_3 \cdots + a_{nn}^1 x_n &= b_n^1 \end{aligned} \tag{8.2}$$

If $a_{11}^1 \neq 0$, we multiply 1st equation by $-a_{i1}^1 / a_{11}^1$ and add it to the ith ($i = 2, 3, \cdots, n$) equation to get

$$\begin{aligned} a_{11}^1 x_1 + a_{12}^1 x_2 + a_{13}^1 x_3 + \cdots + a_{1n}^1 x_n &= b_1^1 \\ a_{22}^2 x_2 + a_{23}^2 x_3 + \cdots + a_{2n}^2 x_n &= b_2^2 \\ \cdots\cdots\cdots\cdots\cdots\cdots\cdots\cdots\cdots & \\ \cdots\cdots\cdots\cdots\cdots\cdots\cdots\cdots\cdots & \\ a_{n2}^2 x_2 + a_{n3}^2 x_3 \cdots + a_{nn}^2 x_n &= b_n^2 \end{aligned}$$

Again, if $a_{22}^2 \neq 0$, we multiply 2nd equation by $-a_{i2}^2 / a_{22}^2$ and add it to the ith ($i = 3, 4, \cdots, n$) equation to get

$$\begin{aligned} a_{11}^1 x_1 + a_{12}^1 x_2 + a_{13}^1 x_3 \cdots + a_{1n}^1 x_n &= b_1^1 \\ a_{22}^2 x_2 + a_{23}^2 x_3 \cdots + a_{2n}^2 x_n &= b_2^2 \\ a_{33}^3 x_3 \cdots + a_{3n}^3 x_n &= b_3^3 \\ \cdots\cdots\cdots\cdots\cdots\cdots\cdots\cdots\cdots & \\ \cdots\cdots\cdots\cdots\cdots\cdots\cdots\cdots\cdots & \\ a_{n3}^3 x_3 + \cdots + a_{nn}^3 x_n &= b_n^3 \end{aligned}$$

Proceeding in this way, we get

$$\begin{aligned} a_{11}^1 x_1 + a_{12}^1 x_2 + a_{13}^1 x_3 + \cdots + a_{1n}^1 x_n &= b_1^1 \\ a_{22}^2 x_2 + x_{23}^2 x_3 + \cdots + a_{2n}^2 x_n &= b_2^2 \\ a_{33}^3 x_3 + \cdots + a_{3n}^3 x_n &= b_3^3 \\ \cdots\cdots\cdots\cdots\cdots\cdots\cdots\cdots\cdots & \\ \cdots\cdots\cdots\cdots\cdots\cdots\cdots\cdots\cdots & \\ a_{nn}^n x_n &= b_n^n \end{aligned} \tag{8.3}$$

where $a_{ij}^{k+1} = a_{ij}^k + m_{ik}a_{kj}^k$; $b_i^{k+1} = b_i^k + m_{ik}b_k^k$ and $m_{ik} = -a_{ik}^k / a_{kk}^k$ and $i, j = k+1, \cdots, n; k = 1, 2, \cdots, n-1$. The coefficient matrix of the reduced system of equation is an upper triangular matrix whose determinant value is equal to $a_{11}^1 \, a_{22}^2 \, a_{33}^3 \ldots a_{nn}^n \neq 0$ as per our assumption. So the above-reduced system of equations has a unique solution. From the last equation of (8.3) we get

$$x_n = \frac{b_n^n}{a_{nn}^{(n)}}.$$

Substituting this value of x_n in the second last equation, one gets

$$x_{n-1} = \frac{b_{n-1}^{n-1} - a_{n-1,n}^{n-1} x_n}{a_{n-1,n-1}^{n-1}}$$

Using x_{n-1}, x_n in the third last equation we get

$$x_{n-2} = \frac{b_{n-2}^{n-2} - a_{n-2,n-1}^{n-2} x_{n-1} - a_{n-2,n}^{n-2} x_n}{a_{n-2,n-2}^{n-2}}.$$

Proceeding in this way, we get

$$x_k = \frac{b_k^k - \sum_{j=k+1}^{n} a_{kj}^k x_j}{a_{kk}^k}, \; k = n, n-1, \cdots, 3, 2, 1.$$

This process of determination of solution of the system of equation (8.3) is known as *back substitution*.

The coefficients $a_{ii}^i, i = 1, 2, \cdots, n$ (assumed non-zero) of the reduced system of equations (8.3) are called *pivots* and the corresponding equations are called *pivotal equations*. If any of the coefficients is zero, the system does not have unique solution. Again, if any of the coefficients is too small, then the factor $1/a_{ii}^i$ will be very large and when the pivotal equation will be multiplied by this factor the round-off and other computational error in the coefficients and the constant will be magnified and the results of elimination will be affected. To illustrate this we consider the following example.

The solution of the system of equations

$$0.0003x_1 + 1.566x_2 = 1.569$$
$$0.3454x_1 - 2.436x_2 = 1.018$$

is $x_1 = 10$, $x_2 = 1$. If we use a four-digit calculator to solve this problem by elimination method, then by eliminating x_1 from the second equation with the help of first one we get

$$0.0003x_1 + 1.566x_2 = 1.569$$
$$-1.804x_2 = -1.805$$

which gives $x_2 = 1.001$ and hence from the first equation $x_1 = 3.333$, which is far away from the exact solution. On the other hand, if we eliminate x_1 from the first equation with the help of second equation, we get

$$0.3454x_1 - 2.436x_2 = 1.018$$
$$1.568x_2 = 1.568$$

which when solved yield $x_2 = 1$ and $x_1 = 10$.

Further, during back substitution round off errors may again occur due to division, multiplication. So in order to get a good approximate result some measure should be taken into account what is called pivoting process, which is described as follows:

First we choose the largest coefficient a_{i1}^1, $i = 1, 2, \cdots, n$ of x_1, say a_{11}^1, in the system of equation (8.2) and the corresponding equation as first pivotal equation. Now eliminating x_1 from the remaining equations with the help of first pivotals equation we get a system of $n-1$ equations in $n-1$ unknowns. Next we choose the largest coefficient a_{i2}^2, $i = 2, 3, \cdots, n$ of x_2, say a_{22}^2, taking the corresponding equation as second pivotal equation we eliminate x_2 from the remaining equations (leaving aside the first pivotal equation) with the help of second pivotal equation we get a system of $n-2$ equations in $n-2$ unknowns. Repreated application of the above process yields the final triangular form. If at any stage the pivotal coefficient vanishes, then the system of equation, has no solution. In an another pivoting process we look for the largest coefficient a_{ij}^1, $i, j = 1, 2, \cdots, n$. Let it be a_{pq}^1, then we interchange, 1st row with pth row and 1st column with qth column. Now we eliminate x_q from the remaining equations with the help of 1st equation. Next leaving aside the first equation we look for largest coefficients a_{ij}^2, $i = 1, 2, \cdots, p-1, p+1, \cdots, n$, $j = 1, 2, \cdots, q-1, q+1, \cdots, n$. If it be a_{rs}^2, then we interchange 2nd row with rth row and 2nd column with sth column and then eliminate x_s from the remaining equations with the help of 2nd equation. The above process is repeated at every subsequent steps. The Gauss's elimination method and pivoting process give us the information whether the system of equation has a solution or not and compute the solution in which case it exists. To count the number of operations required to reduce the coefficient matrix of the system of equation (8.1) to upper triangular form in Gauss's elimination method, we proceed as follows:

To eliminate x_1 from the remaining $n-1$ equations (leaving aside the first equation), we need to determine $n-1$ multipliers and for which we have to perform $n-1$ divisions. Now for each of the $(n-1)$ equations we have to perform $(n-1+1)$ i.e., n multiplications. Hence, $\{n-1+n(n-1)\} = n^2-1$ multiplications and divisions are required to eliminate x_1 from $n-1$ equations. Similarly, to eliminate x_2 from the remaining $n-2$ equations (leaving aside the second equation) we need $(n-1)^2-1$ operations. So the total number of operations required to reduce the system of n equations in n unknowns to upper-triangular form we need

$$\sum_{k=1}^{n}(k^2-1) = \frac{n(n+1)(2n+1)}{6} - n.$$

Next to solve the reduced system in back substitution we require one division to determine x_n from the last equation. Then one multiplication and one division to determine x_{n-1} from the second last equation and so on. Hence after reducing the system to triangular form, we need

$$1 + 2 + \cdots + n = n(n + 1)/2$$

multiplications and divisions to solve the reduced system by back substitution method. So the total number of multiplications and divisions (the operations of addition and subtraction are ignored as computer needs less time to perform such operations) needed in Gauss's elimination method is

$$\frac{n(n+1)(2n+1)}{6} - n + \frac{n(n+1)}{2} = \frac{n^3}{3} + n^2 - \frac{n}{3}.$$

8.3 GAUSS-JORDAN METHOD

Gauss-Jordan method is a modification of Gaussian elimination method. In this method elimination is performed not only in the equations below but also in the equations above. In this way we obtain finally a diagonal (even unit) matrix and we get the solution without further computation. Assurning that the system of equation (8.1) has a unique solution, in this method we finally get

$$\begin{aligned} a_1x_1 + 0.x_2 + 0.x_3 + \cdots + 0.x_n &= c_1 \\ 0.x_1 + a_2x_2 + 0.x_3 + \cdots + 0.x_n &= c_2 \\ 0.x_1 + 0.x_2 + a_3x_3 + \cdots + 0.x_n &= c_3 \\ \cdots\cdots\cdots\cdots\cdots\cdots\cdots\cdots\cdots\cdots \\ 0.x_1 + 0.x_2 + \cdots + 0.x_{n-1} + a_nx_n &= c_n \end{aligned} \tag{8.4}$$

The coefficient matrix of the reduced system of equation is a diagonal matrix whose determinant value is equals to $a_1a_2a_3 \cdots a_n \neq 0$ as per our assumption. So the above-reduced system of equations has a unique solution. Solving the reduced system of equations, we get

$$x_k = \frac{c_k}{a_k}, \; k = 1, 2, \cdots, n.$$

To count the number of operations required to reduce the coefficient matrix of the system of equation (8.1) to diagonal form in Gauss-Jordan method, we start with the kth step. In order to determine the multipliers at this stage, we need $(n-1)$ divisions. So, to calculate the multipliers for all the unknowns we need $n(n-1)$ multiplications. Now to eliminate x_k from the remaining $(n-1)$ equations, all the coefficients (except the first) of pivotal equation and the constant term have to be multiplied by the multiplier, which require $(n-k+1)$ multiplications. For $(n-1)$ multipliers, $(n-1)(n-k+1)$ multiplications are required. So to convert the coefficient matrix into diagonal we need

$$n(n-1) + \sum_{k=1}^{n}(n-1)(n-k+1) = n(n-1) + (n-1)\sum_{k=1}^{n} k = \frac{(n-1)n(n+3)}{2}.$$

To solve the reduced system of equations we need n divisions.

Hence the total number of operations of multiplication and division is $\frac{(n-1)n(n+3)}{2}+n$. So when solving a large system of equations using Gauss-Jordan method the number of operations of multiplication and division is roughly $n^3/2$.

Note: It should be noted that the required number of operations of multiplication and division in Gaussian elimination method is $n^3/3$ whereas that of Gauss-Jordan method is $n^3/2$. Moreover, numerical errors can be controlled more easily in Gaussian elimination method. So the former method should usually be preferred, particularly when solving a large system of equations on computer.

8.4 ITERATIVE METHODS

Any solution of our problem is an n-vector x whose components are required to determine. In the next section we introduce and use norms as a convenient way of measuring the size of vectors. The norm of a vector $x = (x_i)$ is a non-negative real number and defined by

$$\|x\| = \max_{1 \le i \le n} |x_i|$$

which satisfies the following properties:

1. For all n-vectors x, $\|x\| \ge 0$ and $\|x\| = 0$ iff $x = \theta$.
2. For all n-vectors x and all numbers c, $\| cx \| = | c | \, \| x \|$
3. For any two vectors x and y, $\|x + y \| \le \| x \| + \| y \|$.

There are also other ways to define the norm of a vector:

1. $\| x \|_2 = \sqrt{|x_1|^2+|x_2|^2+|x_3|^2+\ldots+|x_n|^2}$
2. $\| x \|_1 = \sum_{i=1}^{n} |x_i|$
3. $\| x \|_{p.w} = \left(\sum_{i=1}^{n} |x_i|^p w_i \right)^{1/p}$, where p is some number between 1 and ∞ and w_i are fixed positive quantities.

If $\{x^k\}$, $k = 1, 2, \cdots$, be a sequence of n-vectors. Then the sequence is said to converge to an n-vector α, if $\| x^k - \alpha \| \to \theta$, i.e., $\| e^k \| \to \theta$ as $k \to \infty$ where $e^k = \alpha - x^k$ denotes the error at the kth iteration. In the next section, we will discuss two iterative methods for solving system of linear equations. In both the cases, we start with an initial approximation to the exact solution and if it converges, we will derive a new better approximation and the process will be repeated until desired accuracy is achieved which implies that like direct method the number of operations is not fixed in iterative methods.

8.5 GAUSS-JACOBI METHOD

In this method we rewrite the system of equations given by (15.1), assuming $a_{ii} \neq 0$, $i = 1, 2, \cdots, n$ as

$$x_1 = \frac{1}{a_{11}}[b_1 - a_{12}x_2 - a_{13}x_3 - a_{14}x_4 \ldots - a_{1n}x_n]$$

$$x_2 = \frac{1}{a_{22}}[b_2 - a_{21}x_1 - a_{23}x_3 - a_{24}x_4 \ldots - a_{2n}x_n]$$

$$x_3 = \frac{1}{a_{33}}[b_3 - a_{31}x_1 - a_{32}x_2 - a_{34}x_4 \ldots - a_{3n}x_n]$$

....

....

$$x_n = \frac{1}{a_{nn}}[b_n - a_{n1}x_1 - a_{n2}x_2 - a_{n3}x_3 - \cdots - a_{n,\,n-1}x_{n-1}]$$

More generally,

$$x_i = \frac{b_i - \sum\limits_{j \neq i} a_{ij}x_j}{a_{ii}}, i, j = 1, 2, \cdots, n. \tag{8.5}$$

If we start with an initial approximation $x_i^0, i = 1, 2, \cdots, n$ then the successive iterations are given by

$$x_i^{k+1} = \frac{b_i - \sum\limits_{j \neq i} a_{ij}x_j^k}{a_{ii}}, i, j = 1, 2, \cdots, n \text{ and } k = 0, 1, 2, \cdots. \tag{8.6}$$

It is clear from the above result that an improved set of components is used to obtain a better set of components.

To test the convergence of this method, we set $\mathbf{e}^k = \mathbf{x} - \mathbf{x}^k$ which denotes the error at the *k*th iteration. From the relation (8.6), we get

$$x_i - x_i^{k+1} = \frac{-\sum\limits_{j \neq i} a_{ij}(x_j - x_j^k)}{a_{ii}}, i, j = 1, 2, \cdots, n \text{ and } k = 0, 1, 2, \cdots$$

$$\Rightarrow \quad e_i^{k+1} = \frac{-\sum\limits_{j \neq i} a_{ij}e_j^k}{a_{ii}}, i, j = 1, 2, \cdots, n \text{ and } k = 0, 1, 2, \cdots$$

$$\Rightarrow \quad \left|e_i^{k+1}\right| \leq \frac{\sum\limits_{j \neq i} |a_{ij}|\,|e_j^k|}{|a_{ii}|} \leq \frac{\sum\limits_{j \neq i} |a_{ij}|\,|\mathrm{e}^k|}{|a_{ii}|} = T\left\|\mathrm{e}^k\right\|, \text{ where } T = \frac{\sum\limits_{j \neq i} |a_{ij}|}{|a_{ii}|}$$

$$\Rightarrow \quad \left\|\mathrm{e}^{k+1}\right\| \leq T\left\|\mathrm{e}^k\right\|.$$

Repeated application of the above formula yields $\|e^{k+1}\| \le T\|e^k\| \le \ldots \le T^{k+1}\|e^0\|$ and this shows that $\|e^k\| \to 0$ as $k \to \infty$, if $T < 1$. Hence, if $\sum_{j \ne i} |a_{ij}| < |a_{ii}|$, i.e., if the coefficient matrix is strictly diagonally dominant, the Gauss-Jacobi method converges for any initial approximation x^0.

8.6 GAUSS-SEIDEL METHOD

This method is also applied to solve a linear system of equations with $a_{ii} \ne 0$, $i = 1, 2, \cdots, n$. If x_i^k represents kth approximation to x_i, then the algorithm under proceed as

$$x_1^{k+1} = \frac{1}{a_{11}}[b_1 - a_{12}x_2^k - a_{13}x_3^k - a_{14}x_4^k \ldots - a_{1n}x_n^k]$$

$$x_2^{k+1} = \frac{1}{a_{22}}[b_2 - a_{21}x_1^{k+1} - a_{23}x_3^k - a_{24}x_4^k \ldots - a_{2n}x_n^k]$$

$$x_3^{k+1} = \frac{1}{a_{33}}[b_3 - a_{31}x_1^{k+1} - a_{32}x_2^{k+1} - a_{34}x_4^k \ldots - a_{3n}x_n^k]$$

....

....

$$x_n^{k+1} = \frac{1}{a_{nn}}[b_n - a_{n1}x_1^{k+1} - a_{n2}x_2^{k+1} - a_{n3}x_3^{k+1} - \cdots - a_{n,n-1}x_{n-1}^{k+1}]$$

More generally, we have for kth iteration to x_i

$$x_i^k = \frac{b_i - \sum_{j<i} a_{ij}x_j^k - \sum_{j>i} a_{ij}x_j^{k-1}}{a_{ii}}, \quad k = 0, 1, 2, \cdots \text{ and } i = 1, 2, \cdots, n. \tag{8.7}$$

The above formula indicates that this method uses an improved component as soon as it is available.

Now using the result (8.7), we get

$$e_i^k = \frac{-\sum_{j<i} a_{ij}e_j^k - \sum_{j>i} a_{ij}e_j^{k-1}}{a_{ii}}, \quad k = 0, 1, 2, \cdots \text{ and } i = 1, 2, \cdots, n.$$

If we define $L = \dfrac{\sum_{j<i} |a_{ij}|}{|a_{ii}|}$, then obviously $0 < L < T < 1$ and the above result gives

$$|e_i^k| \le \frac{\sum_{j<i} |a_{ij}||e_j^k| + \sum_{j>i} |a_{ij}||e_j^{k-1}|}{|a_{ii}|} \le \frac{\sum_{j<i} |a_{ij}|\|e^k\| + \sum_{j>i} |a_{ij}|\|e^{k-1}\|}{|a_{ii}|}$$

$$\Rightarrow \|e^k\| \le L\|e^k\| + (T - L)\|e^{k-1}\|$$

$$\Rightarrow \|e^k\| \le \frac{T-L}{1-L}\|e^{k-1}\| \le T\|e^{k-1}\| \text{ as } T < 1.$$

Again, repeated application of the above formula yield $\| e^k \| \leq \cdots \leq T^k \| e^0 \|$ and this shows that $\| e^k \| \to 0$ when $k \to \infty$ as $T < 1$. Hence, if the coefficient matrix is strictly diagonally dominant, the Gauss-Seidel method converges for any initial approximation x^0. In most of the cases it is found that Gauss-Seidel method converges faster than Gauss-Jacobi method, but we cannot say this a rule as there are the cases in which Gauss-Jacobi method converges quickly than Gauss-Seidel method.

8.7 GAUSS-SEIDEL METHOD IN MATRIX NOTATION

In matrix notation if we write A = L + D + U, where L, U denote the lower and upper triangular matrices with zeros in the diagonal, then the equation (8.7) can be written as

$$\mathrm{D}\mathbf{x}^{(k)} = \mathrm{b} - \mathrm{L}\mathbf{x}^{(k)} - \mathrm{U}\mathbf{x}^{(k-1)}$$

which can be solved for $\mathbf{x}^{(k)}$ as follows

$$\mathbf{x}^{(k)} = \mathrm{D}^{-1}[\mathbf{b} - \mathrm{L}\mathbf{x}^{(k)} - \mathrm{U}\mathbf{x}^{(k-1)}]$$

$$(\mathrm{I} + \mathrm{D}^{-1}\mathrm{L})\mathbf{x}^{(k)} = \mathrm{D}^{-1}\mathbf{b} - \mathrm{D}^{-1}\mathrm{U}\mathbf{x}^{(k-1)}$$

$$\mathbf{x}^{(k)} = (\mathrm{I} + \mathrm{D}^{-1}\mathrm{L})^{-1}[\mathrm{D}^{-1}\mathbf{b} - \mathrm{D}^{-1}\mathrm{U}\mathbf{x}^{(k-1)}]$$

$$= (\mathrm{I} + \mathrm{L}^{-1}\mathrm{D})[\mathrm{D}^{-1}\mathbf{b} - \mathrm{D}^{-1}\mathrm{U}\mathbf{x}^{(k-1)}]$$

$$= \mathrm{D}^{-1}\mathbf{b} - \mathrm{D}^{-1}\mathrm{U}\mathbf{x}^{(k-1)} + \mathrm{L}^{-1}\mathbf{b} - \mathrm{L}^{-1}\mathrm{U}\mathbf{x}^{(k-1)}$$

$$= (\mathrm{D} + \mathrm{L})^{-1}\mathbf{b} - (\mathrm{D} + \mathrm{L})^{-1}\mathrm{U}\mathbf{x}^{(k-1)}$$

Examples

1. Solve the equations using Gauss elimination method:

$$6x_1 + 3x_2 + 2x_3 = 6$$
$$6x_1 + 4x_2 + 3x_3 = 0$$
$$20x_1 + 15x_2 + 12x_3 = 0$$

Solution: Multiplying 3rd equation by –10/3 and adding to 1st and 2nd equations we get

$$20x_1 + 15x_2 + 12x_3 = 0$$
$$-1.5x_2 - 1.6x_3 = 6$$
$$-0.5x_2 - 0.6x_3 = 0.$$

Again, multiplying the second of reduced equations by –1/3 and adding it to the next equation, we get

$$20x_1 + 15x_2 + 12x_3 = 0$$
$$-1.5x_2 - 1.6x_3 = 6$$
$$-0.2x_3/3 = -2.$$

Now solving the above system of equations by back substitution method, we get

$$-0.2x_3/3 = -2 \Rightarrow x_3 = 30,$$
$$-1.5x_2 - 1.6(30) = 6 \Rightarrow x_2 = -36,$$
$$20x_1 + 15(-36) + 12(30) = 0 \Rightarrow x_1 = 9.$$

2. Solve the equations using Gauss elimination method:

$$0.34x_1 - 0.58x_2 + 0.94x_3 = 2$$
$$0.27x_1 + 0.42x_2 + 0.13x_3 = 1.5$$
$$0.2x_1 - 0.51x_2 + 0.54x_3 = 0.8.$$

Solution: To solve the problem in tabular form we proceed as follows

Multiplier	Augmented Matrix				Check Column
	0.34	–0.58	0.94	2.0	2.70
	0.27	0.42	0.13	1.5	2.32
	0.20	–0.51	0.54	0.8	1.03
Step 1	0.34	–0.58	0.94	2.0	
–0.794		0.88	–0.62	–0.088	0.17
–0.588		–0.17	–0.01	–0.38	–0.56
Step 2	0.34	–0.58	0.94	2.0	
		0.88	–0.62	–0.088	
0.193			–0.13	–0.40	–0.53

Back substitution:

$$-0.13x_3 = -0.40 \Rightarrow x_3 = 3.1$$
$$0.88x_2 - 0.62(3.1) = -0.088 \Rightarrow x_2 = 2.1$$
$$0.34x_1 - 0.58(2.1) + 0.94(3.1) = 2.000 \Rightarrow x_1 = 0.89$$

3. Solve the equations using Gauss elimination method

$$x_1 + x_2 + 4x_3 = 12$$
$$4x_1 + 11x_2 - x_3 = 33$$
$$8x_1 - 3x_2 + 2x_3 = 20.$$

Solution: To solve the problem in tabular form we proceed as follows

Multiplier	Augmented Matrix				Check Column
	8	–3	2	20	27
	1	1	4	12	18
	4	11	–1	33	47
Step 1	8	–3	2	20	
–0.125		1.375	3.75	9.5	14.625
–0.5		12.5	–2	23	33.5
Step 2	8	–3	2	20	
		1.375	3.75	9.5	
-9.091			-36.091	-63.365	-99.456

Back substitution:

$$-36.091x_3 = -63.365 \Rightarrow x_3 = 1.756$$
$$1.375x_2 + 3.75(1.756) = 9.5 \Rightarrow x_2 = 2.12$$
$$8x_1 - 3(2.12) + 2(1.756) = 20 \Rightarrow x_1 = 2.856.$$

4. Solve the equations using Gauss-Jordan method:

$$3x_1 + 2x_2 + 3x_3 = 18$$
$$2x_1 + x_2 + x_3 = 10$$
$$x_1 + 4x_2 + 9x_3 = 16.$$

Solution: To solve the problem in tabular form we proceed as follows:

	Multiplier	Augmented Matrix			
		3	2	3	18
		2	1	1	10
		1	4	9	16
Step 1		3	2	3	18
	–2/3	0	–1/3	–1	–2
	–1/3	0	10/3	8	10
Step 2	6	3	0	–3	6
		0	–1/3	–1	–2
	10	0	0	–2	–10
Step 3	–3/2	3	0	0	21
	–1/2	0	–1/3	0	3
		0	0	–2	–10

From the last table we get $-2z = -10 \Rightarrow z = 5; -y/3 = 3 \Rightarrow y = -9$ and $3x = 21 \Rightarrow x = 7$.

5. Solve the equations using Gauss-Jordan method:

$$x_1 + 2x_2 + x_3 = 8$$
$$2x_1 + 3x_2 + 4x_3 = 20$$
$$4x_1 + 3x_2 + 2x_3 = 16$$

Solution: To solve the problem in tabular form we proceed as follows

	Multiplier	Augmented Matrix			
		1	2	1	8
		2	3	4	20
		4	3	2	16
Step 1		1	2	1	8
	–2	0	–1	2	4
	–4	0	–5	–2	–16
Step 2	2	1	0	5	16
		0	–1	2	4
	–5	0	0	–12	–36
Step 3	5/12	1	0	0	1
	1/6	0	–1	0	–2
		0	0	–12	–36

From the last table we get $-12z = -36 \Rightarrow z = 3; -y = -2 \Rightarrow y = 2$ and $x = 1$.

6. Solve the equations:

$$-2x_1 + 10x_2 - x_3 - x_4 = 15$$
$$10x_1 - 2x_2 - x_3 - x_4 = 3$$
$$-x_1 - x_2 - 2x_3 + 10x_4 = -9$$
$$-x_1 - x_2 + 10x_3 - 2x_4 = 27.$$

Solution: To solve the above system of equations using Gauss-Jacobi and Gauss-Seidel methods we rewrite the equation as

$$\begin{aligned} 10x_1 - 2x_2 - x_3 - x_4 &= 3 \\ -2x_1 + 10x_2 - x_3 - x_4 &= 15 \\ -x_1 - x_2 + 10x_3 - 2x_4 &= 27 \\ -x_1 - x_2 - 2x_3 + 10x_4 &= -9. \end{aligned}$$

From these equations we express x_1, x_2, x_3, x_4 as follows:

$$\begin{aligned} x_1 &= (3 + 2x_2 + x_3 + x_4)/10 \\ x_2 &= (15 + 2x_1 + x_3 + x_4)/10 \\ x_3 &= (27 + x_1 + x_2 + 2x_4)/10 \\ x_4 &= (-9 + x_1 + x_2 + 2x_3)/10. \end{aligned}$$

If we start with the initial approximation (0, 0, 0, 0), then Gauss-Jacobi method yields the following iterations

n	x_1	x_2	x_3	x_4
1	0.3000	1.5000	2.7000	–0.9000
2	0.7800	1.7400	2.7000	–0.1800
3	0.9000	1.9080	2.9160	–0.1080
4	0.9624	1.9608	2.9592	–0.0360
5	0.9845	1.9848	2.9851	–0.0158
6	0.9939	1.9938	2.9938	–0.0060
7	0.9975	1.9975	2.9976	–0.0025
8	0.9990	1.9990	2.9990	–0.0010
9	0.9996	1.9996	2.9996	–0.0004
10	0.9998	1.9998	2.9998	–0.0002
11	0.9999	1.9999	2.9999	–0.0001
12	1.0000	2.0000	3.0000	0.0000

Where Gauss-Seidel method yields the following successive iterations:

n	x_1	x_2	x_3	x_4
1	0.3000	1.5600	2.8860	–0.1368
2	0.8869	1.9523	2.9563	–0.0248
3	0.9836	1.9899	2.9924	–0.0042
4	0.9968	1.9982	2.9987	–0.0008
5	0.9994	1.9997	2.9998	–0.0001
6	0.9999	1.9999	3.0000	0.0000
7	1.0000	2.0000	3.0000	0.0000

7. Solve the equations:

$$\begin{aligned} 1.876x_1 + 2.985x_2 - 11.62x_3 &= -0.972 \\ 12.214x_1 + 2.367x_2 + 3.672x_3 &= 7.814 \\ 2.412x_1 + 9.879x_2 + 1.564x_3 &= 4.89. \end{aligned}$$

Solution: To solve the above system of equations using Gauss-Jacobi and Gauss-Seidel methods we rewrite the equation as

$$12.214x_1 + 2.367x_2 + 3.672x_3 = 7.814$$
$$2.412x_1 + 9.879x_2 + 1.564x_3 = 4.89$$
$$1.876x_1 + 2.985x_2 - 11.62x_3 = -0.972.$$

From these equations we express x_1, x_2, x_3 as follows

$$x_1 = (7.814 - 2.367x_2 - 3.672x_3)/12.214$$
$$x_2 = (4.89 - 2.412x_1 - 1.564x_3)/9.879$$
$$x_3 = (0.972 + 1.876x_1 + 2.985x_2)/11.62.$$

If we start with the initial approximation (0, 0, 0), then Gauss-Jacobi method yields the following iterations:

n	x_1	x_2	x_3
1	0.63976	0.49499	0.08365
2	0.51868	0.32555	0.31409
3	0.48224	0.31863	0.25102
4	0.50254	0.33751	0.24336
5	0.50119	0.33376	0.25148
6	0.49936	0.33281	0.25030
7	0.50001	0.33344	0.24976
8	0.50005	0.33337	0.25003
9	0.49998	0.33332	0.25001
10	0.50000	0.33334	0.24999
11	0.50000	0.33333	0.25000

Where Gauss-Seidel method yields the following successive iterations:

n	x_1	x_2	x_3
1	0.63976	0.33879	0.27396
2	0.49174	0.33156	0.24821
3	0.50088	0.33340	0.25016
4	0.49994	0.33332	0.24999
5	0.50000	0.33333	0.25000
6	0.50000	0.33333	0.25000

EXERCISE

1. Describe the following methods for solving a system of simultaneous equations:
   (a) Gaussian elimination method,
   (b) Gauss-Jordan method,
   (c) Gauss-Jacobi method,
   (d) Gauss-Seidel method.

2. What is called pivoting? Why pivoting is necessary to solve a system of equations using Gaussian elimination method?

3. Derive Gauss-Seidel method for solving a system of equations in matrix notation.
4. Show that if the coefficient matrix is diagonally dominant, then
   (a) Gauss-Jacobi method and
   (b) Gauss-Seidel method

   converge for any choice of initial guess.
5. For large values of count the number of operations required to solve the system of n equations in n unknowns using
   (a) Gaussian elimination method,
   (b) Gauss-Jordan method.
6. Solve the following system of equations correct to 3 decimal places:

   (a) $2x + y + 4z = 12$, $8x - 3y + 2z = 20$, $4x + 11y - z = 33$

   (b) $x + 3y + z = 8$, $2x + 3y = 7$, $y - 2z = 0$

   (c) $x + 0.67y + 0.33z = 2$, $0.45x + y + 0.55z = 2$, $0.67x + 0.33y + z = 2$

   (d) $5.091x + 3.455y + 1.091z = 1.276$, $2.818x + 6.455y - 4.273z = 4.654$, $1.273x - 3.091y + 7.545z = 2.187$

   (e) $5.6x + 3.8y + 1.2z = 1.4$, $3.1x + 7.1y - 4.7z = 5.1$, $1.4x - 3.4y + 8.3z = 2.4$

   (f) $1.3x + 4.6y + 3.1z = -1$, $5.6x + 5.8y + 7.9z = 2$, $4.2x + 3.2y + 4.5z = -3$.
7. Solve the following systems of equations using (i) Gauss-Jacobi method and (ii) Gauss-Seidel method:

   (a) $10x + 2y + z = 9$, $2x + 20y - 2z = -44$, $-2x + 3y + 10z = 22$

   (b) $5x - y + 2z = 6$, $x - y + 3z = 3$, $2x + 5y + z = 8$

   (c) $23x + 30y + 3z = 56$, $5x + 14y + 21z = 40$, $17x - 3y + 7z = 21$

   (d) $10x + 2y + z = 13$, $2x + 10y + z = 13$, $2x + y + 10z = 13$.

ANSWERS

6. (a) $x = 3, y = 2, z = 1$; (b) $x = 1, y = 2, z = 1$;
   (c) $x = 1, y = 1, z = 1$; (d) $x = -1.992, y = 2.751, z = 1.753$
   (e) $x = -1.99, y = 2.75, z = 1.75$ (f) $x = -4.3, y = -2.42, z = 5.07$
7. (a) $x = 1, y = -2, z = 3$ (b) $x = 1, y = 1, z = 1$
   (c) $x = 1, y = 1, z = 1$ (d) $x = 1, y = 1, z = 1$.

Chapter 9

NUMERICAL SOLUTION OF ORDINARY DIFFERENTIAL EQUATION

9.1 . INTRODUCTION

Many problems in science and technology are related to the problem of ordinary differential equations satisfying certain given conditions. In most of these cases the analytical methods cannot be applied to solve the equations. That is why solving differential equation is one of the major problems in numerical analysis.

The classical initial value problem is to find a function $y(x)$ which satisfies the first differential equation $y' = f(x, y)$ and take the initial value $y(x_0) = y_0$.

The present chapter is focused on the methods, which yield the solution of the above ordinary differential equation in one of the two forms:

(i) a power series in x, from which the value of the function y can be obtained by direct substitution.

(ii) a set of tabulated values of x and y.

The methods due to Taylor and Picard give the solutions in a power series while Euler, Runge-Kutta, Adams-Basforth, etc. give the solution in the form of a tabulated values in x and y. The methods due to Euler, Runge-Kutta, Adams-Basforth, Adam-Moulton and Milne are called step-by-step methods. These methods are classified into two types, viz., single step method in which the value of y_{i+1} depend on the knowledge of the preceding value y_i only and multi-step methods in which the value of y_{i+1} depends on the knowledge of the preceding values $y_i, y_{i-1}, y_{i-2}, \cdots$. In general a method is called k-th step method if the value of y_{i+1} depends on the knowledge of the preceding values $y_i, y_{i-1}, y_{i-2}, \cdots, y_{i-k+1}$.

Thus in a single step method the initial value changes with every step, the initial value for $(i + 1)$th step is y_i. In a k-th step method, the computation cannot be started until and unless the values of $y_0, y_1, \ldots, y_{k-1}$ are obtained by some other method like any single step method.

9.2 PICARD'S METHOD

Consider the first order differential equation,

$$\frac{dy}{dx} = f(x, y), \tag{9.1}$$

where $y = y_0$, when $x = x_0$.

Integrating (9.1) taking limits y_0 to y and x_0 to x, we get

$$y = y_0 + \int_{x_0}^{x} f(x, y)\, dx. \tag{9.2}$$

For a first approximation y_1 to the solution, we replace y in $f(x, y)$ in (9.2) by y_0 and integrate (9.2). Thus we get

$$y_1 = y_0 + \int_{x_0}^{x} f(x, y_0)\, dx. \tag{9.3}$$

For a second approximation y_2 to the solution, we replace y in $f(x, y)$ in (9.3) by y_1 and integrate (9.3). Thus we get

$$y_2 = y_0 + \int_{x_0}^{x} f(x, y_1)\, dx. \tag{9.4}$$

For a third approximation y_3 to the solution, we replace y in $f(x, y)$ in (9.4) by y_2 and integrate (9.4) to get

$$y_3 = y_0 + \int_{x_0}^{x} f(x, y_2)\, dx. \tag{9.5}$$

Proceeding in this way, we get a sequence of functions of x, viz., $y_1, y_2, y_3, y_4, \cdots$,

$$y_{i+1}(x) = y_0 + \int_{x_0}^{x} f(x, y_i(x))\, dx. \tag{9.6}$$

and it can be shown that if the function $f(x, y)$ is bounded and continuous in some region about the point (x_0, y_0) and if $f(x, y)$ satisfies the Lipschitz condition, viz.,

$$|f(x, y) - f(x, y^*)| \le K\,|y - y^*|, \text{ K being a constant,}$$

then the sequence of functions $y_1, y_2, y_3, y_4, \cdots$ converges uniquely to the solution of (9.1).

From (9.6) we get

$$|y_{i+1}(x)-y_i(x)|=\int_{x_0}^{x}|f(x,y_i(x))-f(x,y_{i-1}(x))|\,dx\le K\int_{x_0}^{x}|y_i(x)-y_{i-1}(x)|\,dx.$$

If we assume $x_0 < x < X,\ X - x_0 = h,\ |f(x, y)| < M$ and $|y_i'(x)| < M$, then

$$|y_2(x)-y_1(x)|=|y_0-y_1(x)+\int_{x_0}^{x}f(x,y_1(x))\,dx|$$

$$\le |y_0-y_1(x)|+|\int_{x_0}^{x}M\,dx|\le 2Mh=N\ \text{(say)}.$$

Proceeding in this way, we get

$$|y_3(x)-y_2(x)|\le K\int_{x_0}^{x}N\,dx\le KNh$$

$$|y_4(x)-y_3(x)|\le K\int_{x_0}^{x}NK(x-x_0)\,dx\le\frac{1}{2}NK^2h^2$$

....

....

$$|y_{n+1}(x)-y_n(x)|\le\frac{K}{(n-2)!}\int_{x_0}^{x}NK^{n-2}(x-x_0)^{n-2}\,dx\le\frac{NK^{n-1}h^{n-1}}{(n-1)!}$$

Thus writing

$$y_{n+1}(x)=y_1(x)+\{y_2(x)-y_1(x)\}+\{y_3(x)-y_2(x)\}+\ldots+\{y_{n+1}(x)-y_n(x)\},\quad (9.7)$$

we see that all the terms of the above series except $y_1(x)$ is less in absolute value than the corresponding terms in the series

$$N+Nkh+\frac{1}{2}NK^2h^2+\cdots+\frac{NK^{n-1}h^{n-1}}{(n-1)!},$$

which converges to Ne^{Kh}. Thus the series (9.7) is absolutely and uniformly convergent to a continuous function $y(x)$. To prove the uniqueness we assume $z = z(x)$ is another solution, then

$$z=y_0+\int_{x_0}^{x}f(x,z(x))\,dx\ \text{and}\ y_{n+1}=y_0+\int_{x_0}^{x}f(x,y_n(x))\,dx.$$

Then

$$|z-y_{n+1}| \le \int_{x_0}^{x} |f(x,z)-f(x,y_n)|\,dx \le K\int_{x_0}^{x} |z-y_n|\,dx.$$

But $|z-y_0| \le Mh = \frac{N}{2}$ and hence $|z-y_1| \le \frac{NK(x-x_0)}{2} \le \frac{NKh}{2}$,

$$|z-y_2| \le \frac{NK^2(x-x_0)^2}{2\times 2!} \le \frac{NK^2h^2}{2\times 2!},$$

and finally

$$|z-y_{n+1}| \le \frac{NK^{n+1}(x-x_0)^{n+1}}{2\times(n+1)!} \le \frac{NK^{n+1}h^{n+1}}{2\times(n+1)!} \to 0 \text{ as } n\to\infty.$$

Hence, we have $z(x) = y(x)$.

9.3 EULER'S METHOD

To solve the first order differential equation (9.1), we expand $y(x)$ near $x = x_0$ in Taylor's series up to $n = 1$, i.e., we take

$$y(x) = y(x_0) + (x-x_0)\,y'(x_0) + \frac{(x-x_0)^2}{2!}y''(\xi), \text{ where } \xi \in (x_0, x_n).$$

If we assume y' is approximately y'_k *for* $x \in (x_{k-1}, x_k)$, then for the equi-spaced points $x_0, x_1, x_2, \cdots, x_n$ in the interval $[x_0, x_n]$, we obtain

$$\begin{aligned} y_1 &= y_0 + hf(x_0, y_0),\\ y_2 &= y_1 + hf(x_1, y_1),\\ y_3 &= y_2 + hf(x_2, y_2),\\ &\cdots\cdots\cdots\\ &\cdots\cdots\cdots\\ y_n &= y_{n-1} + hf(x_{n-1}, y_{n-1}) \end{aligned}$$

where $x_k - x_{k-1} = h$ for $k = 1, 2, \ldots, n$.

Thus the Euler's algorithm for solving the differential equation (9.1) is

$$y_{n+1} = y_n + hf(x_n, y_n),\ n = 0, 1, 2, \cdots, n-1$$

and the local error in y_{n+1} is

$$E = \frac{h^2}{2}y''(\xi), \text{ where } \xi \in (x_0, x_n),$$

which depends on h, so small values of h yield better result. Further, if y' changes rapidly over an interval, then its value at the beginning of the interval gives a poor approximation in comparison of the average value over the interval and the calculated value obtained by this method will differ from actual value.

Computational scheme for Euler's rule

n	x_n	y_n	$f(x_n, y_n)$	$hf(x_n, y_n)$	$y_{n+1} = y_n + hf(x_n, y_n)$
0	x_0	y_0	$f(x_0, y_0)$	$hf(x_0, y_0)$	$y_1 = y_0 + hf(x_0, y_0)$
1	$x_1 = x_0 + h$	y_1	$f(x_1, y_1)$	$hf(x_1, y_1)$	$y_2 = y_1 + hf(x_1, y_1)$
2	$x_2 = x_1 + h$	y_2	$f(x_2, y_2)$	$hf(x_2, y_2)$	$y_3 = y_2 + hf(x_2, y_2)$

....

....

9.4 MODIFIED EULER'S METHOD

Here we start with the initial value y_0, an approximate value of y_1, is calculated from the relation

$$y_1^{(1)} = y_0 + hf(x_0, y_0),$$

where $y_1^{(1)}$ is the first approximation of y at $x = x_1$.

The second approximation $y_1^{(2)}$ is then calculated using the formula

$$y_1^{(2)} = y_0 + \frac{h}{2}[f(x_0, y_0) + f(x_1, y_1^{(1)})].$$

The third approximation $y_1^{(3)}$ is then calculated using the formula

$$y_1^{(3)} = y_0 + \frac{h}{2}[f(x_0, y_0) + f(x_1, y_1^{(2)})].$$

Proceeding in this way, we get

$$y_1^{(n+1)} = y_0 + \frac{h}{2}[f(x_0, y_0) + f(x_1, y_1^{(n)})], n = 1, 2, 3, \cdots.$$

In general,

$$y_{j+1}^{(n+1)} = y_j + \frac{h}{2}[f(x_j, y_j) + f(x_{j+1}, y_{j+1}^{(n)})], n = 1, 2, 3, \cdots; j = 0, 1, 2, \cdots$$

9.5 RUNGE-KUTTA METHOD

Euler's method is less important in practical problems because it requires a very small step size, h for reasonable accuracy, the Runge-Kutta methods are designed to give greater accuracy and at the same time to avoid higher order derivatives. The advantage of the Runge-Kutta method is that it requires the values of the function at some selected points on the subinterval. A formula of the following form is sought for simplest form of the Runge-Kutta methods.

$$y_{n+1} = y_n + ak_1 + bk_2,$$

$$k_1 = hf(x_n, y_n),$$

$$k_2 = hf(x_n + \alpha h, y_n + \beta k_1), \tag{9.8}$$

where a, b, α, β are constants to be determined so that the above equation will agree with the Taylor's algorithm of as higher order as possible. Expanding $y(x_{n+1})$ in a Taylor's series through terms of order h^3, we obtain

$$y(x_{n+1}) = y(x_n) + hy'(x_n) + \frac{h^2}{2!}y''(x_n) + \frac{h^3}{3!}y'''(x_n) + \cdots$$

$$= y(x_n) + hf(x_n, y_n) + \frac{h^2}{2!}(f_x + ff_y)_n + \frac{h^3}{3!}(f_{xx} + 2ff_{xy} + f_{yy}f^2 + f_xf_y + f_y^2 f)_n + O(h^4)\cdots \quad [9.9]$$

where $(\ )_n$ means that the values of the function are evaluated at (x_n, y_n).

Again,

$$\frac{k_2}{h} = f(x_n + \alpha h, y_n + \beta k_1) = f(x_n, y_n) + \alpha h f_x + \beta k_1 f_y + \alpha^2 h^2 f_{xx}/2$$

$$+ \alpha\beta h k_1 f_{xy} + \beta^2 k_1^2 f_{yy}/2 + O(h^3).$$

In this case, the values of the derivatives are also evaluated at (x_n, y_n).

So from (9.8) we get

$$y_{n+1} = y_n + ahf + bh\left[f + \alpha h f_x + \beta h f f_y + \frac{1}{2}\alpha^2 h^2 f_{xx} + \alpha h \beta h f f_{xy} + \frac{1}{2}\beta^2 h^2 f^2 f_{yy} + O(h^3)\right]$$

$$= y_n + (a+b)hf + bh^2(\alpha f_x + \beta f f_y) + bh^3\left[\frac{1}{2}\alpha^2 f_{xx} + \alpha\beta f f_{xy} + \frac{1}{2}\beta^2 f^2 f_{yy}\right] + O(h^4)\cdots \quad [9.10]$$

Comparing the coefficients of h^2 and h^3 in (9.9) and (9.10), we obtain

$$a + b = 1$$

$$b\alpha = b\beta = \frac{1}{2}.$$

One of the simple solution of the above system of equations is

$$a = b = \frac{1}{2}, \quad \alpha = \beta = 1.$$

9.6 ALGORITHM OF RUNGE-KUTTA METHOD OF ORDER 2

For the equation

$$y = f(x, y), \; y(x_0) = y_0$$

generate the sequence of approximations $y_n = y(x_0 + nh)$ and $n = 0, 1, 2, \cdots$ using the formula:

$$y_{n+1} = y_n + \frac{1}{2}(k_1 + k_2),$$

Where
$$k_1 = hf(x_n, y_n)$$

$$k_2 = hf(x_n + h, y_n + k_1)..$$

Total local error of this method is of order h^3.

Formulae of Runge-Kutta methods of any order can be determined using the above method but derivations of higher order Runge-Kutta methods are very complicated. The most commonly used formula of this type is Runge-Kutta method of order 4.

9.7 ALGORITHM OF RUNGE-KUTTA METHOD OF ORDER 4

For the equation

$$y = f(x, y),\ \ y(x_0) = y_0$$

generate the sequence of approximations $y_n = y(x_0 + nh)$ and $n = 0, 1, 2, \cdots$ using the formula:

$$y_{n+1} = y_n + \frac{1}{6}(k_1 + 2k_2 + 2k_3 + k_4),$$

where $k_1 = hf(x_n, y_n)$

$$k_2 = hf(x_n + \frac{1}{2}h, y_n + \frac{1}{2}k_1)$$

$$k_3 = hf(x_n + \frac{1}{2}h, y_n + \frac{1}{2}k_2)$$

$$k_4 = hf(x_n + h, y_n + k_3).$$

The local error of this method is of order h^5.

Examples

1. Using Picard's method of successive approximation obtain a solution up to third approximation of the differential equation

$$\frac{dy}{dx} = x + y^2, \text{ where } y = 0, \text{ when } x = 0.$$

Solution: Here integrating the given equation between the limits $x = 0$ to x and $y = 0$ to y, we get

$$y = \int_0^x (x + y^2)\, dx. \tag{1}$$

First approximation y_1 : Replacing by 0 in $x + y^2$, we get from (1), $y_1 = \frac{1}{2}x^2$.

Second approximation y_2 : Replacing by y_1 in $x + y^2$, we get from (1)

$$y_2 = \int_0^x \left(x + y_1^2\right) dx = \frac{1}{2}x^2 + \frac{1}{20}x^5.$$

Third approximation y_3 : Replacing by y_2 in $x + y^2$, we get form (1)

$$y_3 = \int_0^x \left(x + y_2^2\right) dx = \frac{1}{2}x^2 + \frac{1}{20}x^5 + \frac{1}{160}x^8 + \frac{1}{4400}x^{11},$$

which gives a solution of the given equation up to third approximation.

2. Using Picard's method of successive approximation obtain a solution up to third approximation of the differential equation:

$$\frac{dy}{dx} = 1 - 2xy, \text{ where } y = 0, \text{ when } x = 0.$$

Solution: Here integrating the given equation between the limits $x = 0$ to x and $y = 0$ to y, we get

$$y = \int_0^x (1 - 2xy)\, dx. \tag{1}$$

First approximation y_1 : Replacing y by *0* in $1 - 2xy$, we get from (1), $y_1 = x$.
Second approximation y_2 : Replacing y by y_1 in $1 - 2xy$, we get from (2)

$$y_2 = \int_0^x (1 - 2xy_1)\, dx = x - \frac{2x^3}{3}.$$

Third approximation y_3 : Replacing y by y_2 in $1 - 2xy$ we get from (1)

$$y_3 = \int_0^x (1 - 2xy_2)\, dx = x - \frac{2x^3}{3} + \frac{4x^5}{15},$$

which gives a solution of the given equation up to third approximation.

3. Using Picard's method of successive approximation obtain a solution up to fourth approximation of the differential equation

$$\frac{dy}{dx} = x + y, \text{ where } y = 1, \text{ when } x = 0.$$

Solution: Here integrating the given equation between the limits $x = 0$ to x and $y = 1$ to y, we get

$$y = 1 + \int_0^x (x + y)\, dx. \tag{1}$$

First approximation y_1 = Replacing y by 1 in $x + y$ we get from (1), $y_1 = 1 + x + \frac{1}{2}x^2$
Second approximation y_2 = Replacing y by y_1 in $x + y$, we get from (1)

$$y_2 = 1 + \int_0^x (x + y_1)\, dx = 1 + x + x^2 + \frac{1}{6}x^3.$$

Third approximation y_3 : Replacing by y_2 in $x + y$, we get from (1)

$$y_3 = 1 + \int_0^x (x + y_2)\, dx = 1 + x + x^2 + \frac{1}{3}x^3 + \frac{1}{24}x^4.$$

Fourth approximation y_4 : Replacing y by y_3 in $x + y$, we get from (1)

$$y_4 = 1 + \int_0^x (x + y_3)\, dx = 1 + x + x^2 + \frac{1}{3}x^3 + \frac{1}{12}x^4 + \frac{1}{120}x^5,$$

which gives a solution of the given equation up to fourth approximation.

4. Use Euler's method with $h = 0.1$ to find the solution of

$$\frac{dy}{dx} = x^2 + y^2, \; y(0) = 0 \text{ at } x = 0.6.$$

Solution: Let $x_0 = 0$, then $y_0 = 0$. Now we construct the following table:

n	x_n	y_n	$f(x_n, y_n)$	$hf(x_n, y_n)$	$y_{n+1} = y_n + hf(x_n, y_n)$
0	0	0	0	0	0
1	0.1	0	0.01	0.001	0.001
2	0.2	0.001	0.040001	0.004	0.005
3	0.3	0.005	0.090025	0.009	0.014
4	0.4	0.014	0.160196	0.016	0.030
5	0.5	0.030	0.250900	0.025	0.055

So, $y(0.6) = 0.055$.

5. Use Euler's method with $h = 0.2$ to find the solution of

$$\frac{dy}{dx} = \frac{y - x}{y + x}, \; y(0) = 1 \text{ at } x = 1.$$

Solution: Let $x_0 = 0$, then $y_0 = 0$. Now we construct the following table, taking $h = 0.02$:

n	x_n	y_n	$f(x_n, y_n)$	$hf(x_n, y_n)$	$y_{n+1} = y_n + hf(x_n, y_n)$
0	0.0	1.00000	1.00000	0.20000	1.20000
1	0.2	1.20000	0.71430	0.14286	1.34286
2	0.4	1.34286	0.54098	0.10820	1.45106
3	0.6	1.45106	0.41500	0.08300	1.53406
4	0.8	1.53406	0.31450	0.06290	1.59696

So, $y(1.0) = 1.59696$.

6. Use Euler's method with $h = 0.2$ to find the solution of

$$\frac{dy}{dx} = x + y, \; y(0) = 0 \text{ at } x = 1.$$

Solution: Let $x_0 = 0$, then $y_0 = 0$. Now we construct the following table, taking $h = 0.02$:

n	x_n	y_n	$f(x_n, y_n)$	$hf(x_n, y_n)$	$y_{n+1} = y_n + hf(x_n, y_n)$
0	0.0	0.0000	0.0000	0.0000	0.0000
1	0.2	0.0000	0.2000	0.0400	0.0400
2	0.4	0.0400	0.4400	0.0880	0.1280
3	0.6	0.1280	0.7280	0.1456	0.2736
4	0.8	0.2736	1.0736	0.2147	0.4883

So, $y(1.0) = 0.4883$.

7. Given $\dfrac{dy}{dx} = \log(x+y)$, with the initial condition $y(0) = 1$. Find $y(0.2)$ and $y(0.5)$.

Solution: Let $x_0 = 0$, $x_1 = 0.2$ and $x_2 = 0.5$. Given that $y_0 = 1$, we have to find y_1 and y_2.

Here $f(x_0, y_0) = \log 1 = 0$, so $y_1^1 = y_0 + hf(x_0, y_0) = 1 + (0.2)(0) = 1$.

Now $f(x, y_1^1) = \log(0.2+1) = 0.07918$, so $y_1^2 = y_0 + \dfrac{h}{2}\left[f(x_0, y_0) + f(x_1, y_1^1)\right]$

$$= 1 + \frac{0.2}{2}[0 + 0.07918] = 1.0079.$$

$f(x_1, y_1^2) = \log(0.2 + 1.0079) = \log(1.2079) = 0.08203$.

So $y_1^3 = y_0 + \dfrac{h}{2}\left[f(x_0, y_0) + f(x_1, y_1^2)\right] = 1 + \dfrac{0.2}{2}[0 + 0.8203] = 1.0082.$

Again, $f(x_1, y_1^3) = \log(0.2 + 1.0082) = \log(1.2082) = 0.08213$.

So $y_1^4 = y_0 + \dfrac{h}{2}\left[f(x_0, y_0) + f(x_1, y_1^3)\right] = 1 + \dfrac{0.2}{2}[0 + 0.08213] = 1.0082.$

Hence, we see that $y_1^3 = y_1^4$, so we may take $y_1 = 1.0082$ (corrected to 4D)

To find the value $y_2 = y(0.5)$, we start with $y_1 = 1.0082$ and $h = 0.3$.

So $y_2^1 = y_1 + hf(x_1, y_1) = 1.0082 + (0.3)\log(0.2 + 1.0082) = 1.0328.$

Now $f(x_2, y_2^1) = \log(0.5 + 1.0328) = 0.18549$, so $y_2^2 = y_1 + \dfrac{h}{2}\left[f(x_1, y_1) + f(x_2, y_2^1)\right]$

$$= 1.0082 + \frac{0.3}{2}[0.08213 + 0.18549] = 1.0484.$$

Again, $f(x_2, y_2^2) = \log(0.5 + 1.0484) = 0.18988$, so $y_2^3 = y_1 + \dfrac{h}{2}\left[f(x_1, y_1) + f(x_2, y_2^2)\right]$

$$= 1.0082 + \frac{0.3}{2}[0.08213 + 0.18988] = 1.0490.$$

Again, $f(x_2, y_2^3) = \log(0.5 + 1.0490) = 0.19005$, so $y_2^4 = y_1 + \dfrac{h}{2}\left[f(x_1, y_1) + f(x_2, y_2^3)\right]$

$$= 1.0082 + \frac{0.3}{2}[0.08213 + 0.19005] = 1.0490.$$

Here we see that $y_2^3 = y_2^4$, so we may take $y(0.5) = 1.0490$.

8. Given $\dfrac{dy}{dx} = x + y$, with the initial condition $y(0) = 1$. Find $y(0.05)$.

Solution: Let $x = 0$, $x_1 = h = 0.05$. Given that $y_0 = 1$, we have to find y_1.

Here $f(x_0, y_0) = 1$, so $y_1^1 = y_0 + hf(x_0, y_0) = 1 + (0.05)(1) = 1.05$

Now $f(x_1, y_1^1) = (0.05 + 1.05) = 1.1$, so $y_1^2 = y_0 + \dfrac{h}{2}\left[f(x_0, y_0) + f(x_1, y_1^1)\right]$

$$= 1 + \frac{0.05}{2}[1 + 1.1] = 1.0525.$$

$f(x_1, y_1^2) = 0.05 + 1.0525 = 1.1025,$

So $y_1^3 = y_0 + \frac{h}{2}\left[f(x_0, y_0) + f(x_1, y_1^2)\right] = 1 + \frac{0.05}{2}[1 + 1.1025] = 1.0526.$

Again, $f(x_1, y_1^3) = 0.05 + 1.0526 = 1.1026,$

So $y_1^4 = y_0 + \frac{h}{2}\left[f(x_0, y_0) + f(x_1, y_1^3)\right] = 1 + \frac{0.05}{2}[1 + 1.1026] = 1.0526.$

Hence, we see that $y_1^3 = y_1^4$, so we may take $y_1 = 1.0526$ (corrected to 4D).

9. Use Runge-Kutta method of order two to approximate y, when $x = 0.1$ and 0.2 given that

$$\frac{dy}{dx} = y - x, \; y(0) = 2.$$

Solution: Here we take $x_0 = 0$, $y_0 = 1$ and $h = 0.1$.
Then $f(x_0, y_0) = 2$ and $k_1 = hf(x_0, y_0) = (0.1)\,(2) = 0.2$

$$k_2 = hf(x_0 + h, y_0 + k_1) = (0.1)\,f(0.1, 2.2)$$
$$= (0.1)\,(2.2 - 0.1) = 0.21.$$

So, $y(0.1) = y(0) + \frac{1}{2}[k_1 + k_2] = 2 + \frac{1}{2}(0.2 + 0.21) = 2.205.$

To determine $y(0.2)$ we start with $x_0 = 0.1$, $y_0 = 2.205$ and $h = 0.1$.
Then $f(x_0, y_0) = 2.105$ and $k_1 = hf(x_0, y_0) = (0.1)\,(2.205) = 0.2105$

$$k_2 = hf(x_0 + h, y_0 + k_1) = (0.1)\,f(0.2, 2.4155)$$
$$= (0.1)\,(2.4155 - 0.2) = 0.22155.$$

So, $y(0.1) = y(0) + \frac{1}{2}[k_1 + k_2] = 2.205 + \frac{1}{2}(0.2105 + 0.22155) = 2.4210.$

10. Use Runge-Kutta method of order four to approximate y, when $x = 0.1$, given that

$$\frac{dy}{dx} = x + y, \; y(0) = 1.$$

Solution: Here we take $x_0 = 0$, $y_0 = 1$ and $h = 0.1$.
Then $f(x_0, y_0) = 1$ and $k_1 = hf(x_0, y_0) = (0.1)\,(1) = 0.1$

$$k_2 = hf\left(x_0 + \frac{1}{2}h, y_0 + \frac{1}{2}k_1\right) = (0.1)\,f(0.05, 1.005)$$
$$= (0.1)\,(1.1) = 0.11$$
$$k_3 = hf\left(x_0 + \frac{1}{2}h, y_0 + \frac{1}{2}k_2\right) = (0.1)\,f(0.05, 1.055)$$
$$= (0.1)\,(1.105) = 0.1105$$
$$k_4 = hf(x_0 + h, y_0 + k_3) = (0.1)\,f(0.1, 1.1105)$$
$$= (0.1)\,(1.2105) = 0.12105.$$

So, $y(0.1) = y(0) + \frac{1}{6}[k_1 + 2k_2 + 2k_3 + k_4] = 1 + \frac{1}{6}[0.1 + 2(0.11) + 2(0.1105) + 0.12105]$

$$= 1.11034.$$

11. Use Runge-Kutta method of order four to approximate y, when $x = 0.5$, given that

$$\frac{dy}{dx} = \frac{1}{x+y}, \; y(0) = 1.$$

Solution: Here we take $x_0 = 0$, $y_0 = 1$ and $h = 0.5$.

Then $f(x_0, y_0) = 1$ and $k_1 = hf(x_0, y_0) = (0.5)\,(1) = 0.5$

$$k_2 = hf\left(x_0 + \frac{1}{2}h,\, y_0 + \frac{1}{2}k_1\right) = (0.5)\, f(0.25, 1.25)$$

$$= \frac{0.5}{1.5} = 0.333$$

$$k_3 = hf\left(x_0 + \frac{1}{2}h,\, y_0 + \frac{1}{2}k_2\right) = (0.5)\, f(0.25, 1.167)$$

$$= \frac{0.5}{1.417} = 0.353$$

$$k_4 = hf(x_0 + h, y_0 + k_3) = (0.5)\, f(0.5, 1.353)$$

$$= \frac{0.5}{1.853} = 0.270.$$

So, $y(0.5) = y(0) + \frac{1}{6}[k_1 + 2k_2 + 2k_3 + k_4] = 1 + \frac{1}{6}[0.5 + 2(0.333) + 2(0.353) + 0.270]$

$$= 1.357.$$

12. Use Runge-Kutta method of order four to approximate y, when $x = 0.8$, given that

$$\frac{dy}{dx} = \sqrt{x+y}, \; y(0.4) = 0.41.$$

Solution: Here we take $x_0 = 0.4$, $y_0 = 0.41$ and $h = 0.4$.

Then $f(x_0, y_0) = 0.9$ and $k_1 = hf(x_0, y_0) = (0.4)\,(0.9) = 0.36$

$$k_2 = hf\left(x_0 + \frac{1}{2}h,\, y_0 + \frac{1}{2}k_1\right) = (0.4)\, f(0.6, 0.59)$$

$$= (0.4)\,(1.091) = 0.436$$

$$k_3 = hf\left(x_0 + \frac{1}{2}h,\, y_0 + \frac{1}{2}k_2\right) = (0.4)\, f(0.6, 0.628)$$

$$= (0.4)\,(1.108) = 0.443$$

$$k_4 = hf(x_0 + h, y_0 + k_3) = (0.4)\, f(0.8, 0.853)$$

$$= (0.4)\,(1.286) = 0.514.$$

So, $y(0.8) = y(0.4) + \frac{1}{6}[k_1 + 2k_2 + 2k_3 + k_4]$

$$= 0.41 + \frac{1}{6}[0.36 + 2(0.436) + 2(0.443) + 0.514]$$

$$= 0.848.$$

EXERCISE

1. Prove that if the function $f(x, y)$ is bounded and continuous in some region about the point (x_0, y_0) and if $f(x, y)$ satisfies the Lipschitz condition, then Picard's method converges uniquely.
2. Derive algorithm of Runge-Kutta method of order 2.
3. Define single step and multi-steps methods. Give examples.
4. Discuss the advantages and disadvantages of Runge-Kutta method over Taylor's method.
5. Describe Euler's method for the numerical solution of an ordinary differential equation.
6. Using Picard's method of successive approximation obtain a solution up to third approximation of the differential equation:
$$\frac{dy}{dx} = x - y, \text{ where } y = 1, \text{ when } x = 0.$$
7. Using Picard's method of successive approximation, obtain a solution up to third approximation of the differential equation
$$\frac{dy}{dx} = \frac{2y - x}{x}, \text{ given } y(1) = 2.$$
8. Use Euler's method with $h = 0.1$ to find the solution of
$$\frac{dy}{dx} = x + y, \; y(0) = 0 \text{ at } x = 0.4.$$
9. Use Euler's method with $h = 0.1$ to find the solution of
$$\frac{dy}{dx} = 3x + 2y, \; y(0) = 1 \text{ at } x = 0.3.$$
10. Use Eulers method with $h = 0.1$ to find the solution of
$$\frac{dy}{dx} = x + y^2, \; y(0) = 0 \text{ at } x = 0.5.$$
11. Given: $\frac{dy}{dx} = x + y$, with the initial condition $y(0) = 1$. Find $y(0.1)$.
12. Use modified Euler's method to find $y(0.2)$ from
$$\frac{dy}{dx} = x^2 + y, \; y(0) = 1$$
13. Use Runge-Kutta method of order two to approximate y, when $x = 0.3$ and 0.4, given that
$$\frac{dy}{dx} = y - x, \; y(0) = 2.$$
14. Use Runge-Kutta method of order four to approximate y, when $x = 1.1$, given that
$$\frac{dy}{dx} = 3x + y^2, \; y(1) = 1.2.$$

15. Use Runge-Kutta method of order four to approximate y, when $x = 0.1$, given that

$$\frac{dy}{dx} = x^2 + y, \; y(0) = -1.$$

16. Use Runge-Kutta method of order four to approximate y, when $x = 0.2$, given that

$$\frac{dy}{dx} = \frac{y - x}{y + x}, \; y(0) = 1.$$

ANSWERS

6. $1 - x + x^2 - \frac{x^3}{3} + \frac{x^4}{24}$, **7.** $2x - 2\log x - \frac{1}{3}(\log x)^3$, **8.** 0.0611,

9. 1.824, **10.** 0.1005, **11.** 1.11034,

12. 1.2359, **13.** 2.6492, 2.8909, **14.** 1.727,

15. −1.105, **16.** 1.38.

Chapter 10

A FEW SHORT QUESTIONS ON NUMERICAL ANALYSIS

1. What is numerical analysis and what is algorithm?

❑ The objective of numerical analysis is to solve complex numerical problems using simple operations of arithmetic, to develop and evaluate methods for computing numerical results from given data. The methods of computation is called algorithm.

2. Round the numbers 49.3628, 59.45 to 3 significant digits.

❑ 49.4, 59.4.

3. Round the numbers 0.0056812, 63.85435 to 4 decimal places.

❑ 0.0057, 63.8544

4. If we take $\pi = 3.14$ instead of 3.14159, find the absolute error.

❑ Absolute error = (3.14159 − 3.14) = 0.00159

5. If 0.8333 is taken to be an approximate value of 5/6, find the percentage error.

❑ In this case the absolute error = | 5/6 − 0.8333 | = 0.0000333.

So relative error $= \dfrac{0.0000333}{5/6} = 0.00004.$

Hence, the percentage error = 0.00004 × 100% = 0.004%.

6. Evaluate $\Delta^2 x^3$.

❑ $\Delta^2 x^3 = \Delta(\Delta x^3) = \Delta\,[(x+h)^3 - x^3] = \Delta\,[3x^2h + 3xh^2 + h^3] = 3h\,[(x+h)^2 - x^2] + 3h^3$
$= 6xh^2 + 6h^3.$

7. Show that $(1+\Delta)(1-\nabla) f(x) = f(x)$.

❑ $(1-\nabla) f(x) = f(x) - \nabla f(x) = f(x) - f(x) + f(x-h) = f(x-h)$.

So, $(1+\Delta)(1-\nabla) f(x) = (1+\Delta) f(x-h) = f(x-h) + f(x) - f(x-h) = f(x)$.

8. If V_A and V_T denote the approximate and true value of a number, find the number of significant figures in

(a) $V_A = 0.5642$, given its absolute error as 0.25×10^{-2}.

(b) $V_T = 2.1345$, given its relative error as 0.02×10^{-2}.

❑ (a) In this case absolute error $= 0.25 \times 10^{-2} < 0.50 \times 10^{-2}$, so the number is correct to two decimal places and hence, the given number is correct up to two significant figures.

(b) Absolute error $= 2.1345 \times 0.02 \times 10^{-2} < 0.50 \times 10^{-3}$, so the number is correct to three decimal places and hence the given number is correct up to four significant figures.

9. If $y = \dfrac{0.31x + 2.73}{x + 0.35}$, find the relative error in y, when $x = 0.5 \pm 0.1$

❑ Here $y(0.6) = 3.07$, $y(0.5) = 3.39$ and $y(0.4) = 3.81$. So when x is changed to 0.6 from 0.5, the relative error in y is $(3.39 - 3.07)/3.39 = 0.09$ and when x is changed to 0.4 from 0.5, the relative error in y is $(3.81 - 3.39)/3.39 = 0.12$.

10. Find the relative error in $T = P_1P_2$ with $P_1 = 5.83$, $P_2 = 3.82$, if error in both P_1, P_2 is 0.01.

❑ Relative error in T $= \dfrac{(5.83 + 0.01)(3.82 + 0.01)}{5.83 \times 3.82} - 1 = 0.0045.$

11. Let $u = x^3y^2$, if x_0, y_0 be the approximate values of x, y respectively and Δx_0, Δy_0 be the absolute error in them, determine the error in u.

❑ $u = x^3y^2 \Rightarrow \log u = 3\log x + 2\log y$. So the relative error

$$= \left|\frac{\Delta u_0}{u_0}\right| = \left|3\frac{\Delta x_0}{x_0} + 2\frac{\Delta y_0}{y_0}\right| \le 3\left|\frac{\Delta x_0}{x_0}\right| + 2\left|\frac{\Delta y_0}{y_0}\right|.$$

12. Show that $E^{-1} = 1 - \nabla$.

$(1 - \nabla) f(x) = f(x) - f(x) + f(x - h) = f(x - h) = E^{-1}f(x)$, So $E^{-1} = 1 - \nabla$.

13. Prove that $\Delta - \nabla = \Delta\nabla$.

❑ $(\Delta - \nabla) f(x) = \Delta f(x) - \nabla f(x) = \Delta f(x) - f(x) + f(x - h) = \Delta[f(x) - f(x - h)] = \Delta\nabla f(x)$.

14. Prove that $\nabla^n y_k = \Delta^n y_{k-n}$

❑ L.H.S. $= (1 - E^{-1})^n y_k = \{(E - 1)/E\}^n y_k = (E - 1)^n E^{-n} y_k = (E - 1)^n y_{k-n} = \Delta^n y_{k-n}$.

15. Find a function y_k for which $\Delta^2 y_k = 9y_k$. Can you find two such functions?

❑ Let $y_k = C^k$, then $\Delta y_k = C^{k+1} - C^k = C^k(C - 1)$

$$\Rightarrow \Delta^2 y_k = C^k(C - 1)^2 = 9y_k$$

$$\Rightarrow (C - 1)^2 = 9 \rightarrow C - 1 = \pm 3 \Rightarrow C = 4, -2.$$

Therefore, there exists two such functions and those are 4^k and $(-2)^k$.

16. Express $\nabla^3 f_i$ in terms of f_i.

❑ $$\nabla f_i = f_i - f_{i-1},\ \nabla^2 f_i = \nabla f_i - \nabla f_{i-1} = f_i - f_{i-1} - f_{i-1} + f_{i-2} = f_i - 2f_{i-1} + f_{i-2}.$$

$$\nabla^3 f_i = \nabla(f_i - 2f_{i-1} + f_{i-2}) = f_i - 3f_{i-1} + 3f_{i-2} - f_{i-3}.$$

17. Prove that $\Delta\{\log f(x)\} = \log\left[1+\frac{\Delta f(x)}{f(x)}\right]$.

❑ $\Delta\{\log f(x)\} = \log f(x+h) - \log f(x) = \log\left[\frac{f(x+h)}{f(x)}\right]$

$$= \log\left[\frac{f(x)+f(x+h)-f(x)}{f(x)}\right] = \log\left[1+\frac{\Delta f(x)}{f(x)}\right].$$

18. Show that Δ, ∇, E are linear operators.

❑ $\Delta\{af(x) + bg(x)\} = af(x+h) + bg(x+h) - \{af(x) + bg(x)\}$

$= a\{f(x+h) - f(x)\} + b\{g(x+h) - g(x)\}$

$= a\Delta f(x) + b\Delta g(x)$.

Hence, Δ is a linear operator. Similarly, it can be shown that ∇, E are also linear operators.

19. What is a stable algorithm?

❑ In extended calculations it is likely that many round off will be made. Each of which plays an important role of an input error for the remainder of the computation, and each has an impact on the eventual output. Algorithms, for which the cumulative effect of such errors is limited so that a useful result is generated, are called stable algorithms. But it is found sometimes that the cumulative effect is devastating and the solution is overwhelmed by error. Such algorithms are called unstable.

20. What is called collocation polynomial?

❑ A polynomial $P(x)$ which agrees with the values of a function $f(x)$ at $x = x_i$, $i = 0, 1, 2, \cdots, n$ is called collocation polynomial.

21. Why polynomials are used in numerical analysis?

❑ Approximation of a function by a polynomial is a good idea in numerical analysis, because

(i) polynomials are very easy to use. As per need we can increase or decrease the degree of the polynomials easily.

(ii) their derivatives and integrals are also found without much effort and those are again polynomials and

(iii) roots of a polynomial equation can be easily determined.

22. Prove that there exists exactly one polynomial which can take the specified values y_k at given points x_k, $k = 0, 1, 2, \cdots, n$.

❑ Let if possible there exists two polynomials $P(x)$ and $Q(x)$ of degree n, which can interpolate $f(x)$ at $x = x_k$, $k = 0, 1, 2, \cdots, n$. Then $g(x) = P(x) - Q(x)$ will be a polynomial of degree n or less and $g(x_k) = 0$ for $k = 0, 1, 2, \cdots, n$, which shows that an nth (or less) degree polynomial has $n + 1$ roots and this contradiction implies that $P(x) = Q(x)$.

23. If x_j and x_k are any two of the points x_i $(i = 0, 1, 2, \cdots, n)$, then find the linear interpolating polynomial through these points.

❑ Lagrange polynomial for these two points can be written as

$$p_{jk} = f_k L_k(x) + f_j L_j(x),$$

where $L_k(x) = \dfrac{x - x_j}{x_k - x_j}$ and $L_j(x) = \dfrac{x - x_k}{x_j - x_k}$.

So, $p_{jk} = f_k \dfrac{x - x_j}{x_k - x_j} + f_j \dfrac{x - x_k}{x_j - x_k} = \dfrac{f_k(x - x_j) - f_j(x - x_k)}{x_k - x_j}$.

24. When the use of Stirling's formula or Bessel's formula will be made in case central interpolation?

❑ If the arguments are equi-spaced and even in number, i.e., the difference table ends with odd order differences and the given value of x lies near the middle of the range, the central interpolation formula of choice would be Stirling's formula, and if the arguments are equi-spaced and odd in number, i.e., the difference table ends with even order differences and the given value of x lies near the middle of the range the central interpolation formula of choice would be the Bessel's formula.

25. What are the advantages of Lagrange's interpolation formula?

❑ A major advantage of Lagrange's interpolation formula is that the coefficients of y_i in the expressions of the formula are easily determined and secondly, this formula can be applied to either equally or unequally spaced arguments.

26. A function defined in (0, 1) is such that $f(0) = 0$, $f(1/2) = -1$ and $f(1) = 0$. Find the 2nd degree polynomial $p(x)$ which interpolates $f(x)$ in (0, 1).

❑ Using Lagranges formula we get

$$p(x) = \frac{(x-1/2)(x-1)}{(-1/2)(-1)} f(0) + \frac{(x-0)(x-1)}{(1/2)(-1/2)} f(1/2) + \frac{(x-1/2)(x-0)}{(1/2)(1)} f(1) = 4x(x-1).$$

27. Define order of convergence of an iterative method?

❑ The order of convergence of an iterative method is said to be p, if there exists a constant C > 0 such that

$$\lim_{n\to\infty} \left| \frac{e_{n+1}}{e_n^p} \right| = C, \text{ where } e_n \text{ denotes the error in the } n\text{th step.}$$

28. When may the bisection method be used to find a root of the equation $f(x) = 0$?

❑ Bisection method is applicable only for the roots of $f(x) = 0$ about which $f(x)$ changes sign. So if $f(x) = 0$ has double roots, then $f(x)$ does not change sign about these roots. So, bisection method cannot be applied to determine a double root of $f(x) = 0$.

29. Why bisection method is not applied to evaluate a double root of an equation?

❑ If $f(x) = 0$ has double roots, then $f(x)$ does not change sign about these roots. So, bisection method cannot be applied to determine a double root of $f(x) = 0$.

30. What major advantage has the Newton-Raphson method over other methods?

❑ The order of convergence of Newton-Raphson method is higher than any other iterative methods. So, if Newton-Raphson method converges, then it converges faster than any other iterative methods.

31. State the condition of convergence of Newton-Raphson method.

❑ When the equation $f(x) = 0$ is solved using Newton-Raphson method, the method will converge, if $|f(x)\,f''(x)| < |f'(x)|^2$.

32. When Newton-Raphson method fails? How it can be made convergent in such case?

❑ From the algorithm of this method we see that if $f'(x)$ is closed to zero, then the method diverges and in such case we have to start with a fresh approximation obtained by the method of bisection.

33. Let $f(x)$ be a continuous function in $[a, b]$ such that $f(a)f(b) < 0$. Show that if a number c is calculated using regula-falsi method, then the length of a new bracket is

$$\frac{(a-b)\,f(a)}{f(b)-f(a)}, \quad \text{if } f(a)f(c) < 0$$

and
$$\frac{(b-a)\,f(b)}{f(b)-f(a)}, \quad \text{if } f(b)f(c) < 0.$$

❑ Let $f(a) < 0$ and $f(b) > 0$, then using regula-falsi method, we obtain

$$c = \frac{af(b) - bf(a)}{f(b) - f(a)}.$$

If $f(c) > 0$, then the length of the new interval is

$$c - a = \frac{af(b) - bf(a)}{f(b) - f(a)} - a = \frac{(a-b)\,f(a)}{f(b) - f(a)}.$$

If $f(c) < 0$, then the length of the new interval is

$$b - c = b - \frac{af(b) - bf(a)}{f(b) - f(a)} = \frac{(b-a)\,f(b)}{f(b) - f(a)}.$$

34. State a sufficient condition related to convergence of the method of iteration.

❑ Let I be an interval containing the root of the equation $f(x) = 0$, which can be written as $x = \varphi(x)$, where $\varphi(x)$ is a continuously differentiable function in I. Now, if $|\varphi'(x)| < 1\ \forall x \in I$, then the sequence of iterations defined by $x_{n+1} = \varphi(x_n)$, $n = 0, 1, 2, \cdots$ converges to the root of $f(x) = 0$ for any choice of x_0 in I.

35. When does bisection/regula-falsi method fail?

❑ If the equation $f(x) = 0$ has multiple root $x = \alpha$ of even order, then $f(x)$ does not change signs about $x = \alpha$. Hence bisection/regula-falsi method fails to evaluate the root $x = \alpha$.

36. Why regula-falsi method is called a method of linear interpolation?

❑ In regula-falsi method the linear function

$$y = f(a) + \frac{f(a) - f(b)}{a - b}(x - a),$$

which collocate $f(x)$ at $x = a, b$; replace the curve $y = f(x)$ in the neighborhood of the root. That is why regula-falsi method is called a method of linear interpolation.

37. Show that bisection method converges linearly.

❑ If x_{n+1}, e_{n+1} be the nth iteration and error in the nth iteration in approximating the root α of an equation $f(x) = 0$ using bisection method. Then, for large values of n, we have

$$|e_{n+1}| = |\alpha - x_{n+1}| \approx \frac{1}{2}|b_n - a_n|$$

and $$|e_n| = |\alpha - x_n| \approx \frac{1}{2}|b_{n-1} - a_{n-1}| = |b_n - a_n|.$$

Dividing we get $\left|\dfrac{e_{n+1}}{e_n}\right| \approx \dfrac{1}{2}$. This shows that bisection method converges linearly.

38. Show that regula-falsi method converges linearly.

❑ From algorithm of regula-falsi method we get

$$x_{n+1} = x_n - \frac{b_n - a_n}{f(b_n) - f(a_n)} f(x_n), \text{ since } x_n = a_n \text{ or } b_n.$$

$$\Rightarrow (x_{n+1} - x_n)(f(b_n) - f(a_n)) = -(b_n - a_n)(f(x_n) - f(\alpha)),\ f(\alpha) = 0$$

$$\Rightarrow [(\alpha - x_{n+1}) - (\alpha - x_n)] f'(\xi_n) = (x_n - \alpha) f'(\zeta_n), \text{ where } a_n < \xi_n < b_n, \alpha < \zeta_n < x_n$$

$$\Rightarrow e_{n+1} = \frac{f'(\xi_n) - f'(\zeta_n)}{f'(\xi_n)} e_n$$

Since, all the iterations are within $[a_0, b_0]$ and if we assume that $f'(x)$ is continuous and has the same sign in $[a_0, b_0]$, then there exists two positive numbers m, M such that

$$m \le |f'(x)| \le M \text{ for all } x \text{ in } [a_0, b_0]$$

Hence, for large values of n, we get

$$|e_{n+1}| \approx \frac{M - m}{m}|e_n|.$$

This shows that regula-falsi method converges linearly.

39. Which of the methods, Secant and Newton-Raphson, is more rapidly convergent?

❑ The order of convergence of 'Secant method' is 1.618 and that of 'Newton-Raphson method' is 2. So, obviously 'Newton-Raphson method is more rapidly convergent.

40. State the difference between Direct and Iterative methods?

❑ Direct method are those which, in absence of round off or other errors, will yield the exact solution in a finite number of elementary arithmetic operations. Iterative methods are those which start with an initial approximation and which, by applying a suitably chosen algorithm, leads successively better approximations. In this case if the process converges we can only hope to get an approximate result.

41. Derive $x_{n+1} = \frac{1}{2}\left(x_n + \frac{a}{x_n}\right)$ for determining the square root of $a > 0$ using Newton's method.

❑ To find $\sqrt{a}$, let us take $f(x) = x^2 - a$, then $f'(x) = 2x$.

Using Newton's formula: $x_{n+1} = x_n - \frac{f(x_n)}{f'(x_n)}$, we get $x_{n+1} = \frac{1}{2}\left(x_n + \frac{a}{x_n}\right)$.

42. Show that the equation $x^2 + \ln x = 0$ has exactly one root and the root lies in the literval [1/3, 1].

❑ In this case $f(1/3) < 0$ and $f(1) > 0$, so $f(x) = 0$ has a root in [1/3, 1]. Now $f'(x) = 2x + \frac{1}{x} > 0 \; \forall x \in [1/3, 1]$, i.e:, $f(x)$ does not change sign in [1/3, 1]. So $f(x) = 0$ has only one root in [1/3, 1].

43. The equation $x^2 + ax + b = 0$ has two real roots α, β. Show that the method of iteration

$$x_{k+1} = -\frac{ax_k + b}{x_k} \text{ is convergent near } \alpha, \text{ if } |\alpha| > |\beta|.$$

❑ Given that $\alpha + \beta = -a, \alpha\beta = b$. Now, if we take

$$x = -\frac{ax + b}{x} = \phi(x),$$

then $|\phi'(x)| = \left|\frac{b}{x^2}\right| \Rightarrow |\phi'(\alpha)| = \left|\frac{b}{\alpha^2}\right| = \left|\frac{\beta}{\alpha}\right| < 1$, if $|\alpha| > |\beta|$.

So the iteration $x_{k+1} = -\frac{ax_k + b}{x_k}$ converges to α, if $|\alpha| > |\beta|$.

44. The equation $x^2 + ax + b = 0$ has two real roots α, β. Show that the method of iteration

$$x_{k+1} = -\frac{b}{x_k + a} \text{ is convergent near } \alpha, \text{ if } |\alpha| < |\beta|.$$

❑ Given that $\alpha + \beta = -a, \alpha\beta = b$. Now if we take

$$x = -\frac{b}{x + a} = \phi(x),$$

then $|\phi'(x)| = \left|\frac{b}{(x+a)^2}\right| \Rightarrow |\phi'(\alpha)| = \left|\frac{b}{(\alpha + a)^2}\right| = \left|\frac{b}{\beta^2}\right| = \left|\frac{\alpha}{\beta}\right| < 1$, if $|\alpha| < |\beta|$.

So the iteration $x_{k+1} = -\frac{b}{x_k + a}$ converges to α, if $|\alpha| < |\beta|$.

45. The equation $x^2 + ax + b = 0$ has two real roots α, β. Show that the method of iteration

$$x_{k+1} = -\frac{x_k^2 + b}{a} \text{ is convergent near } \alpha, \text{ if } 2|\alpha| < |\alpha + \beta|.$$

❑ Given that $\alpha+\beta=-a, \alpha\beta=b$. Now if we take

$$x=-\frac{x^2+b}{a}=\phi(x),$$

then $$|\phi'(x)|=\left|\frac{2x}{a}\right| \Rightarrow |\phi'(\alpha)|=\left|\frac{2\alpha}{a}\right|=\left|\frac{2\alpha}{\alpha+\beta}\right|<1, \text{ if } 2|\alpha|<|\alpha+\beta|.$$

So the iteration $x_{k+1}=-\dfrac{x_k^2+b}{a}$ converges to α, if $2|\alpha|<|\alpha+\beta|$.

46. Write down the merits and demerits of Newton's method for computing a simple root of $f(x)=0$.

OR, Compare Newton's method and Secant Method.

❑ In Newton's method, one has to start with an approximation "close enough" to a root of $f(x)=0$ to ensure convergence. In this method one has to calculate $f'(x)$ which is not always an easy task. Further, in each step Newton's method requires two evaluations per step while Secant method requires one evaluation per step. But when Secant method converges, it does not converge as fast as Newton's method. So, for general purpose, the Secant method will be more efficient and should be preferred.

47. Why Newton-Raphson method is called method of tangent?

❑ If the tangent to the curve $f(x)=0$ at (x_0, y_0) cuts the x-axis at $(x_1, 0)$ and makes an angle θ with the positive direction of x-axis, then

$$f'(x_0)=\tan\theta=\frac{f(x_0)-0}{x_0-x_1} \Rightarrow x_1=x_0-\frac{f(x_0)}{f'(x_0)}.$$

Thus the tangent to the curve $f(x)=0$ at the point $x=x_0$ gives the next approximation $x=x_1$ of Newton-Raphson method. That is why this method is also known as method of tangent.

48. Show that the iteration $x_{n+1}=\dfrac{1}{2}\left(x_n+\dfrac{a}{x_n}\right)$ has a second order convergence to $\sqrt{a}$.

❑ Here we see that $x_{n+1}-\sqrt{a}=\dfrac{1}{2}\left(x_n+\dfrac{a}{x_n}\right)-\sqrt{a}=\dfrac{(x_n-\sqrt{a})^2}{2x_n}$,

or, $\dfrac{e_{n+1}}{e_n^2}=\dfrac{1}{2x_n}$, where e_n denotes the error in the nth iteration,

which shows that the given iteration has a second order convergence to $\sqrt{a}$.

49. Suppose the iteration $x_{n+1}=g(x_n)$ produces a sequence of numbers converging to a fixed point ξ. If $g''(x)$ exists, then show that the result

$$g(x_n)=g(\xi)+(x_n-\xi)\,g'(\xi)+\frac{(x_n-\xi)^2}{2!}g''(\zeta),$$

where $\zeta\in(\xi, x_n)$ has second order convergence, if $g'(x)=0$ and $g''(x)\neq 0$.

- Using the results $x_{n+1} = g(x_n)$, $\xi = g(\xi)$, $g'(x) = 0$ and $g''(x) \neq 0$ in the given expression, we get $e_{n+1} = e_n^2 \frac{g''(\zeta)}{2}$, which shows that the given iteration has a second order convergence.

50. On what geometric construction is the method of false position based?

- If we join two points on the curve, say $\{x_0, f(x_0)\}$ and $\{x_1, f(x_1)\}$, such that $f(x_0)f(x_1) < 0$, then the next iteration will be the point, where the chord joining those points cuts the x-axis.

51. Why are numerical methods used in solving non-linear equations?

- Generally, it is difficult to find a root of an equation of degree 3. So, except in special case (e.g., when factorisation is easy), we prefer in practice to use numerical methods to solve polynomial equation of degree > 2. Another class of non-linear equations consists of transcendental functions such as e^x, $\log x$, $\sin x$, $\cos x$. Useful analytic solutions of such equation are rare, so we are forced to use numerical methods.

52. When Newton's interpolation formulae are used?

- The Newton's forward and backward difference formulae are well suited for use at the beginning and at the end of a difference table, i.e., when the value of x, at which the value of the function is to be evaluated, lies near the begining or near the end of the table.

53. If the quadrature rule

$$\int_{-1}^{1} y\,dx = ay_0 + by_1$$

is exact for polynomial of highest possible degree, what is the degree?

- Here we have used two points $x_0 = -1$ and $x_1 = 1$ and there exists exactly one polynomial of degree 2 – 1 = 1, which can interpolate $f(x)$ at these points. So the highest possible degree of the polynomial is one.

54. If T_1, T_2 denote the trapezoidal approximation to

$$\int_a^b f(x)\,dx$$

with 1, 2 subintervals respectively, show that $I - T_2 = \frac{1}{3}(T_2 - T_1)$.

- We know that in trapezoidal rule, error is proportional to h^2. Let $I = T_1 + e_1 = T_2 + e_2$, where $e_1 = C(2h)^2$ and $e_2 = Ch^2$ i.e. $e_1 = 4e_2$.

  Now $T_2 - T_1 = e_1 - e_2 = 3e_2 = 3(I - T_2)$, i.e. $I - T_2 = \frac{1}{3}(T_2 - T_1)$.

55. If S_1, S_2 denote the Simpson's approximations to $I = \int_a^b f(x)\,dx$, with 1, 2 subintervals respectively, then show that $I = \frac{16S_2 - S_1}{15}$.

□ We know that in Simpson's 1/3 rule, the error term is proportional to h^4. Let $I = S_1 + e_1 = S_2 + e_2$, where $e_1 = C(2h)^4$, $e_2 = Ch^4$, i.e., $e_1 = 16e_2$. Now

$$S_2 - S_1 = e_1 - e_2 = 15e_2 = 15(I - S_2)$$

$$\Rightarrow I = \frac{16S_2 - S_1}{15}.$$

56. How big should the spacing (h) be so that the error in computing $\int_{1.0}^{1.3} \sqrt{x}dx$, using trapezoidal rule, is less than or equal to 0.5×10^{-4}.

□ Here $f(x) = \sqrt{x} \Rightarrow f'(x) = \frac{1}{2\sqrt{x}}$; $f''(x) = -\frac{1}{4x^{3/2}}$.

Now the transaction error in trapezoidal rule $= -\frac{(b-a)h^2}{12} f''(\xi)$, $1 \le \xi \le 1.3$.

Hence, $\left| \frac{(b-a)h^2}{12} \max_{1 \le \xi \le 1.3} f''(\xi) \right| \le 0.5 \times 10^{-4}$ gives

$$\frac{0.3h^2}{12} \times \frac{1}{4} \le 0.00005$$

$$\Rightarrow h^2 \le 0.008$$

$$\Rightarrow h \le 0.089.$$

57. How big should the spacing (h) be so that the error in computing $\int_{1.0}^{1.3} \sqrt{x}dx$, using Simpson's 1/3 rule, is less than or equal to 0.5×10^{-4}.

□ Here $f(x) = \sqrt{x} \Rightarrow f'(x) = \frac{1}{2\sqrt{x}}$; $f''(x) = -\frac{1}{4x^{3/2}}$; $f'''(x) = \frac{3}{8x^{5/2}}$; $f^{IV}(x) = -\frac{15}{16x^{7/2}}$.

The truncation error in Simpson's 1/3 rule $= -\frac{(b-a)h^4}{180} f^{IV}(\xi), 1 \le \xi \le 1.3$.

Hence, $\left| \frac{(b-a)h^4}{180} \max_{1 \le \xi \le 1.3} f^{IV}(\xi) \right| \le 0.5 \times 10^{-4}$ gives

$$\frac{0.3h^4}{180} \times \frac{15}{16} \le 0.00005$$

$$\Rightarrow h^4 \le 0.032$$

$$\Rightarrow h \le 0.423.$$

58. Find the number of subintervals and the step size (h) so that the maximum error for the composite Simpson's 1/3 rule is 0.5×10^{-6} for the evaluation of $\int_{1.5}^{2.5} e^{-x}dx$.

❑ Here $f(x)=e^{-x} \Rightarrow f'(x)=-e^{-x}$; $f''(x)=e^{-x}$; $f'''(x)=-e^{-x}$; $f^{IV}(x)=e^{-x}$.

Now the truncation error in Simpson's 1/3 rule $=-\dfrac{(b-a)h^4}{180}f^{IV}(\xi), 1.5 \le \xi \le 2.5$.

Hence $$\left|\frac{(b-a)h^4}{180}\max_{1.5\le\xi\le2.5} f^{IV}(\xi)\right| \le 0.5\times10^{-6} \text{ gives}$$

$$\frac{h^4}{180}\times e^{-1.5} \le 0.0000005$$

$$\Rightarrow h^4 \le 0.0004$$

$$\Rightarrow h \le 0.142.$$

As the numbers of subdivision is even, we take $h=0.1$ and the number of subintervals $=10$.

59. Estimate the truncation error of the trapezoidal rule for the evaluation of

$$\int_{1.0}^{1.3}\sqrt{x}dx \text{ with } h=0.05.$$

❑ Here $f(x)=\sqrt{x} \Rightarrow f'(x)=\dfrac{1}{2\sqrt{x}}$; $f''(x)=-\dfrac{1}{4x^{3/2}}$. With h = 0.05, $b-a$ = 0.3 and $f''(x)=-\dfrac{1}{4x^{3/2}}$, we see that the truncation error in trapezoidal rule is

$$\left|\frac{(b-a)h^2}{12}\max_{1\le\xi\le1.3} f''(\xi)\right|$$

$$=\frac{0.3\times0.0025}{12}\times\frac{1}{4}\approx 0.000016.$$

60. Estimate the truncation error of the Simpson's 1/3 rule for the evaluation of $\int_{1.0}^{1.3}\sqrt{x}dx$ with h = 0.05.

❑ Here $f(x)=\sqrt{x} \Rightarrow f'(x)=\dfrac{1}{2\sqrt{x}}$; $f''(x)=-\dfrac{1}{4x^{3/2}}$; $f'''(x)=\dfrac{3}{8x^{5/2}}$; $f^{IV}(x)=-\dfrac{15}{16x^{7/2}}$.

With h = 0.5, $b-a$ = 0.3 and $f^{IV}(x)=-\dfrac{15}{16x^{7/2}}$, we see that the truncation error in Simpson's 1/3 rule is

$$\left|\frac{(b-a)h^4}{180}\max_{1\le\xi\le1.3} f^{IV}(\xi)\right| = \frac{0.3\times0.0000625}{180}\times\frac{15}{16}=0.000000009.$$

61. Define degree of precision.

❑ A quadrature formula is said to have degree of precision k (a positive integer), if it is exact for any arbitrary polynomial of degree, k but there is a polynomial of degree $k+1$ for which it is not exact. For example the degree of precision of Gaussian quadrature formula is $2k+1$.

62. Prove / verify that the Simpson's 1/3 rule is exact for $\int_a^b x^3 dx$ (or for any algebraic function of degree ≤ 3).

❑ We know that the error term in Simpson's 1/3 formula is given by

$$E(f) = -\frac{f^{IV}(\xi)(b-a)^5}{32 \times 90}, \text{ where } \xi \in [a, b].$$

Now, if $f(x) = x^n$, $n \leq 3$. Obviously, $E(f) = 0$, i.e., the integral value will be exact for any algebraic function of degree 3.

63. What is the disadvantage of integral formulae using equally spaced values of arguments?

❑ Integral formulae based on unequally spaced value of arguments give better accuracy than the formulae based on equally spaced value of arguments. Again, the formulae based on equally spaced value of arguments need the value of the function at the end points of interval of integration and in some cases these formula become usuless to determine the value of a convergent integral where the formulae based on unequally spaced value of arguments do not need the value of the function at the end points of interval of integration causing no problem to evaluate such integral.

64. State the reason why higher order Newton-Cote's formula are not commonly used although larger nodes usually yields better approximation.

❑ It is observed that the Cote's coefficients for $n \geq 8$ are of mixed sign and round off errors in Newton-Cotes formula of order 8 or more become large and unsuitable for use.

65. What is the basic principle of numerical differentiation?

❑ The basic principle of numerical differentiation is "Choose an appropriate interpolating polynomial for the given set of data and also for the point at which the value of the derivative is calculated, and differentiate it as many times as you need.

66. If two functions have same values at a point, then what can you say about their derivatives at that point?

❑ If two functions have same values at a point then their derivatives at that point may or may not be closed as two closed curves may differ considerably in their slopes at that point.

67. When is a matrix said to be diagonally dominant?

❑ A square matrix A = (a_{ij}) is said to be strictly diagonally dominant, if

$$|a_{ii}| > \sum_{j \neq i} |a_{ij}|, \quad i, j = 1, 2, \cdots, n$$

68. State the sufficient condition for convergence of (a) Gauss-Seidel method, (b) Gauss-Jacobi method.

❑ In both cases if the coefficient matrix is diagonally dominant then both the iterations converge for any initial guess.

69. What is the degree of the approximately polynomial corresponding to trapezoidal formula and Simpson's 1/3rd rule?

- The degree of the approximating polynomial corresponding to trapezoidal rule is one and Corresponding to Simpson's 1/3rd rule is 2.

  Justification: In trapezoidal formula consecutive ordinates are joined by straight line where in Simpson's 1/3rd rule ordinates are joined by parabolic chain.

70. For large values of n what is the number of operations required to solve the system of n equations in n unknowns using (a) Gaussian elimination method and (b) Gauss-Jordan method.

- In Gaussian elimination method it is approximately $n^3/3$ and in Gauss-Jordan method it is nearly $n^3/2$.

71. What is partial and complete pivoting?

- In partial pivoting, we select the equation with largest coefficient of x_1 as 1st equation and eliminate x_1 from the remaining equations. Next we select the equation with largest coefficient of x_2 in the reduced system (leaving aside the first equation) and eliminate x_2 from the remaining equations and the process is repeated at every subsequent steps. In complete pivoting we select the largest coefficient of all the coefficients of the matrix and eliminate corresponding unknown from the remaining equations and the process is repeated.

72. Explain Euler's method geometrically.

- Using Euler's method for solving a first order differential equation, we get

$$y(x+h) \approx y(x) + hy'(x).$$

  Geometrically, this interprets that if (x, y) be a point on the solution curve $y = y(x)$, then the next iteration $(x + h, y(x + h))$ lies closely to the tangent drawn at (x, y) to the curve $y = y(x)$.

ELEMENTS OF COMPUTER SCIENCE

Chapter 11

COMPUTER: AN INTRODUCTORY CONCEPT

11.1 INTRODUCTION

Computer may be defined as one of the latest and most sophisticated machines (not an electronic brain as it cannot think and cannot take any decision on its own) developed to process data/calculation quickly and accurately. It works as per the instructions of users and can store information needed by the user as long as its memory units work.

A computer is an electronic device which receives and stores data and after processing the input data it produces output results. Though the term "Computer" has originated from the Latin word "*Computare*" which is derived from *com* 'together' + *putare* 'to settle (an account)'. Now-a-days it is being used highly in various fields like:

(i) different scientific fields,
(ii) commercial and financial fields,
(iii) drawing, designing and publishing fields,
(iv) educational fields,
(v) information and communication fields,
(vi) graphics fields,

and in many areas of our daily life, as it can

(i) calculate anything correctly with high speed,
(ii) communicate and provide information rapidly,
(iii) store huge amount of data and information.

11.2 HISTORY OF COMPUTERS

With the development of the device "Abacus" by the Chinese around 600 B.C. the history of computers started. The device "Abacus" is used to calculate based on arithmetical operations. Though research works carried out for the improvement of calculating devices till the end of 18th century, the conceptual changes were actually made at the beginning of 19th century. During the first decade of 19th century, Jacquard invented an automated loom operated by a mechanism controlled by punched cards. During the same period Charles Babbage developed his differential and analytical engines. This device had provisions for inputting data, storing information, performing arithmetic operations and printing out the results. This was the base of modern computers.

Later on Herman Hollerith and James Power designed a machine, which was capable of possessing census information. Code for processing data for alphabetical and numerical was developed by Hollerith at the end of 19th century.

In the early 1940's a significant machine named 'MARK I' was built utilizing electromagnetic relays and the electronic machine known as ENIAC (Electronic Numerical Integrator and Calculator) was introduced for the first time at the middle of 1940's.

During the period, 1946-1952 John Von Neumann developed a high speed digital machine using vacuum tubes with his team. All the computers of present days work on binary system developed by Dr. Neumann.

11.3 COMPUTER GENERATIONS

The different phases of growth of the computer technology are called computer generations. The enrichment in the field of computer technology till the date is divided into four different categories. These are first generation, second generation, third generation and fourth generation. Research works are being carried out extensively in different parts of the world to produce the fifth generation of computers.

First generation (1942-1955)

The main electronic component of the first generation of computers was the vacuum tube. Though the first generation computers had some utilities during that period, it has some disadvantages and limitations. Those are

(i) it was huge in size and it worked slowly,

(ii) excessive heat was generated during its use,

(iii) it could not work without air conditioning system,

(iv) it was a huge power consuming machine,

(v) cost of it was too high.

Second generation (1955-1964)

Replacement of vacuum tube by the transistor leads the start of the second generation of computer and such replacement caused reduction in its size. In 1958, the first model of second generation computers was introduced by IBM. Though like the computers of first generations it needed air conditioning system and its production cost was too high, it had some advantages:

(i) it became smaller in size as compared to the computers of first generations,

(ii) it was a relatively less heat generator machine,

(iii) its action became more faster than those of first generations.

Third generation (1965-1975)

The era of third generation of computer starts with the invention of integrated circuit (I.C.), which integrate a large number of circuits/components into a very small single chip. Though manufacturer of I.C. faced difficulties, due to development of technology the third generation computers had some advantages:

(i) it became more smaller in size,
(ii) generation of heat reduced further,
(iii) power consumption became lower,
(iv) production cost became cheap,
(v) its action became more faster than those of first and second generations.

Fourth generation (1975 -)

With the introduction of very large scale integrated (VSLI) technology, the era of fourth generation computer starts and till the date it is still continuing. More than 30,000 components are integrated in a single chip used in the computers of fourth generation as compared the chip, integrating only 10-20 components, used in the computers of third generation. Use of microprocessor made in the computers of this generation for the first time increased the facilities to the user and due to this speed of working of the computers became too high. The advantages of this generation of computers are:

(i) it became portable and more smaller in size,
(ii) no air conditioning system is required for running it,
(iii) it became more or less maintenance-free,
(iv) power consumption became too low,
(v) production cost became cheaper.

Extensive research works are in progress in different parts of the world for the earliest launch of computers of fifth generation which are based on knowledge information process system (KIPS) and programming in logic language (PROLOG). This type of computers are expected to work parallely, i.e., will be capable of performing several tasks at a time.

11.4 CLASSIFICATION OF COMPUTERS

Today's computers are of different sizes, shapes and costs.

On the basis of employed principle, computers are classified into three categories. These are

(i) *Digital computer*: These computers operate on binary digits 0 and 1 (which are also called bits). The digital computers carry out operations on distinct data and understand information in 0 and 1 only. Due to this digital computers can perform arithmetic operations rapidly. Each piece of information stored in its memory in some unique combinations of 0's and 1's.

(ii) *Analog computer*: Computers of this category carry out operations on continuous analog signals. Though the accuracy of the results obtained by digital computers is greater than that obtained by analog computers, we are compelled to use analog computer for its own field of application as digital computers fail to work there.

(iii) *Hybrid computer*: In some cases we need both the facilities supplied by digital and analog computers and being a combinations of digital and analog computers, hybrid computers serve those purposes by changing signals from one system to other.

Again, on the basis of sizes and of the speed of performance the digital computers are divided into four groups:

(i) *Micro computer*: Major advancement of technology in the middle of 1970's resulted in the development of microcomputers (also called personal computers). These are small in sizes as these are built entirely of integrated circuits and also less expensive computers. Initially an 8-bits (which is also called one byte) microprocessor was used in microcomputer. Later on 16-bits and 32-bits microprocessors were developed which resulted in the increase in the speed of performance of the microcomputers. In a word a microcomputer is the smallest general purpose processing system and is a self-contained unit designed for single user at a time. For having many facilities microcomputers are widely used in homes, schools, colleges, shops and universities for various purposes.

(ii) *Mini computer*: With the enhancement of technology in the field of microprocessor microcomputer turned into minicomputer. Initially minicomputers were made of 16-bits microprocessor, but at present 32-bits microprocessor based computers are also available. For such transition in technology the performance of computer continues to improve dramatically and their cost continues to drop. The memory capacity of minicomputers has also become high. A minicomputer is a medium sized powerful computer which is designed to serve multiple users simultaneously. Due to these facilities people started to use it for a wide variety of personal, educational, commercial and technical applications.

(iii) *Mainframe computer*: Mainframe computers are big in sizes. The memory capacity is higher than that of minicomputers. It works faster than minicomputer as it is based on 32-bits microprocessor and support a large number of users at a time. These are expensive and maintenance cost of these computers is also high. That is why only universities, large industrial institution and business organization can use mainframe computers.

(iv) *Super computer*: Super computers have highest working speed as these are based on 64-bits microprocessor and extremely large storage capacity. That is why these are used for large scaled numerical problems in scientific disciplines. These are very expensive and maintenance cost of these is too high. Only meteorology department, space research organization, seismic research organization use it.

11.5 COMPUTER CHARACTERISTICS

The digital computers are basically electronic devices which can store, manipulate data/information with high speed. So the characteristics of a computer are (i) memory, (ii) flexibility, (iii) speed and (iv) reliability.

Memory: This is the most important unit of a computer which can store data and retrieve the same at a fast speed, which is needed for several purposes of modern society for different reasons.

Flexibility: This feature of computer helps a lot to the mankind. It can do most for analyzing the personal records of an organization.

Speed: The computer can do any job within a short interval of time. It can perform a large number of operations in a second. That is why it so popular to the scientists, engineers and research workers.

Reliability: A computer not only can perform any job in a second, it can do the job without any error as it cannot make mistakes of their own. So the very high speed of the machine is accompanied by an equally high level of reliability. The responsibility of mistakes or erroneous results lies with the user, not with the computer.

11.6 HARDWARE AND SOFTWARE

For convenience the total computer system is mainly divided into two parts: (i) hardware and (ii) software.

Hardware: A computer consists of different devices constructed by electrical, electronic circuits and mechanical parts. These devices are input and output devices, storage, processing and control units (will be discussed in shorts later on). These physical components like integrated circuits, floppy drive, hard disk drive, compact disk drive, etc. form the hardware part of a computer.

Software: A set of instructions (i.e., a program written in some computer language) written either by user or manufacturer for performing a specified task for data processing activities of the computer is called software. Softwares are again divided into two categories: (i) system software and (ii) application software.

System software: This is a package of a large number of programs which is needed for the functioning of the computer itself. System software used to supervise, control and run the users' programmes. This software includes (i) operating system, (ii) assembler, (iii) compilers, (iv) interpreters and (v) utility programmes.

Application software: This is a set of probrammes developed by the user to perform a specific task. Programmes to calculate wage bill of a company, or programmes written in any computer language to calculate any scientific results are the examples of application software.

11.7 COMPUTER UNITS

A computer is a combination of several electronic components which are connected in such a way that a task can be performed by these completely. Five major components of a computer are

(i) input unit,
(ii) output unit,
(iii) memory unit,
(iv) control unit,
(v) arithmetic logic unit.

The input and output devices allow computer to receive, display and print data and information. Most commonly used input devices are keyboard, mouse, monitor, printer, etc. The remaining three units form the central processing unit (CPU), which is the most important unit of a computer and it is also called the brain of the computer. CPU receives data/information through the input or memory and after performing some assigning task it sends data and information to the output device for displaying/printing or to memory for storing. The main parts of the CPU are

(i) control unit,
(ii) arithmetic logic unit,
(iii) memory.

Control Unit: This unit of CPU coordinates all the actions of different components of the computer. It receives data and information from memory and sends out command and control signals. It selects the sequence of instruction and finally sends the result to the output device.

Arithmetic Logic Unit: The ALU unit of CPU performs all mathematical calculations based on addition, subtraction, multiplication, division and exponentiation and logical operations like 'and', 'or', 'not', etc. including comparisons $\leq, \geq, \neq, =, >, <$.

Memory: The memory unit of a computer store data, result, program, etc. This unit is made of numerous number of cells which are capable of storing data in the form 0's and 1's. These cells are made of electro-magnetic substance and distinguish from one another by a sequence of numbers. These numbers are called addresses.

The memory section of a computer is broadly divided into three categories:

(i) primary memory,
(ii) cache memory,
(iii) secondary memory.

Primary memory: This type of memory stores data and program temporarily. It works as long as power supply is on. Its storage capacity is small (up to 512 MB) but its action is very fast.

Examples

ROM, RAM.

ROM

Its full form is "read only memory" and the information stored on it, generally supplied by the manufacturer, can only be read. It is a non-volatile memory which can be used but cannot be altered by storing new information or erasing anything of it. The ROM stores instruction necessary and ensuring proper functioning of the computer. There are three types of ROM and those are

(a) Programmable Read Only Memory (PROM)
(b) Erasable Programmable Read Only Memory (EPROM)
(c) Ultra Violet Programmable Read Only Memory (UVPROM).

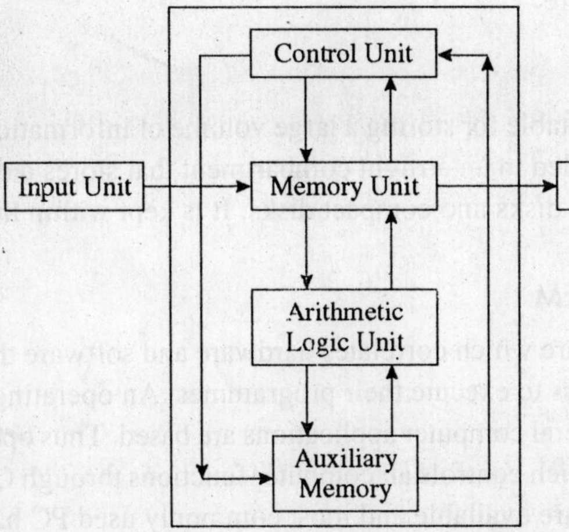

Fig.11.1 Block Diagram of a Digital Computer

RAM

Its full form is "Random Access Memory" and this is the part of the memory which computer uses for performing the user's tasks. It is used by the CPU to load the operating system, software and data as required. It is a volatile memory and the data stored here are changeable and will be lost as soon as the power is off. There are two types of RAM and those are

(a) Static Random Access Memory (SRAM)

(b) Dynamic Random Access Memory (DRAM).

Cache memory: Cache memory can store data with high speed, i.e., in less time but its storage capacity is less than primary memory (up to 4MB). It enhances the speed of CPU and it is expensive.

Secondary memory: Once the data is stored in secondary memory, it will be there permanently as long as the memory works. Its action is slow but storage capacity is too high (up to 120 GB) as compared to other memories.

Examples

Compact disk, Hard disk etc.

Floppy Disk

A floppy disk is made of flexible plastic coated with magnetic oxide and these are the cheaper storage area with small capacities. These are the portable storage unit of a computer.

Compact Disk

It is a high storage capacity media and works faster than floppy disk. It exists for a long time if preserved properly. The chief utilities of CD are that it is portable, its storage capacity is high and different types of data can be stored in it. That is why these are used for storing important software.

Hard Disk

It is a magnetic media suitable for storing a large volume of information/data. This is made of a series of disks sealed in an airtight compartment that stores data faster and at a higher density than floppy disks and compact disks. It is kept within the machine.

11.8 OPERATING SYSTEM

Operating system is a software which correlates hardware and software through a set of programmes which help users to execute their programmes. An operating system is the foundation on which the several computer applications are based. Thus operating system is a group of programmes which controls all computer functions through CPU. Presently different operating systems are available and most commonly used PC based operating systems are MSDOS, LINUX, WINDOWS-95 and onwards, etc.

11.9 PROGRAMMING LANGUAGES

Computer language is a medium of instructing the computer to perform some specific tasks. A computer can understand instructions if those are written in computer languages only. There are three types of programming languages: (i) low level language, (ii) high level language and (iii) fourth generation language. Again, depending on level of syntax, symbols and characters used in preparing computer languages are classified as under:

(i) Machine language: A machine language is represented by sequences of 0's and 1's. Programmes written in machine language is machine dependent, i.e., every computer has its own machine language which the CPU of the computer can understand directly. Machine language programmes are executed with high speed but it is difficult to understand these programmes as these need large numbers of machine codes and the programmer should have clear idea about the machine hardware.

(ii) Assembly language: The sequences of 0's and 1's in the machine language are replaced by symbolic codes (mnemonic codes) to develop assembly language. An instruction of the assembly language consists of operation codes, data and memory addresses. Though it is easier to understand machine language, like machine language (i) it is difficult to write a programme in assembly language, (ii) it is machine dependent and (iii) it needs large number of instruction to write a programme in assembly language.

A programme written in assembly language is translated into machine language through software named assembler. The original programme is called source programme and the translated programme is called object programme.

(iii) High level language: High level languages are problem oriented and machine independent. Programmes written in these languages are easy to use, i.e., correction and modification can be made easily as per need. A programme written in high level language is first translated into machine language by the software compiler or interpreter before the computer executes those instructions.

Examples

BASIC, FORTRAN PASCAL, COBOL, C etc.

(a) ***Compiler***: A compiler is a special programme which translates programmes written in high level language into machine level language so that computer can execute those. It translates the whole programme written in high level language into machine level language before loading it into the computer's memory.

(b) ***Interpreter***: An interpreter is a programme which translates programmes written in high level language line by line, i.e., it translates one instruction at a time into machine level language and execute the same before translating the next instruction. It works slowly than compiler, but due to some teachnical reasons interpreter is used instead of compiler.

EXERCISE

1. Define computer with its area of applications.
2. Write a brief note on the history of computers.
3. What are the different generations of computer?
4. How the computers are classified?
5. Describe different major components of a computer with the aid of block diagram.
6. What is CPU? Write down the functions of CPU. Distinguish between ROM and RAM.
7. What is Software and Hardware of a computer?
8. Write short notes on (i) machine language, (ii) assembly language and (iii) high level language.
9. Write short notes on (i) Compiler and (ii) Interpreter.
10. Write short notes on Operating System.
11. Describe briefly the storage device in a computer.
12. Describe the functions of the major units of a computer.
13. Discuss about the characteristics of a computer.

Chapter 12

NUMBER SYSTEM

12.1 INTRODUCTION

In this chapter we will consider only positional number system which is based on numbers and symbols. Numerical input will usually be in decimal form, since this is the form with which we are most familiar. In this system, any positive integer is expressible as a polynomial of 10 with integral coefficients between 0 to 9. In notation

$$N = (a_n a_{n-1} \ldots a_0)_{10} = a_n 10^n + a_{n-1} 10^{n-1} + \cdots + a_1 10 + a_0,$$

where $0 \le a_i \le 9$ for all i.

A proper fraction in decimal system can be written as

$$(0.a_{-1}a_{-2} \ldots a_{-k} \ldots)_{10} = a_{-1}10^{-1} + a_{-2}10^{-2} + \cdots + a_{-k}10^{-k} + \cdots,$$

where $0 \le a_i \le 9$ for all i. If $a_{-k} = 0$ for some $k \ge m$ (an integer), then the fraction is said to be terminating.

For example, $1/8 = 0.125 = 1.10^{-1} + 2.10^{-2} + 5.10^{-3}$ is a terminating decimal but $1/9 = 0.111111 \cdots = 1.10^{-1} + 1.10^{-2} + 1.10^{-3} + 1.10^{-4} + \cdots$ is a non-terminating decimal.

Modern computers generally find binary representations more convenient as the state of an electrical pulse is either off or on. For positive integers the binary form is

$$N = (d_n d_{n-1} \ldots d_0)_2 = d_n 2^n + d_{n-1} 2^{n-1} + \cdots + d_1 2 + d_0,$$

where $0 \le d_i \le 1$ for all i. This is again a polynomial in 2.

Similarly, in binary system a proper fraction takes the form

$$(0.d_{-1}d_{-2} \cdots d_{-k} \cdots)_{10} = d_{-1}2^{-1} + d_{-2}2^{-2} + \cdots + d_{-k}2^{-k} + \cdots,$$

where $0 \le d_i \le 1$ for all i.

Apart from these two well used system there are also a few other number systems which are as follows

Base	Number system
3	Ternary
8	Octal
12	Duodecimal
16	Hexadecimal

A general expression of a number of any number system is

$$\begin{aligned} \mathrm{N} &= (a_n a_{n-1} \ldots a_0 . a_{-1} a_{-2} a_{-3} \ldots)_K \\ &= a_n K^n + a_{n-1} K^{n-1} + \cdots + a_1 K + a_0 + a_{-1} K^{-1} + a_{-2} K^{-2} + a_{-3} K^{-3} + \cdots, \end{aligned}$$

where $a_{\pm n}$, the digit in the *n*th position left/right calculated from decimal position, is an integer satisfying $0 \le a_{\pm n} \le K - 1$ and K is the base number of the system.

12.2 DECIMAL NUMBER SYSTEM

The most widely used positional number system is decimal number system. The *decimal number system* is also known as base 10 system as only the 10 digits 0, 1, 2, 3, 4, 5, 6, 7, 8, 9 are used to represent a number in this system.

A real number can be represented in this system as under:

$$\begin{aligned} \mathrm{N} &= (a_n a_{n-1} \ldots a_2 a_1 a_0 . a_{-1} a_{-2} a_{-3} \ldots)_{10} \\ &= a_n 10^n + a_{n-1} 10^{n-1} \ldots a_2 10^2 + a_1 10^1 + a_0 10^0 + a_{-1} 10^{-1} + a_{-2} 10^{-2} + a_{-3} 10^{-3} \ldots, \end{aligned}$$

where $0 \le a_{\pm n} \le 9$.

Example

$$(29.4106)_{10} = 2 \times 10^1 + 9 \times 10^0 + 4 \times 10^{-1} + 1 \times 10^{-2} + 0 \times 10^{-3} + 6 \times 10^{-4}.$$

12.3 BINARY NUMBER SYSTEM

Binary number system uses only two digits 0 and 1 to represent a number and the base of this system is 2. Most of the computers work on binary number system.

A real number can be represented in this system as under:

$$\begin{aligned} N &= (a_n a_{n-1} \ldots a_2 a_1 a_0 . a_{-1} a_{-2} a_{-3} \ldots)_2 . \\ &= a_n 2^n + a_{n-1} 2^{n-1} \ldots a_2 2^2 + a_1 2^1 + a_0 2^0 + a_{-1} 2^{-1} + a_{-2} 2^{-2} + a_{-3} 2^{-3} \ldots, \end{aligned}$$

where $a_{\pm n} = 0$ or 1.

Example

$$(110.011)_2 = 1 \times 2^2 + 1 \times 2^1 + 0 \times 2^0 + 0 \times 2^{-1} + 1 \times 2^{-2} + 1 \times 2^{-3}.$$

12.4 OCTAL NUMBER SYSTEM

Some computers work on this system as binary equivalent of a decimal number is large in general. Octal number system uses the digits 0, 1, 2, 3, 4, 5, 6, 7 to represent a number.

A real number can be represented in this system as below:

$$N = (a_n a_{n-1} \dots a_2 a_1 a_0 . a_{-1} a_{-2} a_{-3} \dots)_8$$
$$= a_n 8^n + a_{n-1} 8^{n-1} \dots a_2 8^2 + a_1 8^1 + a_0 8^0 + a_{-1} 8^{-1} + a_{-2} 8^{-2} + a_{-3} 8^{-3} \dots,$$

where $0 \le a_{\pm n} \le 7$.

Example

$$(27.416)_8 = 2 \times 8^1 + 7 \times 8^0 + 4 \times 8^{-1} + 1 \times 8^{-2} + 6 \times 8^{-3}.$$

12.5 HEXADECIMAL NUMBER SYSTEM

Like octal number system, computers also use hexadecimal number system due to same reason. This system uses 10 digits 0, 1, 2, 3, 4, 5, 6, 7, 8, 9 and six symbols A, B, C, D, E and F. Here the symbols A, B, C, D, E, F are equivalent to 10, 11, 12, 13, 14, 15 in decimal system.

A real number can be represented in this system as under:

$$N = (ABa_{n-2} \dots a_2 a_1 a_0 . a_{-1} E a_{-3} F \dots)_{16}$$
$$= A16^n + B16^{n-1} + a_{n-2} 16^{n-2} \dots a_2 16^2 + a_1 16^1 + a_0 16^0 + a_{-1} 16^{-1}$$
$$+ E16^{-2} + a_{-3} 16^{-3} + F16^{-4} \dots,$$

where $0 \le a_{\pm n} \le 9$.

Example

$$(A9.46E)_{16} = A \times 16^1 + 9 \times 16^0 + 4 \times 16^{-1} + 6 \times 16^{-2} + E \times 16^{-3}.$$

12.6 CONVERSION

Many computers use binary system to perform calculations internally while computer users prefer to work in the decimal system. It is, therefore, necessary to have some means of converting one system to other. Suppose the number a is given in a system with base M and we want to convert it into a system with base N. Then we have the equation:

$$a = (a_m a_{m-1} \dots a_0)_M = (b_r b_{r-1} \dots b_0)_N,$$

where the coefficients a_i are known and the coefficients b_r are to be determined. Next we discuss the conversions of decimal system to binary, octal, hexadecimals and conversely. The short cut ways to convert binary numbers to octal and hexadecimal numbers will also be discussed.

1. Binary ↔ decimal conversion

A binary (decimal) number can be converted to decimal (binary) number as follows:

$$(a_n a_{n-1} \dots a_2 a_1 a_0 . a_{-1} a_{-2} \dots)_{10} = (b_k b_{k-1} \dots b_2 b_1 b_0 . b_{-1} b_{-2} \dots)_2,$$

where $b_k^s \left(a_n^s\right)$ are given and $a_n^s \left(b_k^s\right)$ are to be determined.

Example

$$(111)_2 = 1.2^2 + 1.2^1 + 1.2^0 = 7; \ (11011)_2 = 1.2^4 + 1.2^3 + 0.2^2 + 1.2^1 + 1.2^0 = 27.$$

On the other hand, if we want to convert a decimal number, say, 19, to a binary number, we proceed as follows:

2	19	Remainder 1
2	9	Remainder 1
2	4	Remainder 0
2	2	Remainder 0
2	1	Remainder 1
	0	

Now taking the remainders in reverse order, we get $19 = (10011)_2$.

Similarly, when 11 is converted to a binary number, we proceed as follows:

2	11	Remainder 1
2	5	Remainder 1
2	2	Remainder 0
2	1	Remainder 1
	0	

Again taking the remainders in reverse order, we get $11 = (1011)_2$.

Conversion of decimal fraction, say, 0.125, to a binary fraction can be done in the following way:

$2 \times 0.125 = 0.25$ Integral part = 0

$2 \times 0.25 = 0.5$ Integral part = 0

$2 \times 0.5 = 1.0$ Integral part = 1.

Now taking the integral parts we see that $(0.125)_{10} = (0.001)_2$.

Sometimes terminating decimal fraction may be converted to a non-terminating binary fraction. For example, we consider the fraction 0.3 in decimal system, then

$2 \times 0.3 = 0.6$ Integral part = 0

$2 \times 0.6 = 1.2$ Integral part = 1

$2 \times 0.2 = 0.4$ Integral part = 0

$2 \times 0.4 = 0.8$ Integral part = 0

$2 \times 0.8 = 1.6$ Integral part = 1.

So, $(0.3)_{10} = (0.0\ 1001\ 1001\ 1001\ \ldots)_2$.

2. Octal ↔ decimal conversion

Conversion rule from any system to other system is same, so we will illustrate the conversions with examples. Conversion of octal system to decimal system is as follows:

$$(273)_8 = 2.8^2 + 7.8^1 + 3.8^0 = (187)_{10}$$

and for the converse we proceed as follows:

8	187	Remainder 3
8	23	Remainder 7
8	2	Remainder 2
	0	

Now taking the remainders in reverse order, we get $(187)_{10} = (273)_8$.

Conversion of decimal fraction, say, 0.09375 to a octal fraction can be done in the following way:

$$8 \times 0.9375 = 0.7500 \quad \text{Integral part} = 0$$
$$8 \times 0.7500 = 6.0000 \quad \text{Integral part} = 6.$$

Since the fraction part is zero we stop here and write $(0.09375)_{10} = (0.06)_8$.

Again, to convert the decimal fraction 0.1236, to a octal fraction we proceed in the following way:

$$8 \times 0.1236 = 0.9888 \quad \text{Integral part} = 0$$
$$8 \times 0.9888 = 7.9104 \quad \text{Integral part} = 7$$
$$8 \times 0.9104 = 7.2832 \quad \text{Integral part} = 7$$
$$8 \times 0.2832 = 2.2656 \quad \text{Integral part} = 2.$$

Since the given data is corrected up to 4 decimal places in decimal system we carry out the operations 4 times and get $(0.1236)_{10} \approx (0.0772)_8$. To get more accurate result we have to continue the process further.

3. Octal ↔ Binary conversion

Though the binary system is used in modern computers, this system needs more digits to express a big number in decimal system. For example, to represent the number 32 in decimal system it needs 6 digits. To convert a number in the octal system to binary system we proceed as follows:

We convert it into decimal system and then converted number in decimal system is to be converted into binary system. But the octal number presents a kind of adjustment between the binary and decimal system. It is easy to convert the octal system to binary and back. To convert the octal to binary we have to convert octal digits 0 to 7 into binary equivalent as follows

$$000 = 0;\ 001 = 1;\ 010 = 2;\ 011 = 3;\ 100 = 4;\ 101 = 5;\ 110 = 6 \text{ and } 111 = 7.$$

Then we express the octal digits of number written in octal system as above and finally cancel the zeros lying in the left of first 1.

Example

$(123)_8 = (001\ 010\ 011)_2 = (1010011)_2$.

Similarly, $(321)_8 = (011\ 010\ 001)_2 = (11010001)_2$.

Conversely to convert binary to octal we divide the binary numbers into several groups of three digits starting from right and then replace each of the groups by its octal digit.

Example

$(11101001111)_2 = (011\ 101\ 001\ 111)_2 = (3517)_8$.

Similarly, $(10100101)_2 = (010\ 100\ 101)_2 = (245)_8$.

To express an octal fraction we write octal digits as above and finally cancel the zeros lying in the right of the last 1.

Example

$(0.24)_8 = (0.010\ 100)_2 = (0.0101)_2$; $(0.67)_8 = (0.110\ 111)_2$

Conversely to convert binary fraction to octal system we divide the binary numbers into several groups of three digits starting from left and then replace each of the groups by its octal digit.

Example

$(0.10101)_2 = (0.101\ 010)_2 = (0.52)_8$; $(0.100011)_2 = (0.43)_8$.

4. Hexadecimal ↔ Decimal conversion

Conversion of hexadecimal system to decimal system is also like other conversions. The only change in this conversion is that base is now 16.

Example

$(7AE)_{16} = 7 \times 16^2 + 10 \times 16^1 + 14 \times 16^0 = 1792 + 160 + 14 = (1966)_{10}$.

Similarly, to convert $(23717)_{10}$ into hexadecimal system we proceed as follows:

16	23717	Remainder 5
16	1482	Remainder 10 (= A)
16	92	Remainder 12 (= C)
16	5	Remainder 5
	0	

Hence $(23717)_{10} = (5CA5)_{16}$.

To convert decimal fraction 0.003900625 into hexadecimal system we proceed as follows:

$0.003900625 \times 16 = 0.0625$	Integer part 0
$0.062500000 \times 16 = 1.0000$	Integer part 1

So, $(0.003900625)_{10} = (0.01)_{16}$. Again, if we consider the decimal fraction 0.3875, then proceeding as above we obtain

$0.3875 \times 16 = 6.2000$	Integer part 6
$0.2000 \times 16 = 3.2000$	Integer part 3
$0.2000 \times 16 = 3.2000$	Integer part 3
$0.2000 \times 16 = 3.2000$	Integer part 3 and so on.

So we write $(0.3875)_{10} = (0.6333)_{16}$. To get more accurate value of the fraction in hexadecimal we have to proceed more in this way.

Conversely, to convert a hexadecimal fraction 0. AEF into a decimal fraction we calculate as follows:

$$(0.AEF)_{16} = 10 \times 16^{-1} + 14 \times 16^{-2} + 15 \times 16^{-3} = (0.683349609)_{10}.$$

5. Hexadecimal ↔ Binary conversion

The binary number system needs more digits to express a big number in decimal system. To convert a number in the hexadecimal system to binary system we proceed as follows:

Like the octal ↔ binary conversions, hexadecimal number presents a kind of adjustment between the binary and decimal system, which makes the conversion of the numbers in the hexadecimal system to binary system and back easily. Tó convert the hexadecimal to binary we have to convert octal digits 0 to 9 and symbols A, B, C, D, E, and F into binary equivalent as follows:

0000 = 0; 0001 = 1; 0010 = 2; 0011 = 3 0100 = 4; 0101 = 5; 0110 = 6; 0111 = 7
1000 = 8; 1001 = 9; 1010 = A; 1011 = B; 1100 = C 1101 = D; 1110 = E; 1111 = F.

Then we express the hexadecimal digits/symbols of number written in hexadecimal system as above and finally cancel the zeros lying in the left of first 1.

Example

$(D23)_{16} = (1101\ 0010\ 0011)_2 = (110100100011)_2$.

Similarly, $(32A)_{16} = (0011\ 0010\ 1010)_2 = (1100101010)_2$.

Conversely, to convert binary to hexadecimal we divide the binary numbers into several groups of four digits starting from right and then replace each of the groups by its hexadecimal digit.

Example

$(11100100111)_2 = (0111\ 0010\ 0111)_2 = (727)_{16}$.

Similarly, $(10100101)_2 = (10100101)_2 = (A5)_{16}$.

To express an hexadecimal fraction we write hexadecimal digits/symbols as above and finally cancel the zeros lying in the right of the last 1.

Example

$(0.24)_{16} = (0.0010\ 0100)_2 = (0.001001)_2$; $(0.C7)_8 = (0.11000111)_2$.

Conversely, to convert binary fraction to hexadecimal system we divide the binary numbers into several groups of four digits starting from left and then replace each of the groups by its hexadecimal digit/symbol.

Example

$(0.10101)_2 = (0.1010\ 1000)_2 = (0.A8)_{16}$; $(0.1001\ 1011)_2 = (0.9B)_{16}$.

12.7 BINARY ARITHMETIC

Binary numbers can be added, subtracted, multiplied and divided like decimal numbers which are discussed below with examples:

1. Addition of binary numbers

Binary numbers are added according to the following rules:

0 + 0 = 0; 0 + 1 = 1; 1 + 0 = 1 and 1 + 1 = 0 and carry 1 for next column.

Examples

(i) Add 110101.110 and 101.001.

$$\begin{array}{r} 110101.110 \\ +\quad 101.001 \\ \hline 111010.111 \end{array}$$

(ii) Add 1100.101 and 101.101

```
   1101.101
+   101.101
-----------
  10011.010
```

2. Subtraction of binary numbers

Binary numbers are subtracted according to the following rules:

0 – 0 = 0; 1 – 1 = 0; 1 – 0 = 1 and 0 – 1 = 1 and borrow 1 from next column.

Examples

(i) Subtract 1101.001 from 110101.110.

```
  110101.110
–   1101.001
------------
  101000.101
```

(ii) Subtract 101.11 from 1111.101.

```
   1111.101
–   101.110
-----------
   1001.111
```

3. Multiplication of binary numbers

Binary numbers are multiplied just like decimal numbers, i.e.,

0 × 0 = 0; 1 × 0 = 0; 0 × 1 = 0 and 1 × 1 = 1

Examples

(i) Multiply 111 by 11.

```
   111
    11
------
   111
  111×
------
 10101
```

(ii) Multiply 11.01 by 101.1

```
    11.01
    101.1
---------
     1101
    1101×
---------
   100111
    0000
---------
   100111
   1101
---------
 1001.111
```

4. Division of binary numbers

Binary numbers are divided just like decimal numbers, i.e.,

$$0 \div 1 = 0 \text{ and } 1 \div 1 = 1$$

Examples

(i) Divide 1111 by 101.

```
101) 1111 (11
     101
     ----
      101
      101
      ----
        ×
```

(ii) Divide 11100 by 101

```
101) 11100 (101
     101
     -----
       1000
        101
     -----
         11
```

12.8 1'S COMPLIMENT AND 2'S COMPLIMENT METHODS

For any number system there are two types of compliment methods for subtraction. One is the base compliment and other is one less than base compliment. Here we shall discuss the two compliment methods, viz., (i) 1's compliment method and (ii) 2's compliment method for subtractions of binary number systems.

(a) 1's compliment method

The 1's compliment of a binary number is obtained by changing 1 by 0 and 0 by 1. To subtract X from Y in this method, we write the number Y and add to it the 1's compliment of X. If the left-most digit is 1 (which indicates that the result of subtraction is positive), add it to the rest to get the result and if the left-most digit is 0, i.e., if there is no carry (which indicates that the result of subtraction is negative) then we take the 1's compliment of the rest to get the result.

Examples

(1) Subtract 1001 from 1101 using 1's compliment method.

```
                              1101
1's compliment of 1001 is     0110
                             -----
                             10011.
```

The left-most digit is 1, indicating that the result is positive and to get the result we add 1 to 0011, i.e.,

```
  0011
+    1
------
   100
```

Hence, $(1101)_2 - (1001)_2 = (100)_2$.

(ii) Subtract 1111 from 1101 using 1's compliment method.

$$\begin{array}{r} 1101 \\ \underline{0000} \\ 01101 \end{array}$$

1's compliment of 1111 is 0000

The left most digit is 0, indicating that the result is negative and the result is –(1's compliment of 1101).

Hence, $(1101)_2 - (1111)_2 = -(10)_2$.

(b) 2's compliment method

The 2's compliment of a binary number is obtained by adding 1 to its 1's compliment. To subtract X from Y in this method, we write the number Y and add to it the 2's compliment of X. If the left-most digit is 1 (which indicates that the result of subtraction is positive), then the rest is the result and if the left-most digit is 0, i.e., if there is no carry (which indicates that the result of subtraction is negative), then we take the 2's compliment of the rest to get the result.

Example

(i) Subtract 1001 from 1101 using 2's compliment method.

$$\begin{array}{r} 1101 \\ \underline{0111} \\ 10100. \end{array}$$

2's compliment of 1001 is 0111

The left-most digit is 1, indicating that the result is positive and the result is 100.

Hence, $(1101)_2 - (1001)_2 = (100)_2$

(ii) Subtract 1111 from 1101 using 2's compliment method.

$$\begin{array}{r} 1101 \\ \underline{0001} \\ 01110 \end{array}$$

2's compliment of 1111 is 0001

The left-most digit is 0, indicating that the result is negative and the result is –(2's compliment of 1110).

Hence $(1101)_2 - (1111)_2 = -(10)_2$.

EXERCISE

1. Write short notes on different number systems.
2. Write short notes on 1's compliment and 2's compliment methods.
3. Convert the following binary numbers into decimal numbers:
   (i) 110011, (ii) 10101, (iii) 1011011, (iv) 110011,
   (v) 11000, (vi) 11.011, (vii) 1.0101, (viii) 11.101.
4. Convert the following decimal numbers into binary numbers:
   (i) 135, (ii) 101, (iii) 10.375, (iv) 214.42.
5. Convert the following binary numbers into octal numbers:
   (i) 110011, (ii) 10101, (iii) 1011011, (iv) 1100.101, (v) 110.01.

6. Convert the following octal numbers into binary numbers:
   (i) 135, (ii) 101, (iii) 1037.15, (iv) 2124.02.
7. Convert the following binary numbers into hexadecimal numbers:
   (i) 110011, (ii) 10101, (iii) 1011011, (iv) 110011,
   (v) 11000, (vi) 11.011, (vii) 1.0101, (viii) 11.101.
8. Convert the following hexadecimal numbers into binary numbers:
   (i) A35, (ii) 1CD, (iii) 10.AB5, (iv) 1001.
9. Convert the following octal numbers into decimal numbers:
   (i) 135, (ii) 101, (iii) 1037.15, (iv) 2124.02.
10. Convert the following decimal numbers into octal numbers:
   (i) 135, (ii) 101, (iii) 1037.15, (iv) 2124.02.
11. Add the following binary numbers:
   i. 11011, 10001, 11101
   ii. 11,111,1111
   iii. 11001,111,1100.
12. Evaluate the subtractions in binary system:
   i. 11011−10001
   ii. 11001−1111
   iii. 11001−1100.
13. Evaluate the following in binary system:
   i. 11011×101
   ii. 1001×1111
   iii. 1101×1100.
14. Evaluate the following in binary system:
   i. 11011÷101
   ii. 11001÷111
   iii. 1101÷11.
15. Evaluate the subtractions in binary system using (i) 1's compliment and 2's compliment methods:
   i. 11011−10001
   ii. 11001−1111
   iii. 11001−1100
   iv. 1111−10001
   v. 1001−1011
   vi. 1001−10111.

Chapter

13 ALGORITHM AND FLOWCHART

13.1 INTRODUCTION

We have already mentioned that a computer works according to the instructions supplied to it by the users. So the users must have adequate knowledge in writing the programme so that he or she can prepare the instructions properly. The method of computations, which is a finite set of instructions supplied to the computer for performing a specific task, is called an *algorithm.* An algorithm must satisfy the following criteria:

(i) A precise input data are required to supply to get the desired output accurately and the algorithm should be such that small error in input produces a small error in output only.

(ii) Instructions must be non-ambiguous and can be carried out with high speed.

As the computer programme are generally complex and are written after preparing the corresponding *algorithm*, an intermediate step is necessary to plan, design and write a programme which is a diagrammatic representation of algorithm. This diagrammatic representation of algorithm is called *flowchart* and it gives the different steps of an algorithm and also transfers control between various steps. Different types of operations are indicated by geometrical figures of different shapes as given below:

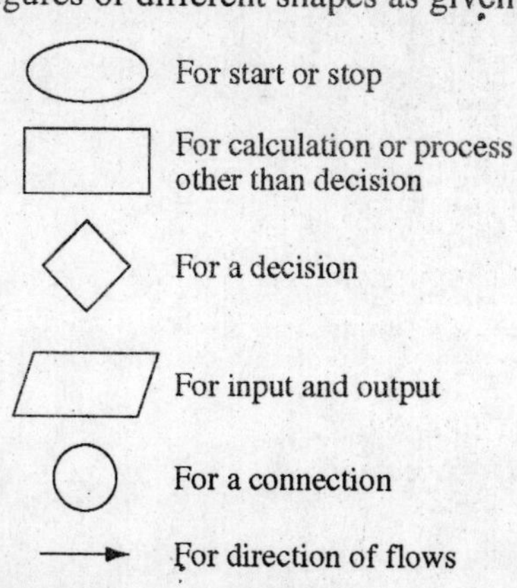

Examples

1. Write an algorithm and draw the flowchart to find the area and perimeter of a circle of radius r.

Algorithm:

Step 1: Start

Step 2: Read Pi, r

Step 3: Calculate A = Pi × r^2 and C = 2 × Pi × r

Step 4: Print A, C.

Step 5: Stop.

Flowchart:

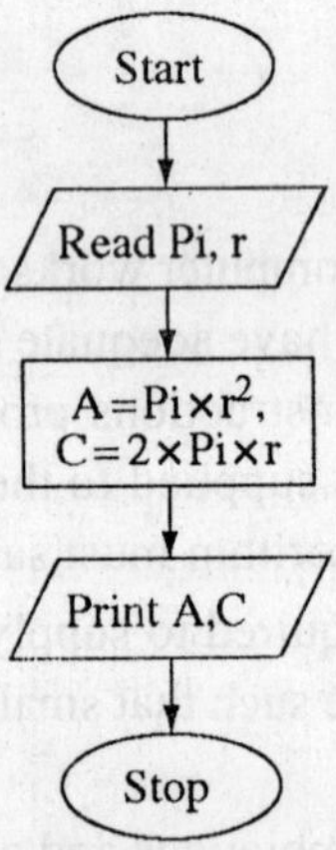

2. Write an algorithm and draw the flowchart to find the sum of all positive odd integers up to 100.

Algorithm:

Step 1: Start

Step 2: Set $S = 0$ and $I = 1$

Step 3: $S = S + I$

Step 4: $I = I + 2$

Step 5: If $I \leq 100$, go to step 3

Step 6: Print S

Step 7: Stop.

Flowchart:

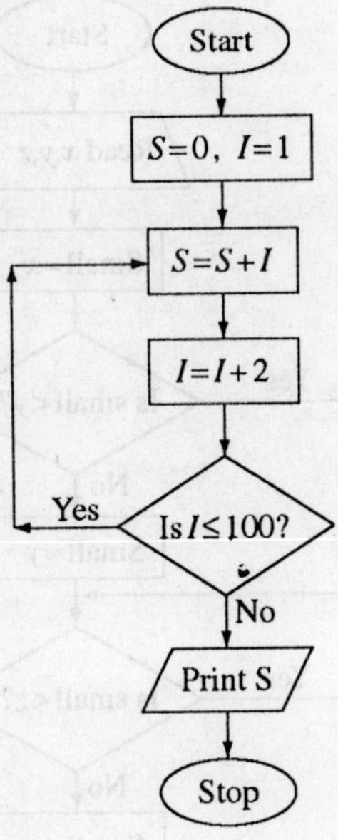

3. Write an algorithm and draw the flowchart to find the smallest of the three numbers x, y, z.

Algorithm:

Step 1: Start
Step 2: Read x, y, z
Step 3: Set Small = x
Step 4: If Small < y, go to step 6
Step 5: Set Small = y
Step 6: If Small < z, go to step 8
Step 7: Small = z
Step 8: Print Small
Step 9: Stop.

Flowchart:

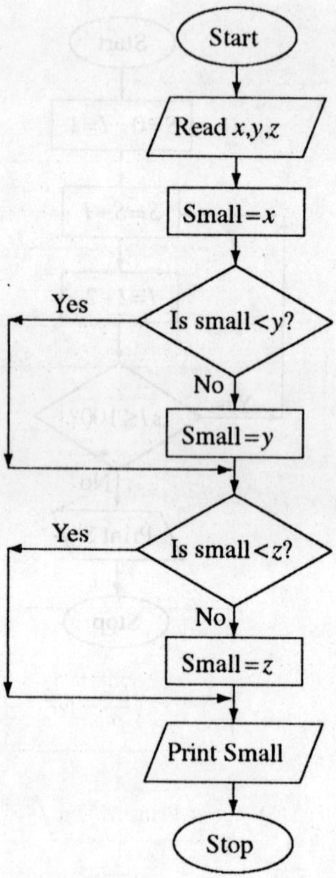

4. Write an algorithm and draw the flowchart to find the mean and s.d. of n numbers.

Algorithm:

Step 1: Start

Step 2: Read x_k; $1 \leq k \leq n$

Step 3: Set $S1 = 0$; $S2 = 0$ and $k = 1$

Step 4: $S1 = S1 + x_k$; $S2 = S2 + x_k^2$

Step 5: $k = k + 1$.

Step 6: If $k \leq n$, go to step 4

Step 7: Mean $= \frac{S1}{n}$; s.d.$= \sqrt{\frac{S2}{n} - (\text{Mean})^2}$

Step 8: Print Mean, s.d.

Step 9: Stop.

Flowchart:

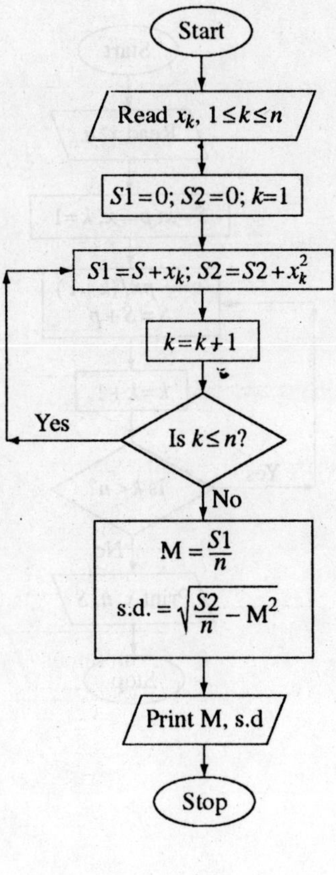

5. Write an algorithm and draw the flowchart to evaluate the sum

$$x - \frac{x^2}{2!} + \frac{x^3}{3!} - \cdots + (-1)^n \frac{x^n}{n!}.$$

Algorithm:

Step 1: Start

Step 2: Read n, x

Step 3: Set $S = x$; $p = x$ and $k = 1$

Step 4: $p = -p \times \frac{x}{k+1}$; $S = S + p$

Step 5: $k = k + 1$

Step 6: If $k < n$, go to Step 4

Step 7: Print n, x, S

Step 8: Stop.

Flowchart:

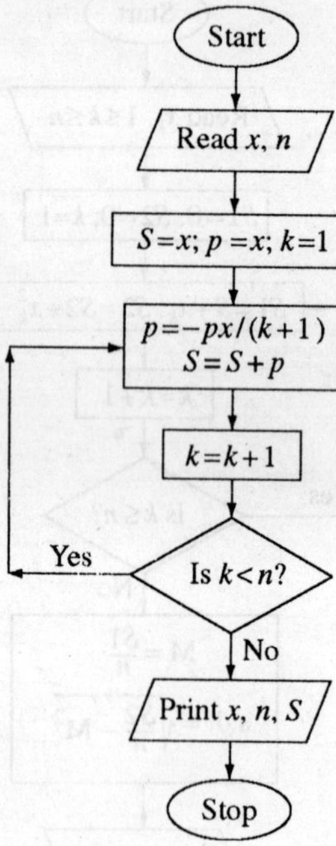

6. A function is defined by

$$f(x) = x^2 + 3x, \; x < 2$$
$$= x^2, \quad x = 2$$
$$= x^2 - 5, \; x > 2.$$

Write an algorithm and draw the flowchart to print the function.

Algorithm:

Step 1: Start

Step 2: Read x

Step 3: If $x < 2$, set $f(x) = x^2 + 3x$ and go to Step 6

Step 4: If $x = 2$, set $f(x) = x^2$ and go to Step 6

Step 5: If $x > 2$, set $f(x) = x^2 – 5$ and go to Step 6

Step 6: Print $f(x)$

Step 7: Stop.

Flowchart:

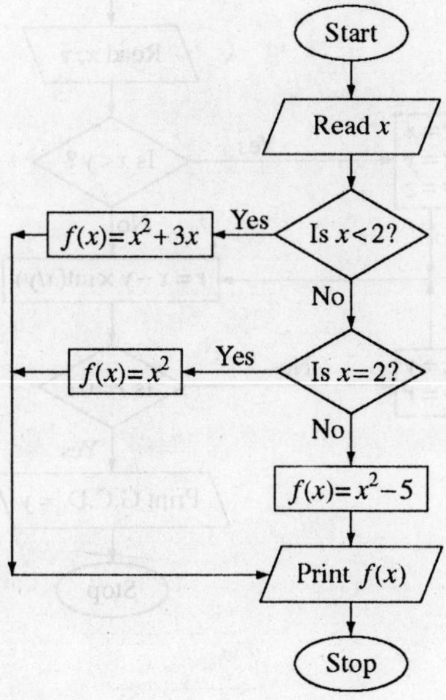

7. Write an algorithm and draw the flowchart to find the g.c.d. of two positive integers.

Algorithm:

Step 1: Start

Step 2: Read x, y

Step 3: If $x > y$, go to Step 5

Step 4: Interchange x and y

Step 5: Divide x by y to get the remainder r

Step 6: If $r = 0$ go to Step 8

Step 7: Set $x = y$ and $y = r$. Go to Step 5

Step 8: Print G.C.D. = y

Step 9: Stop.

Flowchart:

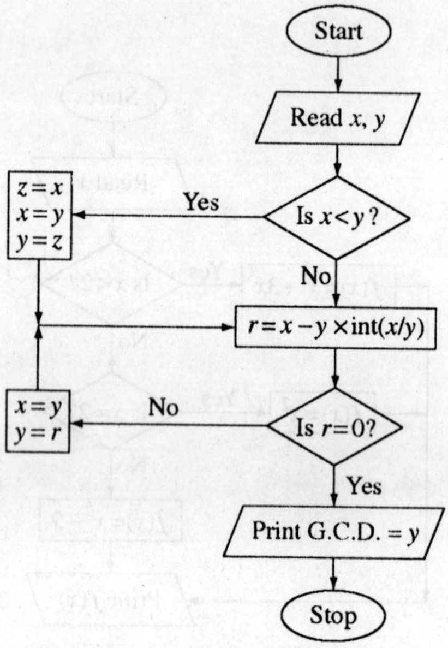

8. Write an algorithm and draw the flowchart to find the real root of the equation $ax^2 + bx + c = 0$.

Algorithm:

Step 1: Start

Step 2: Read a, b, c

Step 3: Set $D = b^2 - 4ac$

Step 4: If $D < 0$, go to Step 6

Step 5: Calculate: $x_1 = \dfrac{-b+\sqrt{D}}{2a}$; $x_2 = \dfrac{-b-\sqrt{D}}{2a}$. Go to Step 7

Step 6: Print "Equation has no real root". Go to Step 8

Step 7: Print x_1 and x_2

Step 8: Stop.

Flowchart:

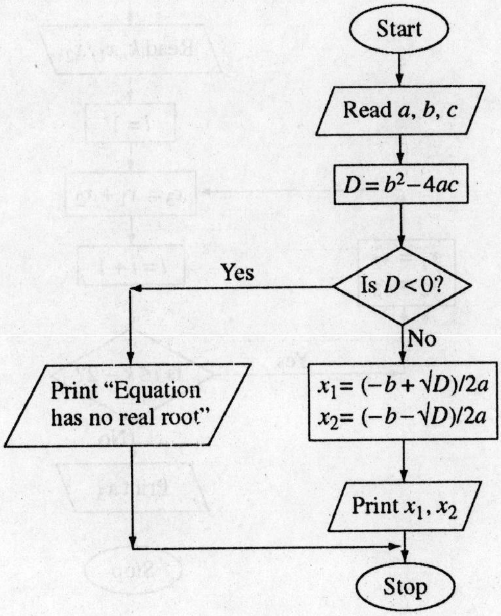

9. Write an algorithm and draw the flowchart to find kth element of Fibonacci sequence satisfying $x_{n+2} = x_{n+1} + x_n, n \geq 1$.

Algorithm:

Step 1: Start

Step 2: Read k, x_1 and x_2

Step 3: Set $i = 1$

Step 4: $x_3 = x_1 + x_2$

Step 5: $i = i + 1$

Step 6: If $i \leq k-2$. Set $x_1 = x_2$, $x_2 = x_3$. Go to Step 4

Step 7: Print x_3

Step 8: Stop.

Flowchart:

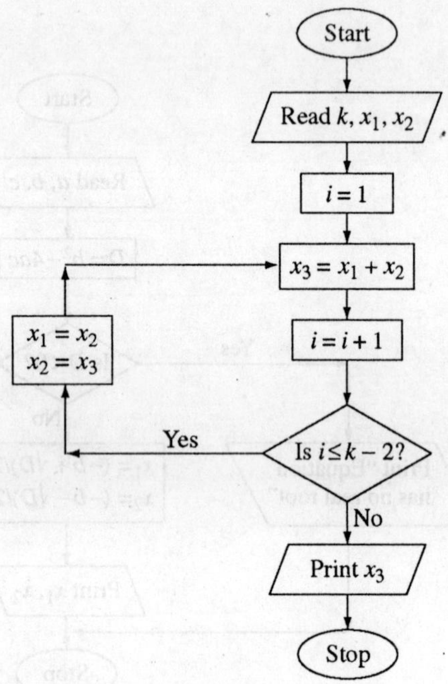

10. Write an algorithm and draw the flowchart to evaluate the value of an *n*th degree polynomial:

$$a_0 + a_1x + a_2x^2 + \ldots + a_nx^n .$$

Algorithm:

Step 1: Start

Step 2: Read x, n, a_k, $0 \le k \le n$

Step 3: Set $P = a_n$, $i = n - 1$

Step 4: $P = P \times x + a_i$

Step 5: $i = i - 1$

Step 6: If $i \ge 0$, go to Step 4

Step 7: Print P

Step 8: Stop.

Flowchart:

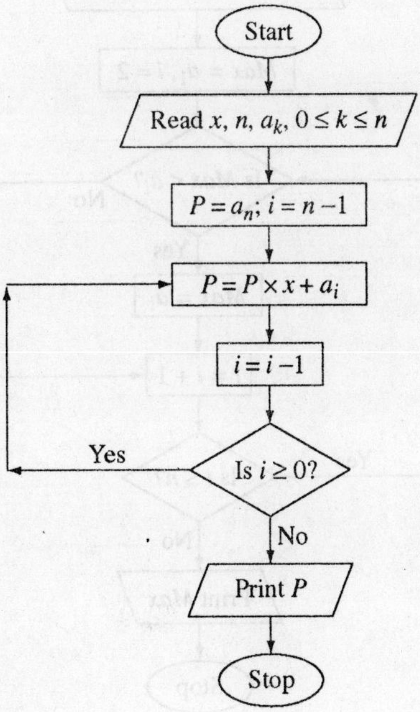

11. Write an algorithm and draw the flowchart to find the largest of n given integers.

Algorithm:

Step 1: Start

Step 2: Read n, a_k, $0 \le k \le n$

Step 3: Set $Max = a_1$ and $i = 2$

Step 4: If $Max < a_i$, set $Max = a_i$

Step 5: Set $i = i + 1$

Step 6: If $i \le n$ go to Step 4

Step 7: Print Max

Step 8: Stop.

Flowchart:

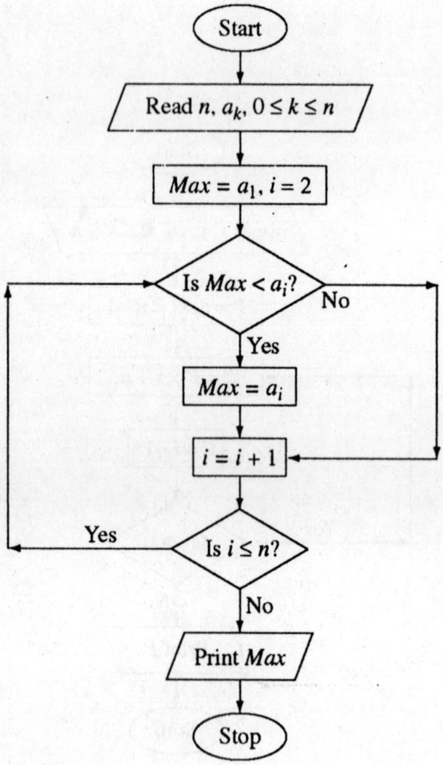

12. Write an algorithm and draw the flowchart to write n numbers in descending order.

Algorithm:

Step 1: Start

Step 2: Read x_k, $0 \le k \le n$

Step 3: Set $i = 1$

Step 4: Set $j = i + 1$

Step 5: If $x_i < x_j$, go to Step 7.

Step 6: Interchange x_i and x_j

Step 7: If $j < n$, set $j = j + 1$ and go to Step 5

Step 8: If $i < n - 1$, set $i = i + 1$ and go to Step 4

Step 9: Print x_i

Step 10: Stop.

Flowchart:

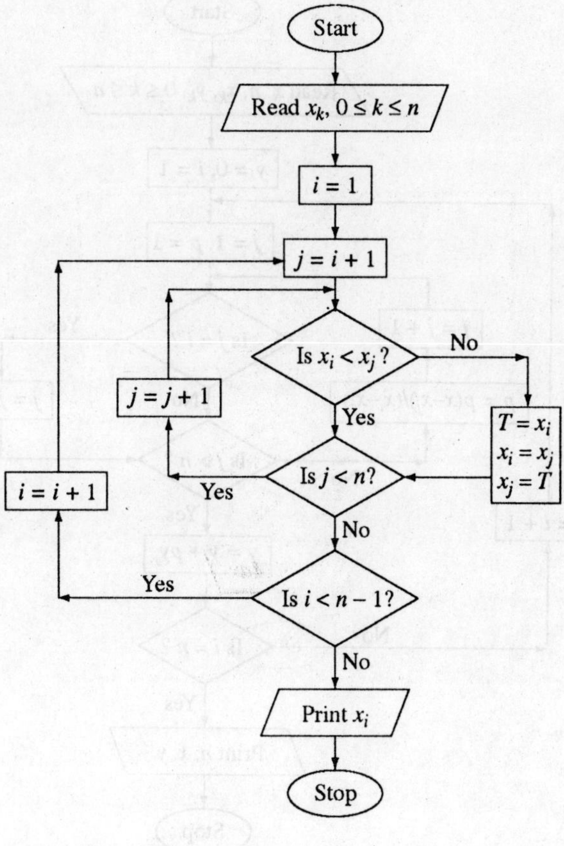

13. **Write an algorithm and draw the flowchart of Lagrange's interpolation:**

Algorithm:

Step 1: Start

Step 2: Calculate: $w(x) = (x - x_0)\,(x - x_1)\,(x - x_2)\, \ldots\, (x - x_n)$

Step 3: Calculate: $D_k = (x-x_k)(x_k-x_0)\, \ldots\, (x_k-x_{k-1})(x_k-x_{k+1})\, \ldots\, (x_k-x_n)$ for $0 \le k \le n$

Step 4: Calculate $\dfrac{y_k}{D_k}$, for $0 \le k \le n$

Step 5: Use above values in the following Lagrange's interpolation formula:

$$L(x) = w(x) \sum_{k=0}^{n} \frac{y_k}{D_k}.$$

Step 6: Stop.

Flowchart:

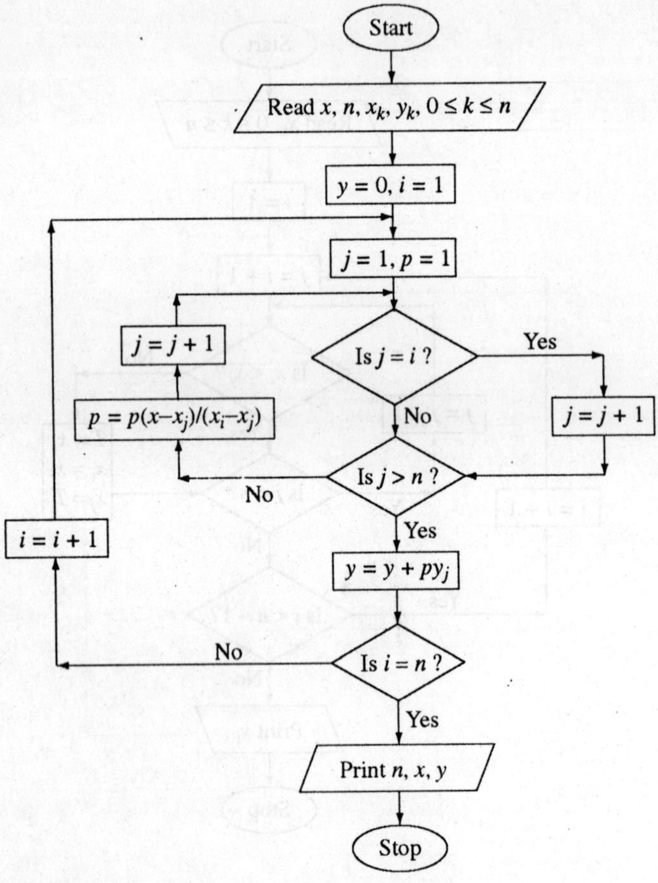

14. Write an algorithm and draw the flowchart of bisection method to find a real root of $f(x) = 0$.

Algorithm:

Step 1: Start

Step 2: Select an interval $[a, b]$ using tabulation method such that $f(a) < 0$ (or > 0) and $f(b) > 0$ (or < 0)

Step 3: Compute $x = \dfrac{a+b}{2}$

Step 4: If $|b-a| < e$, a pre-assigned small number (depends on given level of accuracy) then go to Step 6

Step 5: Compute $f(x)$. If $f(x)\,f(b) > 0$, then set $b = x$, otherwise set $a = x$ and go to step 3.

Step 6: Write 'The value of the required root correct to given decimal places is x (after rounding it)"

Step 7: Stop.

Flowchart:

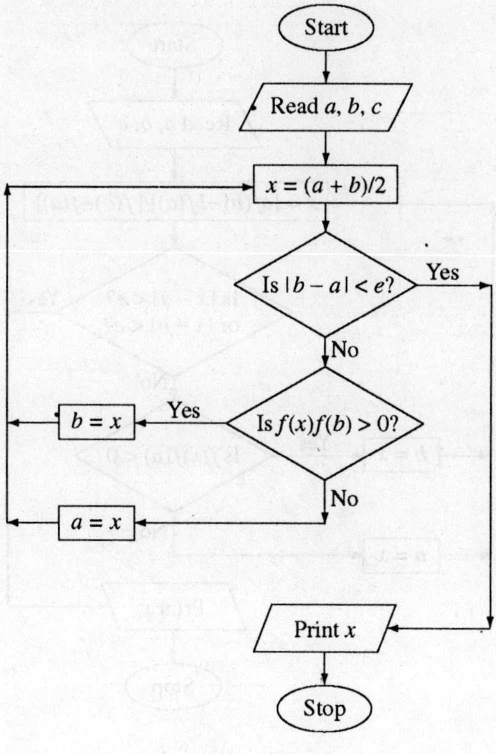

15. Write an algorithm and draw the flowchart of regula falsi method to find a real root of $f(x) = 0$.

Algorithm:

Step 1: Start

Step 2: Select an interval $[a, b]$ using tabulation method such that $f(a) < 0$ (or > 0) and $f(b) > 0$ (or > 0)

Step 3: Compute $x = \dfrac{af(b) - bf(a)}{f(b) - f(a)}$

Step 4: If $|x - a| < e$ or $|x - b| < e$, a pre-assigned small number (depends on given level of accuracy), then go to Step 6

Step 5: Compute $f(x)$. If $f(x)\,f(a) < 0$, then set $b = x$, otherwise set $a = x$. and go to step 3

Step 6: Write 'The value of the required root correct to given decimal places is x (after rounding it)"

Step 7: Stop.

Flowchart:

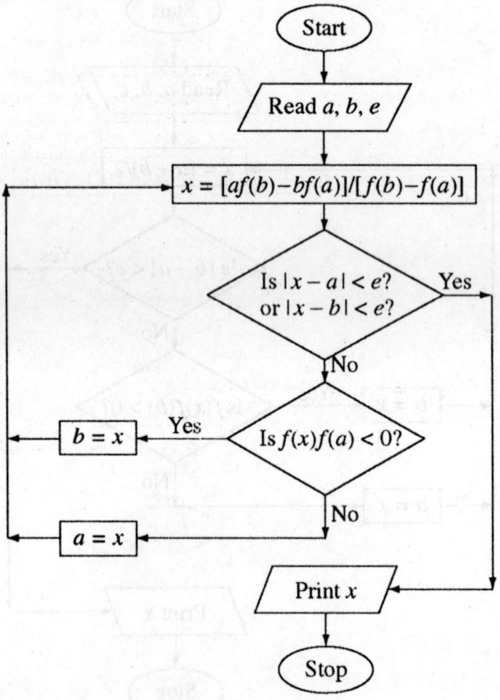

16. Write an algorithm and draw the flowchart of secant method to find a real root of $f(x) = 0$.

Algorithm:

Step 1: Start

Step 2: Select an interval $[a, b]$ (not necessarily containing the root of $f(x) = 0$

Step 3: Compute $x = \dfrac{af(b) - bf(a)}{f(b) - f(a)}$

Step 4: If $|x - a| < e$ or $|b - x| < e$, a pre-assigned small number (depends on given level of accuracy), then go to Step 6

Step 5: Set $b = x$, $a = b$ and go to step 3

Step 6: Write "The value of the required root correct to given decimal places is x (after rounding it)"

Step 7: Stop.

Flowchart:

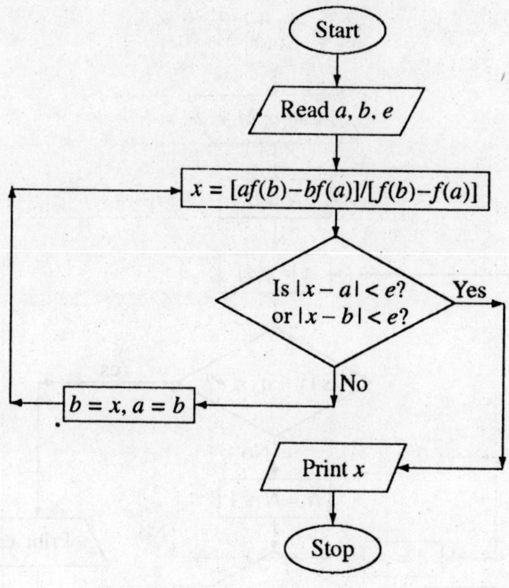

17. Write an algorithm and draw the flowchart of method of iteration to find a real root of $f(x) = 0$.

Algorithm:

Step 1: Start

Step 2: Choose an initial approximation a of the root of $f(x) = 0$ and a positive integer m

Step 3: Express $f(x) = 0$ as $x = g(x)$ such that $|g'(a)| < 1$

Step 4: Set $n = 0$

Step 5: Compute $x = g(a)$

Step 6: If $|x - a| < e$, a pre-assigned small number (depends on given level of accuracy) then to to Step 9

Step 7: Set $n = n + 1$. If n is less than m, go to Step 5

Step 8: Write "The method does not converge after m iterations". Go to step 10

Step 9: Write "The value of required root correct to given decimal places is x (after rounding it)"

Step 10: Stop.

Flowchart:

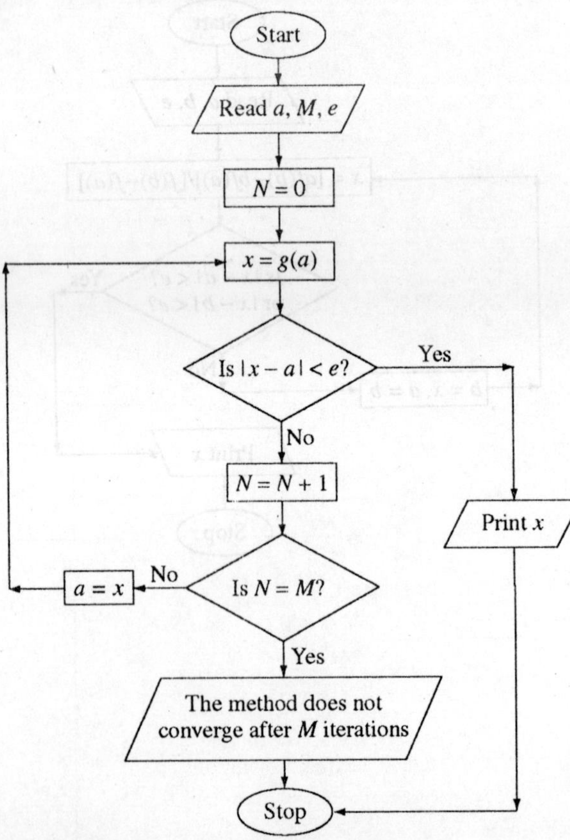

18. Write an algorithm and draw the flowchart of Newton-Raphson's method to find a real root of $f(x) = 0$.

Algorithm:

Step 1: Start

Step 2: Read $f(x)$ and evaluate $f'(x)$

Step 3: Choose an initial approximation a of the root of $f(x) = 0$ and a positive integer m

Step 4: Set $n = 0$

Step 5: Compute $x = a - \dfrac{f(a)}{f'(a)}$

Step 6: If $|x - a| < e$, a pre-assigned small number (depends on given level of accuracy), then go to Step 9

Step 7: $n = n + 1$. If n is less than m, go to Step 5

Step 8: Write "The method does not converge after m iterations". Go to step 10

Step 9: Write "The value of required the root correct to given decimal places is x (after rounding it)"

Step 10: Stop.

Flowchart:

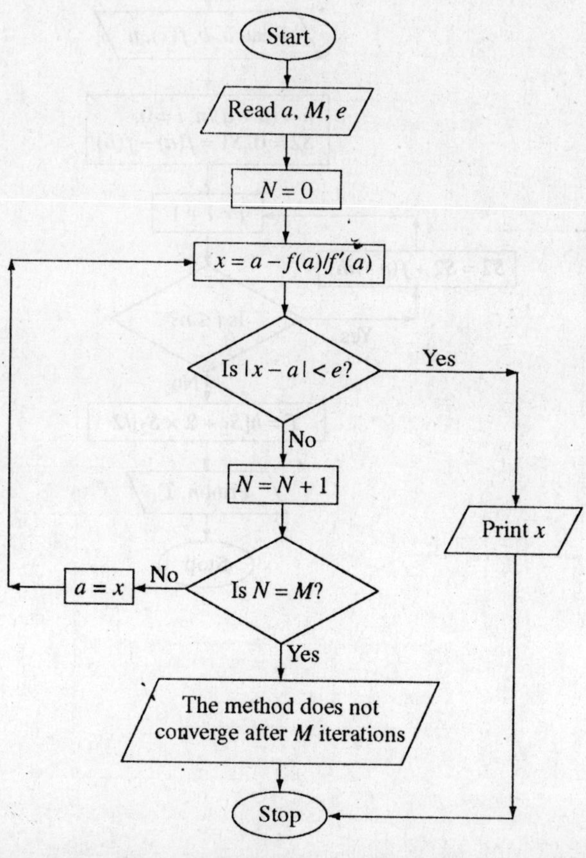

19. Write an algorithm and draw the flowchart of trapezoidal rule for numerical integration.

Algorithm:

Step 1: Start

Step 2: Read a, b, $f(x)$ and select a positive integer n

Step 3: Compute $h = \dfrac{b-a}{n}$

Step 4: Compute $x_i = a + ih$ and $y_i = f(x_i)$, $0 \le i \le n$

Step 5: Calculate the value of the integral for n divisions of $[a, b]$ using the following formula:

$$I_n = \frac{h}{2}\left[y_0 - y_n + 2(y_1 + y_2 + \cdots + y_{n-1} + y_n)\right]$$

Step 6: Write "The value of integral for n divisions of $[a, b]$ is I_n (after rounding it)"

Step 7: Stop.

Flowchart:

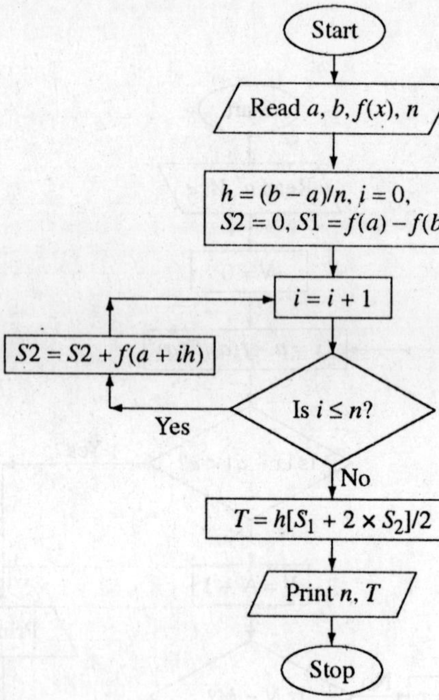

20. Write an algorithm and draw the flowchart of Simpson's 1/3 rd rule for numerical integration.

Algorithm:

Step 1: Start

Step 2: Read a, b, n

Step 3: Compute $h = \dfrac{b-a}{2n}$

Step 4: Compute $x_i = x_0 + ih$ and $y_i = f(x_i)$, $0 \le i \le 2n$.

Step 5: Calculate the value of the integral for n divisions of $[a, b]$ using the following formula:

$$I_n = \frac{h}{3}[y_0 - y_{2n} + 4(y_1 + y_3 + \ldots + y_{2n-1}) + 2(y_2 + y_4 + \ldots + y_{2n-2} + y_{2n})]$$

Step 6: Write "The value of integral for n divisions of $[a, b]$ is I_n (after rounding it)"

Step 7: Stop.

Flowchart:

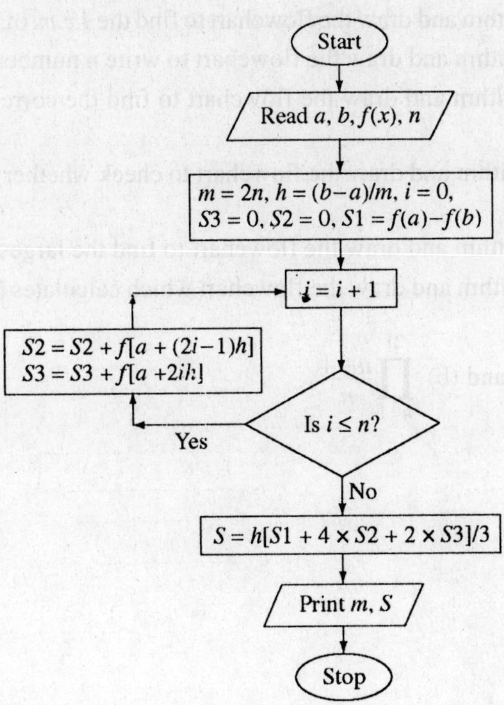

EXERCISE

1. Write an algorithm and draw the flowchart to find the area and perimeter of a triangle whose sides are a, b, c.
2. Write an algorithm and draw the flowchart to find the smallest of three distinct numbers x, y, z.
3. Write an algorithm and draw the flowchart to find the sum of all positive even integers up to 100.
4. Write an algorithm and draw the flowchart to find the factorial of a positive integer n.
5. Write an algorithm and draw the flowchart to evaluate the sum

$$x - \frac{x^3}{3!} + \frac{x^5}{5!} - \ldots + (-1)^n \frac{x^{2n+1}}{(2n+1)!}.$$

6. A function is defined by

$$\begin{aligned} f(x) &= x, & x &< 1 \\ &= x^2, & 1 \le x &\le 2 \\ &= x^2 - 5x, & x &> 2. \end{aligned}$$

   Write an algorithm and draw the flowchart to evaluate the function.
7. Write an algorithm and draw the flowchart to find the *l.c.m.* of two positive integers.
8. Write an algorithm and draw the flowchart to write n numbers in ascending order.
9. Write an algorithm and draw the flowchart to find the correlation coefficients of two variables.
10. Write an algorithm and draw the flowchart to check whether a number is prime or not.
11. Write an algorithm and draw the flowchart to find the largest of n given integers.
12. Write an algorithm and draw the flowchart which calculates (to 4 decimal places):

$$\text{(a) } \sum_{n=1}^{10} \frac{1}{2n+1} \text{ and (b) } \prod_{n=1}^{21} \frac{n+1}{n}.$$

Chapter

14 INTRODUCTION TO FORTRAN 77

14.1 INTRODUCTION

FORTRAN, which stands for formula translation, is the first high level language which was accepted throughout the world. Though FORTRAN was first developed in 1957, after several modifications FORTRAN 77 was developed in 1977 which and whose latest versions are now being used by large number of scientists, statisticians, and mathematicians. In this section we will discuss the features of FORTRAN 77 only.

14.2 CHARACTERS

Like other computer languages, FORTRAN 77 has its own set of characters as follows:

Digit: 0 1 2 3 4 5 6 7 8 9

Alphabetic characters: A B C D E F G H I J K L M
N O P Q R S T U V W X Y Z

Symbols: + − * / . , " ( ) ; =

A blank space is also treated as character.

14.3 CONSTANTS AND VARIABLES

In FORTRAN there are two distinct types of constants—(i) the integer constant and (ii) the real constant. An integer constant is a signed or unsigned whole number, consisting of at most 9 digits, without a decimal point.

Example

A few valid integer constants are

18, −4, 0, 71234, −9867.

The following are the invalid integer constants:

−12.	(decimal point is not allowed)
4.0	(decimal point is not allowed)
3,456	(comma is not allowed)
1000000000	(contains more than 9 digits)

Real constants are written in two different forms—(i) the decimal form and (ii) the exponential form. A decimal point is used in both forms to make the difference between real and integer constants.

Decimal form: A real constant, in decimal form, is written as a signed or unsigned digit(s) (generally up to 8 significant digits) with a decimal point only.

Example

A few valid real constants in decimal form are

18.0, −4.3, 0.7, 71234, −98.672.

The following are the invalid real constants in decimal form:

−40	(decimal point is not given)
3,456.0	(comma is not allowed)
1000000000.0	(contains more than 8 significant digits).

Exponential form: In exponential form a real constant is written as a combination of two parts: (i) a decimal form and (ii) a part starting with E and a signed or unsigned integer constant of maximum two digits.

Example

A few valid real constants in exponential form are

18.1E−02, −4.0E + 2, 0.7E05.

The following are the invalid real constants in exponential form:

−6E − 2	(decimal point is not given)
3,456.0E + 2	(comma is not allowed).

Like constants there are two distinct types of variables—(i) the integer variable and (ii) the real variable. An integer variable in FORTRAN starts with any one of the characters i, j, k, l, m, n and consists of generally at most 6 characters/digits. If the variable name starts with any one of the remaining alphabets, then it is called real variable. But a few words are there, such as Read, Write, Real, Integer, Stop, End, Pause, which are the parts of FORTRAN language and cannot be used as a variable.

Example

A few valid real variables are

XY, C1, BM2.

The following are the invalid real variables:

2AZ	(starts with digit)
A2*C	(symbol is not allowed)
ASDFGHJ	(consisting of more than 6 characters)
IABC	(starts with I).

A few valid integer variables are

LXY, K1, MM.

The following are the invalid integer variables:

2L1	(starts with digit)
K2*C	(symbol is not allowed)
LLMNAHJ	(consisting of more than 6 characters)
AL12	(starts with A).

In FORTRAN, complex variables are written as (a, b), where a, b are real positive or negative constants

Example

A few valid complex constants are

(4., 3.), (1.24, −3.2), (−0.3, 2.), (0.5E2, 3.76).

The following are the invalid integer variables

−(4.3)	(Sign outside the bracket is not allowed)
(4.6)	(Decimal point is not given)
(23, *a*.)	(*a* is not a real constant).

Complex variables used in FORTRAN are declared as

Complex C1, C2, ...

and assume only complex constants.

Example

```
Complex C1, C2
C1 = (1., 2.)
C2 = (1.2., -2.5).
```

14.4 ARITHMETIC OPERATIONS AND EXPRESSIONS

The arithmetic operations in FORTRAN are as follows:

Operations	*Symbols*
Addition	+
Subtraction	−
Multiplication	*
Division	/
Exponentiation	**

The order of execution of the above parenthesis free operations is as follows: (i) Exponentiation; (ii) Multiplication and division and (iii) Addition and subtraction.

If there be two consecutive operations of same precedence, then the leftmost one is preformed first. We have already discussed that there are two types of constants, real and integer, so there are two types of arithmetic, real arithmetic and integer arithmetic.

The real/integer arithmetic is performed if the operands are real/integer and yields real/integer value. For example, 6. + 2., 6. − 2., 6. * 2. yield 8., 4., 12. in real arithmetic and 6 + 2, 6 − 2, 6*2 yield 8, 4, 12. in integer arithmetic. Integers division gives the integer part of the quotient, i.e., the fractional part is deleted but in case of division of reals we get the result like ordinary division. For examples, 6/5 = 1, 3/5 = 0 but 6./5. = 1.2 and 3/5 = 0.6.

The above arithmetics are meaningful when the FORTRAN compilers do not allow mixed mode operations. But now-a-days, most of the compilers allow mixed mode operations and in such cases both variables/constants can be used in a single arithmetic expression. When both real and integer constants are used in an expression, the compiler first converts the integer into real and after performing real arithmetic it gives the real value.

Arithmetic expressions are the expressions formed by the combination of variables and constants. Though an integer expression is a combination of integer variables and integer constants and a real expression is a combination of real variables and real constants, now-days both variables/constants can be used in a single arithmetic expression as most of the compilers allow mixed mode operations.

For example, $\frac{my}{3}$ can be expressed either as Float(m)*y/3. or as m*Ifix(y)/3 and even as m*y/3, if compiler allows.

Before going to write an expression in FORTRAN, one should have clear information about the library functions, which will be required frequently at the time of writing.

14.5 BUILT-IN-FUNCTIONS

FORTRAN provides a library of functions to the programmer for writing a program. The most commonly used set of functions are given below:

Function	**Mathematical form**	**FORTRAN expression**
Square root of x	$\sqrt{x}$	sqrt(x)
Absolute value of x	$\lvert x \rvert$	abs(x)
Exponential of x	e^x	exp(x)
Sine of x	$\sin x$	sin(x)
Cosine of x	$\cos x$	cos(x)
Tangent of x	$\tan x$	tan(x)
Log of x	$\log_{10} x$	a log10(x)
Natural log of x	$\log_e x$	a log(x)
Conversion of integer I into real		float(I)
Integer part of a real x	$[x]$	ifix(x)
Inverse of sin x	$\sin^{-1} x$	a sin(x)
Inverse of cos x	$\cos^{-1} x$	a cos(x)
Inverse of tan x	$\tan^{-1} x$	a tan(x)
Sin hyperbolic of x	$\sinh x$	sinh(x)
Cos hyperbolic of x	$\cosh x$	cosh(x)
Tan hyperbolic of x	$\tanh x$	tanh(x)

14.6 EXAMPLES OF ARITHMETIC EXPRESSIONS

Mathematical form	Fortran expression
1. $a + bc/d$	1. a + (b*c)/d
2. $a + b.c - d^2$	2. a + b*c – d**2
3. $a \sin x + \sqrt{b \cos x + c \log_e x}$	3. a*sin(x) + sqrt(b*cos(x) + c*alog(x))
4. $e^{\sin x} + \log_{10} x + \|x\|$	4. exp(sin(x)) + alog10(x) + abs(x)
5. $\sin^{-1} x + \frac{x^3+5}{\sqrt{x^4+3x}}$	5. asin(x) + (x**3 + 5)/sqrt(x**4 + 3*x)
6. $\sinh x + b\log_{10}x + ax^2 + bx + c$	6. sinh(x) + b*alog10(x) + a*x**2 + b*x + c
7. $\sqrt{\frac{a^2}{b+c}}$	7. Sqrt(a**2/(b + c))

An arithmetic expression is evaluated as follows:

(i) Expressions lying within a parenthesis is evaluated first. In case of nested parenthesis the expressions of the innermost parenthesis is evaluated first,

(ii) Evaluation of library functions used in an expression,

(iii) Exponentiation,

(iv) Division and multiplication,

(v) Addition and subtraction.

14.7 RELATIONAL EXPRESSION

The six mathematical relations between arithmetic expressions are represented by relational operators. A relational expression consists of two expressions connected by one of the six relational operators and represents a condition which is true or false. The general form of the relational operation is

expression1.operator.expression2

Mathematical relation	Equivalent Fortran expressions
Less than	.LT.
Less than or equal to	.LE.
Equal	.EQ.
Not equal	.NE.
Greater than	.GT.
Greater than or equal to	.GE.

Examples

Mathematical relation	Equivalent Fortran expressions
$I \geq J$	I.GE.J
$(A+B) < 5.2$	(A+B).LT.5.2
$A = B + 2$	A.EQ.(B+2.0)
$S \neq G$	S.NE.G
$B^2 \leq 4AG$	B**2.LE.4*A*C
$X > Y$	X.GT.Y

In addition to the relational expressions, which are also called as logical expressions, FORTRAN has three more logical expressions: .NOT., .AND., .OR. Logical expressions around NOT is evaluated first then logical expressions around AND is evaluated and finally logical expressions around OR is evaluated. Two logical expressions le1 and le2 can be compounded using logical operators as follows:

le1 .AND. le2

le1 .OR. le2

.NOT. le1

The operator AND results true when both logical expressions are true, otherwise it results false. The operator OR results false when both logical expressions are false, otherwise it results true and the operator NOT results in change of the values of a logical expression from true to false or false to true as can be seen from following table:

le1	le2	le1 . AND . le2	le1 . OR . le2
T	T	T	T
T	F	F	T
F	T	F	T
F	F	F	F

le1	.NOT . le1
T	F
F	T

Example

(i) (X.LT.5.2).AND.(Y.GE.3.2). This expression is true, if both $X < 5.2$ and $Y \geq 3.2$.

(ii) (J.LE.5).OR.(K.GE.3). This expression is true, if $J \leq 5$ or $K \geq 3$.

(iii) .NOT.(K.GE.3). This expression is true, if $K < 3$.

14.8 FORTRAN STATEMENTS

There are two types of FORTRAN statements: (i) executable statement and (ii) non-executable statement. The statements which result in machine language instruction and which specify action are called executable statement and the statements which only provides information to the computer and do not specify any action are called non-executable statements:

Executable statements: (i) Assignment statement,
(ii) Control statement,
(iii) Input/Output (I/O) statement.

Non-executable statements: (i) Specification statement,
(ii) Data statement,
(iii) Format statement,
(iv) Function defining statement,
(v) Function/Subroutine statement.

14.8.1 Type Statement

The variables used in FORTRAN can be declared by the Type statement which has the form:

Real list of variables separated by comma
Integer list of variables separated by comma
Complex list of variables separated by comma

Example

Real *a*, *b*, *c*, I1
Integer *k*, *m*, *n*, *b*2
Complex C1, C2, val

It tells the compiler that throughout the program the type of the variables declared will remain same. Being a non-executable statement, it is to be placed before first executable statement of a computer program written in FORTRAN.

14.8.2 Data Statement

The general form of DATA statement is

$$\text{DATA } x_1, x_2, \cdots, x_n \,/\, a_1, a_2, \cdots, a_n \,/ ,$$

which tells the computer to assign the values $a_1, a_2, \cdots, a_n$ to the variables $x_1, x_2, \cdots, x_n$ respectively and hence it is an executable statement. If some of a_i's are equal, $a_1 = a_2 = \cdots = a_k = a$, then above statement can be written as

$$\text{DATA } x_1, x_2, \cdots, x_n \,/\, k{*}a, a_{k+1}, \cdots, a_n \,/ ,$$

Example

(i) DATA *x*, *y*, *z*/1.2, 3.4, 5.6/, which tells the computer to assign 1.2 to *x*, 3.4 to *y* and 5.6 to *z*.

(ii) DATA *x*, *y*, *z*/3*1.2/, which tells the computer to assign 1.2 to all the three variables *x*, *y* and *z*.

(iii) DATA *x*, *y*, *k*/2*3.4,5/, which tells the computer to assign 3.4 to both *x* and *y* and assign 5 to the variable *k*.

14.8.3 Parameter Statement

The general form of PARAMETER statement is

$$\text{PARAMETER } (p_1 = a_1, p_2 = a_2, \ldots, p_n = a_n),$$

which tells the computer to assign the values $a_1, a_2, \ldots, a_n$ to the parameters/symbolic constants $p_1, p_2, \cdots, p_n$, hence it is an executable statement.

Example

(i) Parameter (Pi = 3.14), which tells the computer to assign 3.14 to the parameter Pi.

14.8.4 Arithmetic Assignment Statement

The general form of an arithmetic assignment statement is

Variable name = Arithmetic expression

which means that the arithmetic expression in the right-hand side is evaluated first and then it is stored in the storage location of the variable used in the left, i.e., the "=" sign stands for replacement.

Example

(i) $a = b*c$

(ii) $a = a + 2$

(iii) Area = sqrt(s*(*s*–*a*)*(*s*–*b*)*(*s*–*c*))

(iv) Volume = 3.14*r**2*h.

14.8.5 Read and Write Statements

In FORTRAN computer reads input data and prints output data by READ and WRITE statements. Though the READ and WRITE statements (which are also called I/O statements) are associated with FORMAT statement which provides the computer the desired format of the data, i.e., real, integer or exponential. Format-free I/O statements are also used. The general form of the format-free I/O statements are

Read*, variable list (separated by commas)

and Print*, variable list (separated by commas),

whereas the formatted I/O statements are

Read(m, n) variable list (separated by commas)

and Write (m, n) variable list (separated by commas),

where the integer constant m is the input device reference number and the integer constant n denotes the statement level of the format statement. Instead of m in READ statement we can use * which denotes that the data to the computer will be supplied through keyboard and the same in WRITE statement denotes that the output data will be printed in the screen of the monitor, whereas in both the cases * in the place of n implies that the data are format-free.

14.8.6 FORMAT Statement

The FORMAT statement is a non-executable statement and can be placed anywhere within the program. This informs the computer about the desired format of the I/O data, i.e., how the input will be found and how the output data will be printed. The general form of the FORMAT statement is

n Format (*ak*, *bl*, *cm*, ···, *dj*),

where *n* is the statement level number, *ak*, *bl*, *cm*, ··· are format specifications. Next we discuss different format specifications for write statement only as it is wise to use format-free READ statement. It is to be noted that in all the cases data will be printed in right justified positions.

(i) **The I-field:** The general form of the I-field specification is Iw, which is used for integer constant. Here I denotes that the data is of integer type and w is an unsigned integer denoting the field width.

Example:

```
      Write(*, 12) Jac
12    Format(I5)
```

(ii) **The F-field:** The general form of the F-field specification is fw.d, which is used for real constant. Here *f* denotes that the data is of real type and w,d are unsigned integers giving the field width and number of decimal digits respectively.

Example:

```
      Write(*, 12) x
12    Format(f9.6)
```

(iii) **The E-field:** The general form of the E-field specification is ew.d, which is used for real constant in exponential form. Here *e* denotes that the data is of real type and the constant is in exponential form and w,d are unsigned integers giving the field and number of decimal digits respectively. In this field specification it is to be noted that if a data is desired to be printed correct to 'd' decimal places then the field width w should satisfy $w \geq d + 7$ as can be seen from the following figure.

±	0	·	d figures	E	±	2 figures

Example:

```
      Write(*, 12) x
12    Format(e12.5)
```

(iv) **The X-field:** The general form of the X-field is wX, where w denotes the field width. This field is used for separating the constants on the same line giving blank spaces between them. Thus wX generates w blank spaces in the output.

Example:

```
      Write(*, 12) x, y
12    Format(f9. 5, 4x, f9.5).
```

which tells the computer to print 4 blank spaces between the values of x and y.

(v) **The Literal field:** This field is used for printing some message or some message with numerical value.

Example 1:

```
      Write(*, 12)
12    Format(1x, 'Apurba Narayan Das')
```

Example 2:

```
      Write(*, 12) a, i
12    Format(1x, 'value of A = ', f9.5, 2x, 'value of I = ', I5)
```

Note: Apart from the above fields of specification we have more options to get the output in different fashions. For example, one may use/(slash) at the end of field which implies that the next output data will be printed in the next line. Use of n slasiies at the end of one record implies that n–1 lines will be skipped. Again, if same specification is required for n variables, then instead of writing that field specification, say f9.5, n times we can write nf9.5.

Example 1:
```
        Write(*, 12) AVE, SD
12      Format (2x,'AVERAGE = ', f9.4/'S.D. = ', f10.4)
```

Example 2:
```
        Write(*, 13) A1, A2, J1, J2, J3
13      Format (1x, 2f9.5, 3x, 3I4)
```

14.8.7 Control Statements

Usually a computer follows the sequence of statements line by line. But there are cases where the execution order of the statements may have to alter or a few statements may have to skip depending on some decision. This is accomplished by using control statements. Different control statements of FORTRAN are given below:

(a) Logical If statement: The general form of Logical If statement is

If (logical expression) an executable statement

It tells the computer to execute the statement given in the right side if the logical expression is true, otherwise execute the statement next to it.

Example 1:
```
If (x.gt.y) m = m + 1
m = m – 1
```

This program segment tells the computer increases m by 1 if x > y, otherwise decrease m by 1.

Example 2:
```
    I = 1
5   I = I + 1
    Write(*,*) I
    If (I . It . 100) Go to 5
```

The above program segment tells the computer to print 2 to 100.

(b) Unconditional GO TO statement: The general form of the Unconditional GO TO statement is

GO TO n

where n is an statement level number to which the control will be transferred.

Example:
```
10    Read (*,*) a, b, c,
      D = a + b + c
      Av = D/3.0
      Write(*,*) Av
      Go to 10
```

The above program segment tells the computer to read different sets of three values and calculate their average value.

(c) Computed GO TO statement: The general form of the Computed GO TO statement is

$$\text{GO TO } (n_1, n_2, \cdots, n_k), \text{I}$$

where $n_1, n_2, \cdots, n_k$ are the statement levels and I is an integer variable.

It tells the computer to transfer the control to the statement level number n_p if I assume the value p, $1 \le p \le k$. For example the statement GO TO (10, 20, 30), I is equivalent to

```
If (I.eq.1) go to 10
If (I.eq.2) go to 20
If (I.eq.3) go to 30
```

(d) Arithmetic If statement: The general form of Arithmetic If statement is

If (arithmetic expression) *l*, *m*, *n*

where *l*, *m*, *n* are statement numbers. It tells the computer to transfer the control to the statement level number *l* or *m* or *n* according as the expression is <, =, >, 0. For example the statement If (x) 10, 20, 30 is equivalent to

```
If (x.lt.0) go to 10
If (x.eq.0) go to 20
If (x.gt.0) go to 30
```

(e) If-Then-Endif statement: The general form of If-Then-Endif statement is:

```
If (logical expression) then
Sequence of statements
Endif
```

It tells the computer to execute the sequence of statements given within the block if the logical expression is true, otherwise transfer the control to the statement next to endif.

Example:

```
If (D.lt.0) then
Write(*,*) "The equation has no real root"
Endif
Write(*,*) "The equation has two real roots"
```

The above program segment tells the computer to print

The equation has no real root

If D < 0, otherwise print

The equation has two real roots

(f) If-Then-Else-Endif statement: The general form of If-Then-Else-Endif statement is:

```
If (logical expression) then
1st set of statements
Else
2nd set of statements
Endif
```

It tells the computer to execute the 1st set of statements, if the logical expression is true, otherwise execute the 2nd set of statements and transfer the control to the statement next to endif.

Example:
```
If (D.lt.0) then
Write (*,*) "The equation has no real root"
Else
Write (*,*) "The equation has two real roots"
Endif
```

The above program segment tells the computer to print

The equation has no real root

if D < 0, otherwise print

The equation has two real roots

(g) If-Then-Elseif-Else-Endif statement: The general form of If-Then-Elseif-Else-Endif statement is:

```
If (logical expression 1) then
1st set of statements
Elseif (logical expression 2) then
2nd set of statements
Elseif (logical expression 3) then
3rd set of statements
Else
4th set of statements
Endif
```

It tells the computer to execute the 1st or 2nd or 3rd set of statement if the logical expression 1 or logical expression 2 or logical expression 3 is true, otherwise execute the 4th set of statements and transfer the control to the statement next to endif.

Example:
```
If (D.lt.0) then
Write (*,*) "The equation has no real root"
Elseif (D.eq.0) then
Write (*,*) "The equation has two equal real roots"
Else
Write (*,*) "The equation has two unequal real roots"
Endif
```

The above program segment tells the computer to print

The equation has no real root

if D < 0, otherwise print

The equation has two equal real roots

if D = 0, otherwise print

The equation has two unequal real roots

14.8.8 The Do-Continue Statement

We have already shown that to execute a set of instructions repeatedly logical If statement can be used. The same job can be done using the Do-Continue statement which has the following form:

```
        Do N k = l, m, n
        Body of the loop
   N    Continue
```

Where the symbols N, *l*, *m*, *n* have the following meaning:

(i) The integer N is the statement level number of the last executable statement of the loop, which is generally a Continue statement,

(ii) k represents an integer variable,

(iii) *l*, *m*, *n* are positive (may be all negative) integer constants and denote the initial value, final value and increment of the variable k. If the increment is 1, then it may be discarded.

When the integer variable exceeds, the final value control will be transferred to the executable next to the loop and this type of exit from the loop is called normal exit. On the other hand, if there be a logical if statement within the loop, then exit from the loop may take place, if the logical expression is true, even if the integer variable does not reach to the final value and such type of exit from the loop is called abnormal exit.

Example 1:

```
        Do 10k = 1, 5, 2
   10   Write (*,*) k
```

This program segment will print 1,3,5.

Example 2:

```
        Do 10k = –1, –5, –2
   10   Write (*,*) k
```

This program segment will print –1, –3, –5.

Example 3:

```
        Do 10k = 10, 5, 2
   10   Write (*,*) k
```

This program segment will print 10, as the initial value exceeds the final value it will be executed once only.

All the above examples are of normal exit from the loop and the following example shows the case of abnormal exit.

Example 4:

```
        J = 1
        Do 10k = 1, 5, 2
        If (k.gt.j) go to 20
   10   Write (*,*) k
   20   k = j + 1
        Write (*,*) k
```

This segment of program will print 1, 2.

Use of a set of Do-loops is possible in FORTRAN. In such cases one Do-loop (called inner Do-loop) is to lie completely within another Do-loop (called outer Do-loop), i.e.,

Do-loops cannot overlap and such set of Do-loops with different integer variables is called nested Do-loops. Control can not be transferred from outside of a Do-loop to inside of the Do-loop but from inside of a Do-loop control can be transferred to the outside of it and in case of nested Do-loops control can be transferred from inner Do-loop to an outer Do-loop. The statement level of the last executable statement of the loops may or may not be same as can be seen from following examples:

Example 1:

```
      Do 10k = 1, 5
      Do 20j = 1, 5, 2
      * * * * * * * * *
      * * * * * * * * *
   20 Continue
      * * * * * * * * *
      * * * * * * * * *
   10 Continue
```

Example 2:

```
      Do 10k = 1, 5, 2
      * * * * * * * * *
      * * * * * * * * *
      Do 10j = 1, 5
      * * * * * * * * *
      * * * * * * * * *
   10 Continue
```

14.8.9 The Dimension Statement

A memory cell is required to store a single value, say x. This type of variable is called unsubscripted variable and different cells are required to store different values with different names. But sometimes we have to work with $x_1, x_2, \cdots, x_n$ or $y_{11}, y_{12}, \cdots, y_{nn}$ etc, which need different cells in computer memory with same name. These type of variables are called subscripted variables. To store subscripted variables in the memory we have to use the Dimension statement which has the form:

Dimension $x(10)$, $y(10, 10)$

The subscripts can also be variables, i.e., they may be simple arithmetic expressions of the form

$$\text{N}, \quad \text{k}, \quad \text{N} + \text{k}, \quad \text{k*N}, \quad \text{k*N} \pm \text{m}$$

where k, m are integer constants and N is an integer variables. The DIMENSION statement is a non-executable statement and it specifies the reserved space in the memory for subscripted variables. The above program segment tells the computer for the variable x, 10 memory locations are to be allocated, and for the variable y, $10 \times 10 = 100$ memory locations are to be allocated, and all of which may not be used by those variables but under any circumstances the number of allocations cannot be less than the number of used variables.

14.8.10 Implied Do-loop

The READ or WRITE a set of subscripted variables we use Implied Do-loop, which has the form

Read (*,*) (x(i), i = 1, 5)

It is so called because it causes a looping process within the READ and WRITE statements. The above statement tells the computer to read the values of $x(1)$, $x(2)$, $x(3)$, $x(4)$, $x(5)$. Like Do-loops, implied Do-loops can also be nested. Let y be a m × n array, then we use two Do-loops one inside the other as can be seen from the following example:

```
      Write (*,14) ((y(i, j), i = 1, 5), j = 1, 5)
14    Format (5(f9.6, 3x))
```

In the above case j is the integer variable of the outer Do-loop and i is integer variable of inner Do-loop. For each value of j of the outer Do-loop, the inner Do-loop will be executed. So the above statement tells the computer to write the values of y_{ij} in the following form:

y(1,1)	y(2,1)	y(3,1)	y(4,1)	y(5,1)
y(1,2)	y(2,2)	y(3,2)	y(4,2)	y(5,2)
y(1,3)	y(2,3)	y(3,3)	y(4,3)	y(5,3)
y(1,4)	y(2,4)	y(3,4)	y(4,4)	y(5,4)
y(1,5)	y(2,5)	y(3,5)	y(4,5)	y(5,5)

14.8.11 The STOP Statement

The STOP statement is an executable statement which tells the computer to stop from executing the next executable statement of the program. The general form of this statement is

STOP

Example 1:

```
      Read (*,*) x, y
      If (x.gt.y) stop
      Write (*,*) x
      end
```

which tells the computer if x > y, then terminate the program, otherwise print the value of x.

Example 2:

```
      If (x.gt.y) go to 10
      Write (*,*) x
10    Stop
      End
```

which tells the computer if x > y, then control is transferred to the statement level number 10, i.e., terminate the program otherwise print the value of x.

14.8.12 The END Statement

The END statement is a non-executable statement which tells the computer to end the program and no other statement will be executed after it. The general form of this statement is

END

14.9 SUBPROGRAM

In FORTRAN complex programs can be divided into smaller programs which are called subprograms. A subprogram is a complete and independent program which is executed separetely and called by the main program as and when required. It receives values from main program and after performing calculations it returns the result(s) to the main program. We have already discussed about some in-built functions of FORTRAN library which are also subprograms. There are two types of subprograms in FORTRAN: (i) FUNCTION subprograms which usually returns one result and (ii) SUBROUTINE subprograms which returns more than one results.

14.9.1 FUNCTION Subprogram

The general form of the FUNCTION subprogram is

```
FUNCTION name (x1, x2, ..., xn)
* * * * * * * * * * * * * * * * * * *
name = expression
* * * * * * * * * * * * * * * * * * *
Return
End
```

In the above subprogram name is the name of the function of at least one of dummy variables ($x_1, x_2, \cdots, x_n$) and more than one return statement can be used depending on the number of expressions assigned to the name function. Control is transferred from the main program to the FUNCTION subprogram when function name is encountered in the main program. After the execution of subprograms the results are supplied to the main program through the RETURN statement of the subprogram.

Example: Calculate the sum $1-\frac{1}{2!}+\frac{1}{3!}-\frac{1}{4!}+\frac{1}{5!}$.

```
C   MAIN PROGRAM
    S = 0
    DO 5 I = 1, 5
5   S = S + (–1)**(I–1)/FACT(I)
    WRITE (*,*) 'SUM = ',S
    STOP
    END
C   SUBPROGRAM
    FUNCTION FACT(K)
    FACT = 1
    DO 10 L = 1, K
10  FACT = FACT*L
    RETURN
    END
```

14.9.2 SUBROUTINE Subprogram

The general form of the SUBROUTINE subprogram is

SUBROUTINE name $(x_1, x_2, \ldots, x_n)$

```
* * * * * * * * * * * *
name = expression
* * * * * * * * * * * *
Return
End
```

Like FUNCTION subprogram, the above subprogram name is the name of the subroutine and dummy variables are $(x_1, x_2, \ldots, x_n)$. If no dummy variable is needed then the parenthesis is of no use. The values of the SUBROUTINE subprogram is called by the main program by an executable statement named CALL statement which has the form:

CALL name $(x_1, x_2, \ldots, x_n)$

Control is transferred from the main program to the SUBROUTINE subprogram when it is called by the main program. After the execution of subprograms the results are supplied to the main program through the RETURN statement of the subprogram.

Example: Write a program which prints 4 numbers in descending form.

```
C   MAIN PROGRAM
    DIMENSION A(4)
    READ(*,*) (A(I), I = 1, 4)
    DO 10 I = 1, 3
    DO 10 J = 1, 3
    IF (A(J).LT.A(J + 1)) THEN
    CALL EXCH (T, J, A)
    ENDIF
10  CONTINUE
    WRITE (*,30) (A(I), I = 1, 4)
30  FORMAT (4(3X, F6.3))
    STOP
    END
C   SUBPROGRAM
    SUBROUTINE EXCH (T, J, A)
    DIMENSION A(4)
    T = A(J)
    A (J + 1) = A(J)
    A(J + 1) = T
    RETURN
    END
```

14.10 PROGRAMMING TECHNIQUE

After going through all the statements of FORTRAN mentioned above one can write a program in FORTRAN but the programmer should have knowledge about the FORTRAN text editor. Here we shall discuss about the editor.

FORTRAN text editor provides 80 columns for typing a line. Out of which columns 1 to 5 are kept for statement level number and out of which first column is for the use of the characters c and *. The column number 6 is used for continuation of a statement and columns 7 to 72 are used for writing FORTRAN statements. FORTRAN compilers do not read anything written in the columns 73 to 80. To identify the program these columns can be used by the user.

Column Numbers

1 2 3 4 5	6	7 8 9 .... 71 72	73 74 80
Statement level number	Continuation column	Fortran Statements	Ignored by Compilers

Statement numbers can of atmost five digits i.e. from 1 to 99999 and no two different levels can be numbered same as these serve as an identifier and helps to transfer control in a program. A line starting with C or * in the first column indicates that as a comment line and always ignored by the compiler. If any statement crosses the column number 72 then we can continue it to the next line writing 1 in the column 6. If it further crosses the second line then we can continue it to the next line writing 2 in the column 6 and so on.

Example:

12345	6	7 8 9 .... 71 72	73 74 ... 80
C		Fortran Program for calculating area	Program No. 1
	1	and perimeter of a triangle.	
		Read (*.*) a, b, c	
		S = (a + b + c)/2.0	
		Area = SQRT (S*(S–a)*(S–b)*(S–c))	
		AP = 2.0*S	
		Write (*,10) Area, AP	
10		FORMAT (2x,'AREA = ',f9.4/ 'Perimeter = ', f.8.3)	
		STOP	
		END	

14.11 MISCELLANEOUS EXAMPLES

1. Write a program for finding the area and perimeter of a circle of radius r.

```
C       AREA AND PERIMETER OF A CIRCLE FOR RADIUS R
        READ (*,*) R,PI
        AREA  = PI*R*R
        PERI = 2.0*PI*R
        WRITE (*,12) AREA, PERI
   12   FORMAT (2X,'AREA = ',F8.3/'PERI = ',F 8.3)
        STOP
        END
```

2. Write a program to find the sum of all positive odd integers up to 100.

```
C     PROGRAM OF FINDING SUM OF ALL POSITIVE ODD INTEGERS
      UP TO 100
      SUM = 0
      I =1
10    SUM = SUM +I
      I = I+2
      IF(I .LT.100)GO TO 10
      WRITE(*,12) SUM
12    FORMAT(2X,'SUM = ', I5)
      STOP
      END
```

3. Write a program to find the smallest of the three numbers x, y, z.

```
C     POGRAM TO FIND THE SMALLEST OF THREE NUMBERS
      READ(*,*)X,Y,Z
      SMALL = X
      IF(SMALL . LT . Y)GO TO 6
      SMALL = Y
6     IF(SMALL . LT . Z)GO TO 8
      SMALL = Z
8     WRITE(*,12) SMALL
12    FORMAT(3X,'SMALL = ',F8.3)
      STOP
      END
```

4. Write a program to find the mean and s.d. of 100 numbers.

```
C     PROGRAM FOR MEAN AND S.D. OF 100 NUMBERS
      DIMENSION X(100)
      READ(*,*)(X(I),I =1,100)
      S1 = 0
      S2 = 0
      DO 10 I = 1,100
      S1 = S1+ X(I)
10    S2 = S2 + X(I)**2
      AVE = S1/100.0
      SD = SQRT(S2/100.0-AVE**2)
      WRITE(*,12)AVE,SD
12    FORMAT(4X,'MEAN = ',F9.6//'S.D.=',F9.6)
      STOP
      END
```

5. Write a program to evaluate the sum

$$x-\frac{x^2}{2!}+\frac{x^3}{3!}-\cdots+(-1)^n\frac{x^n}{n!}$$

```
C       PROGRAM FOR EVALUATION OF A SUM
        READ(*,*)N,X
        S = X
        P = X
        K = 1
   10   P = -P*X/(K+1)
        S = S+P
        K= K+1
        IF (K .LE . N) GO TO 10
        WRITE(*,12)X,N,S
   12   FORMAT(3X,'FOR X =',F5.2,3X,'AND N=',I3,4X,'THE SUM IS =',S)
        STOP
        END
```

6. Write a program to find the g.c.d. of two positive integers.

```
C       PROGRAM FOR G.C.D. OF TWO POSITIVE INTEGERS
        READ(*,*)X,Y
        IF (X .GT . Y) GO TO 5
        T = X
        X = Y
        Y = T
    5   R = X-Y*IFIX(X/Y)
        IF (R . EQ . 0) GO TO 6
        X = Y
        Y = R
        GO TO 5
    6   WRITE(*,7)Y
    7   FORMAT(3X,'G.C.D. =',Y)
        STOP
        END
```

7. Write a program to find the l.c.m. of two positive integers.

```
C       PROGRAM FOR L.C.M.. OF TWO POSITIVE INTEGERS
        READ(*,*)X,Y
        PROD = X*Y
        IF (X .GT . Y) GO TO 5
        T = X
        X = Y
        Y = T
```

```
5     R = X-Y*IFIX(X/Y)
      IF (R . EQ . 0) GO TO 6
      X = Y
      Y = R
      GO TO 5
6     LCM = PROD/Y
      WRITE(*,7)LCM
7     FORMAT(3X,'L.C.M. =',I4)
      STOP
      END
```

8. Write a program to find the real root of the equation $ax^2 + bx + c = 0$.

```
C     PROGRAM FOR FINDING THE REAL ROOT OF AN EQUATION
      READ(*,*)A,B,C
      D = B**2-4.0*A*C
      IF (D. LT . 0) THEN
      WRITE(*,*)' THE EQUATION HAS NO REAL ROOT'
      GO TO 13
      ENDIF
      X1 = (–B+SQRT(D))/(2.0*A)
      X2 = (–B-SQRT(D))/(2.0*A)
      WRITE(*,11)
11    FORMAT(1X,'REAL ROOTS ARE')
      WRITE(*,12) X1,X2
12    FORMAT(2X,2(3X,F9.3))
13    STOP
      END
```

9. Write a program to find the *k*th element of Fibonnaci sequence satisfying

$$x_{n+2} = x_{n+1} + x_n,\ n \geq 1.$$

```
C     PROGRAM FOR FINDING THE K-TH ELEMENT OF FIBONACCI
      SEQUENCE.
      READ(*,*)K,X1,X2
      I = 1
10    X3 = X1 +X2
      I = I+1
      IF(I.LE.K–2)THEN
      X1 = X2
      X2 = X3
      GO TO 10
      ENDIF
      WRITE(*,11)X3
```

```
11    FORMAT(3X,I4)
      STOP
      END
```

10. Write a program to evaluate the value of an nth degree polynomial:

$$a_0 + a_1x + a_2x^2 + \dots\dots\dots + a_nx^n$$

```
C     PROGRAM FOR EVALUATING THE VALUE OF AN N-TH  DEGREE
      POLYNOMIAL
      DIMENSION A(100)
      READ(*,*)X,N,(A(I),I=0,N)
      AP = A(N)
      I = N-1
5     AP = AP*X+A(I)
      I = I-1
      IF (I. GE . 0) GO TO 5
      WRITE(*,15)X,N,AP
15    FORMAT (3X,'FOR X=',F5.3,3X,'AND FOR N=',I2,3X,'THE SUM =',F9.4)
      STOP
      END
```

11. U and V are linear arrays, each with n elements. Write a program in FORTRAN to compute

$$\frac{\sqrt{u_1^2 + u_2^2 + \cdots + u_n^2}\sqrt{v_1^2 + v_2^2 + \cdots + v_n^2}}{\sqrt{u_1v_1 + u_2v_2 + \cdots + u_nv_n}}.$$

```
      DIMENSION U(100),V(100)
      READ(*,*)N,(U(I),I=1,N),(V(I),I=1,N)
      S1=0
      S2=0
      S3=0
      DO 10 I =1,N
      S1=S1+U(I)*U(I)
      S2=S2+V(I)*V(I)
10    S3=S3+U(I)*V(I)
      S=SQRT(S1)*SQRT(S2)/SQRT(S3)
      WRITE(*,15)S
15    FORMAT(3X,F10.4)
      STOP
      END
```

12. Suppose the Frictional force acting on a particle is given by the equations

$$f(y) = 0 \text{ for } |y| \leq 0.5$$
$$= 2y^2 \text{ for } |y| > 0.5$$

Frictional force = $(mv/t) - f(v)^2 + at$

Write a program in FORTRAN to evaluate the frictional force for given values of m, v, a and t.

```
      READ(*,*)M,V,A,T
      IF((Y.LE.0.5).AND.(Y.GE.-0.5)) F(Y) = 0
      F(Y) = 2*Y**2
      FF =M*V/T-(F(V))**2+A*T
      WRITE(*,10) FF
10    FORMAT(3X,F10.5)
      STOP
      END
```

13. Write a program in FORTRAN which reads two integers *M* and *N* and finds the sum of all odd integers between *M* and *N* (excluding *M* and *N*).

```
      READ(*,*)M,N
      S=0
      M=M+2
      IF(M/2*2.EQ.M) M=M–1
      N=N-1
      DO 10 I = M,2,N
10    S = S + I
      WRITE(*,15)S
15    FORMAT(2X,I5)
      STOP
      END
```

14. Eccentricity of a conic is given. Write a FORTRAN program by using Arithmetic IF statement.

```
      READ(*,*)E
      E = E–1
      IF(E)10,20,30
10    WRITE(*,*)'THE CONIC IS AN ELLIPSE'
      GO TO 40
20    WRITE(*,*)'THE CONIC IS  A PARABOLA'
      GO TO 40
30    WRITE(*,*)'THE CONIC IS  A HYPERBOLA'
40    STOP
      END
```

15. Write a FORTRAN program to determine only complex root of a quadratic equation.

```
      COMPLEX X1,X2
      READ(*,*)A,B,C
      D = B*B– 4*A*C
      IF(D.GE.0) STOP
      D1 = SQRT(-D)
      B1= –B/(2*A)
      D2=D1/(2*A)
      X1 = CMPLX(B1,D2)
      X2 = CMPLX(B1, –D2)
      WRITE(*,*) X1,X2
      END
```

16. Write a program in FORTRAN which reads three positive numbers *A,B,C* and determine whether *A,B,C* can from the sides of a triangle and if so, determine whether the triangle will be a right angled triangle.

```
      READ(*,*)A,B,C
      IF(((A+B).GT.C).OR.((A+C).GT.B).OR.((B+C).GT.A))THEN
      WRITE(*,*) 'A,B,C FORM THE SIDES OF A TRAINGLE'
      GO TO 10
      ENDIF
      GO TO 15
10    H1 = SQRT(A**2+B**2)
      H2 = SQRT(B**2+C**2)
      H3 = SQRT(A**2+C**2)
      IF((C.EQ.H1).OR.(A.EQ.H2).OR.(B.EQ.H3))THEN
      WRITE(*,*)'THE TRIANGLE IS RIGHT ANGLED TRIANGLE'
      ENDIF
      GO TO 16
15    WRITE(*,*) ' A,B,C  DO NOT FORM A TRIANGLE'
16    STOP
      END
```

17. Write a FORTRAN program segment to find the sum of the numbers which are multiple of three and lie within (0,*N*) (both are inclusive).

```
      READ(*,*)N
      S = 0
      DO 10  I = 1,N
      IF (I/3*3.EQ.I)THEN
      S = S + I
      ENDIF
```

```
10      CONTINUE
        WRITE(*,15)S
15      FORMAT(2X,I4)
        STOP
        END
```

18. Write a program to write *n* numbers in ascending order.

```
C       PROGRAM FOR WRITNIG N NUMBERS IN ASCENDING ORDER.
        DIMENSION X(100)
        READ(*,*)N,(X(I),I = 1,N)
        DO 10 I = 1,N-1
        DO 10 J = I+1,N
        IF(X(J) . GT . X(I))THEN
        T = X(I)
        X(I) = X(J)
        X(J) = T
        ENDIF
10      CONTINUE
        WRITE(*,11)(X(I),I = 1,N)
11      FORMAT (3X,5(2X,F5.2))
        STOP
        END
```

19. Write a program which checks whether a number is prime or not.

```
C       PROGRAM TO CHECK A NUMBER
        READ(*,*)N
        K = 2
10      IF(N/K*K . EQ . N) GO TO 12
        K = K+1
        IF( K .LE. N/2) GO TO 10
        WRITE(*,11) N
11      FORMAT(2X,I4,2X,'IS A PRIME NUMBER')
        STOP
12      WRITE(*,13) N
13      FORMAT(2X,I4,2X,'IS NOT A PRIME NUMBER')
        END
```

20. Write a program to find the largest of given numbers.

```
C       PROGRAM TO FIND THE LARGEST OF N NUMBERS
        DIMENSION A(100)
        READ(*,*)N,(A(K),K = 1,N)
        MAX = A(1)
        DO 10 J = 2,N
```

```
      IF(MAX . GE . A(J)) GO TO 10
      MAX = A(J)
10    CONTINUE
      WRITE(*,*)'LARGEST NUMBER IS =', MAX
      STOP
      END
```

21. Write a program in FORTRAN-77, which prints all odd positive integers less than 100, omitting those divisible by 7.

```
      DO 10 I = 1,99,2
      IF(I/7*7.EQ.I) GO TO10
      WRITE(*,15) I
10    CONTINUE
15    FORMAT(3X,I2)
      STOP
      END
```

22. Write a FORTRAN program to classify the conic represented by

$$ax^2 + 2hxy + by^2 + 2gx + 2fy + c = 0$$

```
      READ(*,*)A,B,H
      E = H*H-A*B
      IF(E)10,20,30
10    WRITE(*,*)'THE CONIC IS AN ELLIPSE'
      GO TO 40
20    WRITE(*,*)'THE CONIC IS  A PARABOLA'
      GO TO 40
30    WRITE(*,*)'THE CONIC IS  A HYPERBOLA'
40    STOP
      END
```

23. Income tax rates for different income groups are given below. Write a FORTRAN program to calculate an individual's income tax.

Net Income Range	Rate of Income Tax
Up to 50,000/-	Nil
Rs. 50,001/- to 60,000/-	10% of the amount by which total income exceeds Rs. 50,000/-
Rs. 60,001/- to 1,50,000/-	Rs. 1000/- +20% of the amount by which total income exceeds Rs. 60,000/-
Above 1,50,000/-	Rs. 19000/- +30% of the amount by which total income exceeds Rs. 1,50,000/-

```
      READ(*,*)I
      IF(I.GT.150000)THEN
      TAX=19000 + (I-150000)*0.3
```

```
      ELSEIF((I.GT.60000).AND.(I.LE.150000))THEN
      TAX=1000 + (I-60000)*0.2
      ELSEIF((I.GT.50000).AND.(I.LE.60000))THEN
      TAX = (I-50000)*0.1
      ELSE
      TAX = 0
      ENDIF
      WRITE(*,10) TAX
   10 FORMAT(3X,I6)
      STOP
      END
```

24. Write a FORTRAN program to read a 5×7 matrix and then to find its maximum and minimum elements.

```
      DIMENSION A(5,7)
      READ(*,*)((A(I,J),I=1,5),J=1,7)
      MAX =A(1,1)
      MIN = A(1,1)
      DO 10 I =1,5
      DO 10 J=1,7
      IF(MAX.LT.A(I,J))MAX=A(I,J)
      IF(MIN.GT.A(I,J))MIN=A(I,J)
   10 CONTINUE
      WRITE(*,*)'THE LARGEST COEFFICIENT =',MAX
      WRITE(*,*)'THE SMALLEST COEFFICIENT =',MIN
      STOP
      END
```

25. Write a FORTRAN program to compute ${}^nC_r$

```
      DIMENSION FACT(100)
      READ(*,*) N,I
      FACT(1) =1
      DO 10 L = 2,N
   10 FACT(L) = FACT(L-1)*L
      NCR = FACT(N)/(FACT(I)*FACT(N-I))
      WRITE(*,12)NCR
   12 FORMAT(I5)
      STOP
      END
```

26. Find a root of the equation $x^3 - 4x - 9 = 0$ using bisection method corrected to 4D.

```
 C       SOLUTION OF A NON-LINEAR EQUATION BY BISECTION METHOD
         F(X) = X**3-4.0*X-9.0
         READ(*,*)A,B,N,E
         DO 10 I =1,N
         X = (A+B)/2.0
         IF((ABS(X-A).LT.E).OR.(ABS(X-B).LT.E)GO TO 11
         IF(F(X)*F(B).LT.0)THEN
         A = X
         ELSE
         B = X
         ENDIF
    10   CONTINUE
         WRITE(*,*)'ITERATION DOES NOT CONVERGE AFTER ',N,'
         ITERATIONS'
         GO TO 20
    11   WRITE(*,12) X
    12   FORMAT('THE REQUIRED ROOT CORRECTED TO 4D IS =',F9.4)
    20   STOP
         END
```

27. Find a root of the equation $x^3 - 4x - 9 = 0$ using regula falsi method corrected to 5D.

```
 C       SOLUTION OF A NON-LINEAR EQUATION BY REGULA FALSI METHOD
         F(X) = X**3–4.0*X–9.0
         READ(*,*)A,B,N,E
         DO 10 I = 1,N
         X = (A*F(B)–B*F(A))/(F(B)–F(A))
         IF((ABS(X-A).LT.E).OR.(ABS(X-B).LT.E)GO TO 11
         IF(F(X)*F(B).LT.0)THEN
         A = X
         ELSE
         B = X
         ENDIF
    10   CONTINUE
         WRITE(*,*)'ITERATION DOES NOT CONVERGE AFTER
         ',N,'ITERATIONS'
         GO TO 20
    11   WRITE(*,12) X
    12   FORMAT('THE REQUIRED ROOT CORRECTED TO 5D IS =',F9.5)
    20   STOP
         END
```

28. Find a root of the equation $x^3 - 4x - 9 = 0$ using Newton-Raphson method corrected to 6D.

```
C      SOLUTION OF A NON-LINEAR EQUATION BY NR METHOD.
       DIMENSION Y(100)
       F(X) = X**3-4.0*X-9.0
       DF(X) = 3.0*X**2-4.0
       READ(*,*)Y(1),N,E
       DO 10 I =1,N
       Y(I+1)= Y(I)-F(Y(I))/DF(Y(I))
       IF(ABS(Y(I+1)-Y(I)).LT.E)GO TO 11
 10    CONTINUE
       WRITE(*,*)'ITERATION DOES NOT CONVERGE AFTER ',N,'
       ITERATIONS'
       GO TO 20
 11    WRITE(*,12) Y(I+1)
 12    FORMAT('THE REQUIRED ROOT CORRECTED TO 6D IS =',F9.6)
 20    STOP
       END
```

29. Evaluate the integral $\int_0^1 e^x dx$ by trapezoidal rule and Simpson's $\frac{1}{3}$ rd rule taking $n = 12$.

```
C      PRGRAM FOR EVALUATION OF AN INTEGRAL
       F(X) = EXP(X)
       READ(*,*) A,B,N
       H= (B-A)/N
       S1= F(A)-F(B)
       S3 = 0
       S4 = 0
       DO 10 I = 1,N/2
       S3 = S3 + F(A+(2*J-1)*H)
 10    S4 = S4 +F(A+2*J*H)
       T = H*(S1+2*(S3+S4))/2.0
       S = H*(S1+4*S1+2*S4)/3.0
       WRITE(*,*)'VALUE OF THE INTEGRAL BY TRAPEZOIDAL RULE =',T
       WRITE(*,*)'VALUE OF THE INTEGRAL BY SIMPSON'S RULE = ',S
       STOP
       END
```

Examples

1. Which of the followings are acceptable real constants?

   (a) 3,234.0 (b) 12.345 (c) 12 (d) 1000000000.0

   (e) 3E-02 (f) 3.0E2.3 (g) E + 05 (h) 1.0E123

Answers

(a) Unacceptable: comma is not allowed.

(b) Acceptable.

(c) Unacceptable: contains no decimal point.

(d) Most computers allow real constants up to 9 digits.

(e) Unacceptable: before E the number must be in the form of real constant.

(f) Most computer allow only integer exponent.

(g) Unacceptable: an exponent cannot appear alone.

(h) Unacceptable: exponent part allows only two integers.

2. Which of the followings are acceptable real variables?

   (a) 3XY (b) E-COM (c) MAXMIN (d) QUESTION

   (e) ANSWER (f) X123 (g) A+B (h) KLINK

Answers

(a) Unacceptable: first character is not a letter.

(b) Unacceptable: only letter or digit can be used and (–) is not allowed.

(c) Unacceptable: first letter cannot be I,J,K,L,M,N.

(d) Unacceptable: contains more than six characters.

(e) Acceptable.

(f) Acceptable.

(g) Unacceptable: only letter or digit can be used and (+) is not allowed.

(h) Same as (c).

3. Which of the followings are acceptable integer constants?

   (a) 3234.0 (b) 12,345 (c) 12 (d) 1230000000 (e) 3E–02

Answers

(a) Unacceptable: decimal point is not allowed.

(b) Unacceptable: comma is not allowed.

(c) Acceptable.

(d) Most computers allow real constants up to 9 digits.

(e) Unacceptable: integers cannot be written in exponential form.

4. Which of the followings are acceptable integer variables?

   (a) 3J (b) J+123 (c) MINMAX (d) LAMBDA

   (e) ALPHA (f) L.P.P. (g) J(2)L (h) KLINK

Answers

(a) Unacceptable: first character is not a letter.
(b) Unacceptable: only letter or digit can be used and (+) is not allowed.
(c) Acceptable.
(d) Acceptable.
(e) Unacceptable: first letter can only be one of I,J,K,L,M,N.
(f) Unacceptable: decimal point is not allowed.
(g) Unacceptable: only letter or digit can be used and () is not allowed.
(h) Acceptable.

5. Write the following mathematical expressions in FORTRAN:

(a) $a+c-\frac{b}{de^2}$; (b) $\sqrt{x_1^2+x_2^2}$; (c) $\log_{10} x+\sin x+2y$; (d) $\frac{1}{|xy|}+e^z$;

(e) $\left(\frac{x}{y}\right)^{n+m}$; (f) $\sin(\log_e(3x+2y))$; (g) $\frac{a^2+b^2}{c-d}$; (h) $\left(1+\frac{x}{y}\right)^{y/x}$.

Answers

(a) $a+c-b/(d*c**2)$
(b) $sqrt(x(1)**2+x(2)**2)$
(c) $a\log 10(x)+\sin(x)+2.0*y$
(d) $1/abs(x*y)+\exp(x)$
(e) $(x/y)**(n+m)$
(f) $\sin(a\log(3.0*x+2.0*y))$
(g) $(a**2+b**2)/(c-d)$
(h) $(1+x/y)**(y/x)$.

6. Which of the following are acceptable FORTRAN statements?
(a) A = BC; (b) A = B + C; (c) A = B*C; (d) A + B = C; (e) A*B = C.

Answers

(a) Acceptable. (b) Acceptable. (c) Acceptable.
(d) Unacceptable: A+B is not a variable name.
(e) Unacceptable: A*B is not a variable name.

7. Write the following statements in FORTRAN:
(a) if $a-b=x^2$, go to the statement level no. 31.
(b) if $i \neq j$, stop.

Answers

(a) IF ((A – B) . EQ . X**2) GO TO 31.
(b) IF(I . NE . J) STOP.

8. Find error, if any, in the following computed GO TO statements.

(a) GO TO (0,1,2), J (b) GO TO (1,2,3,4), RAM

(c) GO TO (4,3,2,1) JAC (d) GO TO (4,5,3,2), JJJ

Answers

(a) '0' cannot be a statement number.

(b) The variable must be an integer variable.

(c) There must be a comma before JAC.

(d) No error.

9. Write the equivalent (a) Logical IF statement and (b) Arithmetic IF statement of the computed GO TO statement: GO TO (4,5,3), J

Answers

(a) IF (J. EQ . 1) GO TO 4

IF (J. EQ . 2) GO TO 5

IF (J. EQ . 3) GO TO 3

(b) IF (J-2) 4,5,3

10. Find error , if any, in the following DO statements

(a) DO 10, I= 3,41,2 (b) DO 12 P =1,4 (c) DO 11 JACK= 6,91,3,

(d) DO 11 JACK= JILL1,JILL2,JILL3 (e) D0 12 J = 1,2*N

Answers

(a) Comma is not allowed after 10.

(b) The variable(or index) must be an integer variable.

(c) Comma is not allowed after 3.

(d) No error if JILL1,JILL2,JILL3 are defined before it.

(e) No arithmetic operation is allowed within the DO –loop.

11. Find the errors, if any , in the following Implied DO –loops:

(a) READ(*,*),(A(K),K=1,10) (b) WRITE(*,12) A(K),K=1,10

(c) WRITE(*,12) ((A(K+1),K=1,10) (d) READ(*,*)(A(I,J),I=1,5,J=1,5)

Answers

(a) comma is not allowed after READ.

(b) variable list must be given within a parenthesis.

(c) parameter should be an integer variable, not an expression.

(d) In case of nested DO-loops, variables are to be kept within a parenthesis and the parentheses are to be separated by comma.

12. Find the errors, if any , in the following DIMENSION statements:

(a) DIMENSION, A(100) (b) DIMENSION X(2,0,3) (c) DIMENSION X(2,–3)

(d) DIMENSION A(10) B(12) C(23) (e) DIMENSION A(10.0)

Answers

(a) Comma is not allowed after DIMENSION.
(b) '0' can not be a subscript.
(c) Negative subscripts are not allowed.
(d) Variables must be separated by comma.
(e) Subscripts cannot be real constant.

13. Find errors , if any, in each of the following subprogram statements:
(a) FUNCTION, RAM(A,B) (b) FUNCTION RAM(A*B,A,B)
(c) FUNCTION RAM A,B (d) SUBROUTINE RAM(X(I),X(J))
(e) SUBROUTINE (X,Y) (f) SUBROUTINE RAM(T,X,X)

Answers

(a) No comma is allowed after FUNCTION.
(b) A*B can not be a parameter.
(c) Parameters must be given within a parenthesis.
(d) Parameters must be un-subscripted variable.
(e) No function name is given.
(f) A parameter can appear once only.

EXERCISE

1. Which of the followings are acceptable real constants?
(a) 334.0 (b) 45 (c) 12E-2 (d) 1234567890.0
(e) 3.25E 02 (f) 3.0E2.3 (g) E-05 (h) 21.0E-123.

2. Which of the followings are acceptable real variables?
(a) 3*XYZ (b) R.COM (c) MABABA (d) TYPHOID.

3. Which of the followings are acceptable integer constants?
(a) 3.24 (b) 2,35 (c) 125 (d) 1234567890 (e) 56E12.

4. Which of the followings are acceptable integer variables?
(a) IJK (b) JJ+13 (c) MINIMAXI (d) MU
(e) THETA (f) NEXT (g) J*L (h) BETA.

5. Write the following mathematical expressions in FORTRAN:

(a) $2a^2 + 3c^2 - \dfrac{|b|}{d - e^2}$; (b) $\sqrt{|a^2 - b^2|}$; (c) $\log_{10} x + 2y^3 + \dfrac{1}{|xy|} + e^z$;

(d) $\cos(\log_{10}|3x - 2y|)$; (e) $\dfrac{a^2 + b^2}{c^3 - d^3} + 3x^2y - 2xy^3$.

6. Which of the following are acceptable FORTRAN statements?
(a) A**2 = BC; (b) A = B**2*C ; (c) BC= A**2 ; (d) A/B = C.

7. Write the following statements in FORTRAN:
(a) if $A^2 > B^2$, go to the statement level no.11,
(b) if $B^2 - 4AC < 0$, stop.

8. Find error, if any, in the following computed GO TO statements.
   (a) GO TO (1,2,3,4,5) J (b) GO TO (1.0,2.0,3.0,4.0),JJJ
   (c) GO TO (40,15,30,20), JAC (d) GO TO (40,15,30,20), J*J.
9. Find error ,if any, in the following DO statements
   (a) DO I = 3,41,2 (b) DO 12 K =1,4, (c) DO 11 JILL= 6,91,3 (d) D0 12 J = 1,5+N.
10. Find the error, if any , in the following Implied DO -loops:
   (a) READ(*,*) (A(K) K=1,10) (b) WRITE(*,12),(A(K),K=1,10)
   (c) WRITE(*,12) ((A(3*K),K=1,10) (d) READ(*,*)((A(I,J),I=1,5)J=1,5).
11. Find the errors, if any , in the following DIMENSION statements:
   (a) DIMENSION A(2*K-3) (b) DIMENSION X(2*K)
   (c) DIMENSION X(–3) (d) DIMENSION A(10.0),I(12).
12. If scores of three students are given, write a program which finds and prints the number of these scores ≥ 80.
13. The commissions of a salesman on total sales is as follows:
   (i) If sale < 500, then there is no commission.
   (ii) If $500 \leq$ sales ≤ 1000, the commission = 10% of total sale.
   (iii) If sales > 1000, then commission is = 100 + 15% of total sale.
   Write a program which reads the total sales and calculate the commission.
14. Write a program which reads three positive numbers and checks whether they can form a triangle or not.
15. Health insurance premium of the employees of a company is deducted from their salary according as:

$$\text{Premium} = \begin{cases} 200, \text{ if unmarried,} \\ 350, \text{ if married without child} \\ 450, \text{ if married with child.} \end{cases}$$

   Write a program which reads the employees salaries and calculate their net salaries.
16. Write a program to find the largest of the three numbers x,y,z.
17. Write a program which reads and prints N numbers in descending form.
18. Write a program which calculates the sums correct to 4 decimal places.

$$\text{(i) } 1+\frac{1}{3}+\frac{1}{5}+\cdots+\frac{1}{21}, \text{ (ii) } 1-\frac{1}{2^2}+\frac{1}{3^2}-\cdots-\frac{1}{10^2} \text{ and (iii) } \frac{2\,4\,6}{1\,3\,5}\cdots\frac{20}{19}.$$

19. Write a program which writes the numbers 1 to 100 in reverse order.
20. Write a program prints all odd positive integers less than 100 leaving aside those which are multiple of 5.
21. Write a program to find the smallest of given N numbers.
22. Write a program to find the smallest coefficient of a matrix.

Chapter 15

BOOLEAN ALGEBRA AND LOGIC GATES

15.1 DEFINITION

Boolean algebra is an algebraic structure defined on a set of elements together with a set of operators and a number of postulates.

A non-empty set B with two binary operators, usually written as (+), (.) and a unary operator $'$ is called Boolean algebra if the following postulates, due to E.V.Huntington (1904), are satisfied:

P1: The operations (+) and (.) are commutative, i.e.,

$$x + y = y + x \quad \text{and} \quad x.y = y.x \quad \text{for all } x,y \text{ in } B.$$

P2: There exist two elements 0,1 in B such that

$$x + 0 = 0 + x \text{ and } x.1 = 1.x \quad \text{for all } x \text{ in } B.$$

P3: The operations (+) and (.) are distributive over one another, i.e.,

$$x.(y + z) = x.y + x.z \text{ and } x + y.z = (x + y).(x + z) \quad \text{for all } x, y, z \text{ in } B.$$

P4: For each x in B, there exists an element x' in B such that

$$x + x' = x' + x = 1 \quad \text{(identity element for .)}$$

and $$x.x' = x'.x = 0 \quad \text{(identity element for +)}$$

Note:

(i) Huntington postulates do not include the associative laws but this law holds for Boolean algebra.

(ii) Boolean algebra has no additive or multiplicative inverses implying that there is no concept of subtraction and division in Boolean algebra.

(iii) Apart from two-element Boolean algebra, the elements of a Boolean algebra are not defined properly.

EXAMPLES

(a) The power set $P(A)$ of the non-empty set A is a Boolean algebra with respect to (i) $\cup$ (union), (ii) $\cap$ (intersection) and (iii) $'$ (complementation).

(b) If we define $B = \{1,2,4,5,8,10,20,40\}$, the set of all positive divisors of 40.

$$x + y = \text{l.c.m. of } x,y,$$
$$xy = \text{h.c.f. of } x,y$$
$$\text{and} \quad x' = 40/x.$$

Then $\{B,+,.,'\}$ is not a Boolean algebra as P4 is not satisfied.

Similarities of set theory and Boolean algebra are shown in the following table.

Set Algebra	Boolean Algebra
$A \cup B$	$a + b$
$A \cap B$	$a.b$
A^C	a'
S	1
ϕ	0
$A' \cup A = S$	$a' + a = 1$
$A' \cap A = \phi$	$a'.a = 0$
$A \cup A = A$	$a + a = a$
$A \cap A = A$	$a.a = a$

Depending on the number of elements of the set B, Boolean algebra is classified as follows:

(i) *One-element Boolean algebra*: If the set B of the Boolean algebra $\{B, +, ., '\}$ contains only one-element, then it is called one-element Boolean algebra. In this case the zero element and unit element are same. This algebra is called degenerate algebra. Example: $\{\phi, +, ., '\}$ is a one-element Boolean algebra.

(ii) *Two-element Boolean algebra*: If the set A contains one element only, then its power set will contain two elements ϕ, A, i.e., the Boolean algebra $\{P(A),\cup,\cap,'\}$ contains two elements, viz., zero element and unit element. This is called two–element Boolean algebra. Example: $\{(0,1),+, ., '\}$ is a two element Boolean algebra.

(iii) *Infinite element Boolean algebra*: Let X be an infinite set (or real numbers), then $\{P(X),\cup,\cap,'\}$ is an infinite Boolean algebra in which ϕ (null set) is the zero element and the set X is the unit element of the Boolean algebra.

15.2 PRINCIPLE OF DUALITY

The dual of a proposition in a Boolean algebra is a proposition obtained by interchanging the operations (+),(.) and interchanging the identities 0,1 and any one of these are called dual of other.

For example, dual of $x + y.z = (x + y).(x + z)$ is $x.(y + z) = x.y + x.z$.

15.3 FUNDAMENTAL THEOREMS ON BOOLEAN ALGEBRA

Theorem 1: The identity elements of a Boolean algebra for the binary operations are separately unique.

Proof: Let 0 and 0* be the identity elements for the operation + in the Boolean algebra B. Then $a + 0 = a$ and $a + 0^* = a$ for all a in B. Since $0, 0^* \in B$, we have

$0 + 0^* = 0$, since 0* is an identity element for the operation +

and $0 + 0^* = 0^*$, since 0 is an identity element for the operation + .

Hence, $0 = 0^*$.

Again, let 1 and 1* be the identity elements for the operation . in the Boolean algebra B. Then $a \,.\, 1 = a$ and $a \,.\, 1^* = a$ for all a in B. Since $1, 1^* \in B$, we have

$1.1^* = 1$, since 1* is an identity element for the operation

and $1.1^* = 1^*$, since 1 is an identity element for the operation.

Hence, $1 = 1^*$.

Theorem 2: The complement of any element in a Boolean algebra is unique.

Proof: Let 0,1 are the identity elements for the operations (+), (.) respectively. Let a', a'' be the complements of a in the Boolean algebra B. Then we have

$a + a' = a' + a = 1$ and $a.a' = 0$

and $a + a'' = a'' + a = 1$ and $a.a'' = 0$.

Now
$$\begin{aligned} a' &= a'.1 \\ &= a'.(a + a'') \\ &= a'.a + a'.a'' \\ &= a'.a'' \\ &= a.a'' + a'.a'' \\ &= (a+ a').a'' \\ &= 1.a'' = .a''. \end{aligned}$$

Theorem 3: For all a in B, (i) $a + a = a$ and $a.a = a$, (ii) $a + 1 = 1$ and $a.0 = 0$.

Proof: (i)
$$\begin{aligned} a &= a + 0 \\ &= a + a.a', \text{ where } a' \text{ is the complement of } a \text{ in } B. \\ &= (a+a).(a+a') \quad \text{[distributive law]} \\ &= (a+ a).1 \\ &= a + a. \end{aligned}$$

Again, we have
$$\begin{aligned} a &= a.1 \\ &= a.(a + a') \\ &= a.a + a.a' \quad \text{[distributive law]} \\ &= a.a + 0 \\ &= a.a. \end{aligned}$$

Thus $a + a = a$ and $a.a = a$ for all a in B. These are known as *idempotent* laws.

(ii)
$$\begin{aligned} 1 &= a+a' \\ &= a+a'.1 \\ &= (a+a').(a + 1) \\ &= 1.(a + 1) \\ &= a + 1. \end{aligned}$$

Again, we have

$$\begin{aligned} 0 &= a.a' \\ &= a.(a' + 0) \\ &= a.a' + a.0 \\ &= 0 + a.0 \\ &= a.0. \end{aligned}$$

Thus $a + 1 = 1$ and $a.0 = 0$ for all a in B.

Theorem 4: In a Boolean algebra $a + a.b = a$ and $a.(a + b) = a$ for a, b in B.

Proof: $a + a.b$

$$\begin{aligned} &= a.1 + a.b \\ &= a.(1 + b) \\ &= a.1 \\ &= a. \end{aligned}$$

Again, from this result we see that

$$\begin{aligned} a &= a + a.b \\ &= a.a + a.b \\ &= a.(a + b) \end{aligned}$$

Thus $a + a.b = a$ and $a.(a + b) = a$ for a, b in B. These are known as *absorption* laws.

Theorem 5: In a Boolean algebra (i) $a + x = a + y$ and $a' + x = a' + y$ imply $x = y$;
(ii) $a.x = a.\ y$ and $a'.\ x = a'.\ y$ imply $x = y$,

for a,x,y in B.

Proof: (i) From the given results we can write

$$(a + x).(a' + x) = (a + y).(a' + y)$$

or, $a.a' + x = a.a' + y$ [distributive law]

or, $0 + x = 0 + y$

or, $x = y$.

(ii) From the given results we can write

$$a.x + a'.x = a.y + a'.y$$

or, $(a + a').x = (a + a').y$ [distributive law]

or, $1.x = 1.y$

or, $x = y$.

Theorem 6: In a Boolean algebra (i) $a+(b+c) = (a+b)+c$, (ii) $a.(b.c) = (a.b).c$ for a,b,c in B.

Proof: Let $l = a + (b + c)$ and $r = (a + b) + c$. Now

$$\begin{aligned} a.l &= a.[a + (b + c)] \\ &= a.a + a.(b + c) \quad \text{[distributive law]} \\ &= a + a \quad \text{[idempotent and absorption laws]} \\ &= a. \end{aligned}$$

Similarly, $a.r = a.[(a+b)+c]$

$= a.(a+b) + a.c$ [distributive law]

$= a + a.c$ [absorption law]

$= a.$ [absorption law]

So $a.l = a.r$ (1)

Again, $a'.l = a'.[a+(b+c)]$

$= a'.a + a'.(b+c)$ [distributive law]

$= 0 + a'.(b+c)$

$= a'.(b+c).$

Similarly, $a'.r = a'.[(a+b)+c]$

$= a'.(a+b) + a'.c$ [distributive law]

$= a'.a + a'.b + a.c$ [distributive law]

$= 0 + a'.(b+c)$ [distributive law]

$= a'.(b+c).$

Thus $a'.l = a'.r.$ (2)

From (1) and (2) we see that $l = r$, i.e., $a + (b+c) = (a+b) + c$ for a,b,c in B.

Again let $l = a.(b.c)$ and $r = (a.b).c$. Now

$a + l = a + a.(b.c)$

$= (a+a).(a+b.c)$ [distribution law]

$= a.(a+b.c)$ [idempotent law]

$= a.$ [absorption law]

Similarly, $a + r = a + (a.b).c.$

$= (a + a.b).(a+c)$ [distributive law]

$= a.(a+c)$ [absorption law]

$= a.$ [absorption law]

Thus, $a + l = a + r.$ (3)

Again, $a' + l = a' + a.(b.c)$

$= (a'+a).(a'+b.c)$ [distribution law]

$= 1.(a'+b.c)$

$= a' + b.c.$

Similarly, $a' + r = a' + (a.b).c$

$= (a' + a.b).(a'+c)$ [distributive law]

$= (a'+a).(a'+b).(a'+c)$

$= 1.(a'+b.c)$

$= a' + b.c$

Thus, $a' + l = a' + r.$ (4)

From (3) and (4), we see that $l = r$, i.e., $a.(b.c) = (a.b).c$ for a,b,c in B. The given laws are known as *associative* laws.

Theorem 7: In a Boolean algebra B, $a+x=1, a.x=0$ and $a+y=1, a.y=0 \Rightarrow x=y$.

Proof: $x = x.1$

$= (a+y).x$

$= a.x + y.x$ [distributive law]

$= 0 + x.y$ [commutative law]

$= x.y + a.y$

$= (x+a).y$ [distributive law]

$= (a+x).y$

$= 1.y = y.$

Theorem 8: In a Boolean algebra B, $(a')' = a$, for all a in B.

Proof: $(a')' = (a')' + a.a'$

$= ((a')' + a).((a')' + a')$ [distributive law]

$= ((a')' + a).1$

$= ((a')' + a).(a' + a)$

$= (a')'\, a' + a$

$= 0 + a$

$= a.$

Thus, $(a')' = a$, for all a in B. This is known as *involution* law.

Theorem 9: In a Boolean algebra B, $(a+b)' = a'.b'$ and $(a.b)' = a' + b'$, for all a,b in B.

Proof: $a + b + a'.b' = (a+b+a').(a+b+b')$ [distributive law]

$= (b+a+a').(a+b+b')$ [associative law]

$= (b+1).(a+1)$

$= 1.1$

$= 1.$ (1)

Similarly, $(a+b).(a'.b') = a.(a'.b') + b.(a'.b')$ [distributive law]

$= (a.a').b' + a'.(b.b')$ [associative law]

$= 0.b' + a'.0$

$= 0 + 0$

$= 0.$ (2)

From (1) and (2), we see that $(a+b)' = a'.b'$ for all a,b in B.

Again, $a' + b' + a.b = (a' + b' + a).(a' + b' + b)$ [distributive law]

$= (b' + a' + a).(a' + b' + b)$ [commutative law]

$= (1 + b').(1 + a')$

$= 1.1$

$= 1.$ (3)

Similarly, $(a.b).(a' + b') = (a.b).\, a' + (a.b).\, b'$ [distributive law]

$= b.(a.a') + a.(b.b')$ [commutative and associative laws]

$= b.0 + a.0$

$= 0 + 0$

$= 0.$ (4)

From (3) and (4) we see that $(a.b)' = a' + b'$, for all a,b in B.

These laws are known as *De Morgan's* law.

Theorem 10. In a Boolean algebra B, $a + b' = 1$, iff $a + b = a$ and $a.b' = 0$, iff $a.b = a$, for all a, b in B.

Proof:
$$\begin{aligned} a + b &= (a + b).1 \\ &= (a + b).(a + b') \\ &= a + b.b' && \text{[distributive law]} \\ &= a + 0 = a. \end{aligned}$$

Conversely,
$$\begin{aligned} a + b' &= 1.(a + b') \\ &= (a + a').(a + b') \\ &= a + a'.b' && \text{[distributive law]} \\ &= a + (a + b)' && \text{[De Morgan's law]} \\ &= a + a' = 1. \end{aligned}$$

Again,
$$\begin{aligned} a.b &= a.b + 0 \\ &= a.b + a.b' \\ &= a.(b + b') && \text{[distributive law]} \\ &= a.1 = a \end{aligned}$$

Conversely,
$$\begin{aligned} ab' &= 0 + ab' \\ &= aa' + ab' \\ &= a\,(a' + b') && \text{[distributive law]} \\ &= a\,(ab)' && \text{[De Morgan's law]} \\ &= aa' = 0. \end{aligned}$$

Theorem 11. In a Boolean algebra B, $a + b = b$, iff $a.b = a$, for all a,b in B.

Proof:
$$\begin{aligned} & a + b = b \\ \Rightarrow\ & a.b = a.(a + b) \\ \Rightarrow\ & a.b = a && \text{[absorption law]} \end{aligned}$$

Again,
$$\begin{aligned} & a.b = a \\ \Rightarrow\ & a + b = a.b + b \\ \Rightarrow\ & a + b = b + b.a && \text{[commutative law]} \\ \Rightarrow\ & a + b = b. && \text{[absorption law]} \end{aligned}$$

Note:

After going through all the laws, theorems discussed above it is now clear that Boolean algebra has similarities and dissimilarities between the algebra of real numbers. Here we shall point out the dissimilarities between the above algebras:

(i) The distributive law of (+) over (.) is valid for Boolean algebra but not for algebra of real numbers.

(ii) Algebra of real numbers includes subtraction and division but Boolean algebra has no subtraction and division operation.

(iii) Algebra of real numbers does not include complement operation like Boolean algebra.

(iv) Idempotent laws, absorption laws, involution law and De Morgan's law hold in Boolean algebra, not for the elements of algebra of real numbers.

(v) Principle of duality holds in Boolean algebra only, not for the elements of algebra of real numbers.

15.4 BOOLEAN FUNCTION

In Boolean algebra, the term constant means any symbol which represents a specific element such as 0 and 1, while a variable refers to a literal symbol such as x, y, z, etc., used to represent an arbitrary element or specified element of a Boolean algebra. In Boolean algebra, a single element or an expression of more elements formed with the aid of operation (.) and with or without prime notation (′) is called a monomial and an expression formed with several monomials, which are connected by the operation (+) is called a polynomial, and the monomials are called its terms. Henceforth, for simplicity $x.y$ will be written as xy.

Example

$a + ab' + abc'$ is a polynomial where its terms a, ab', abc' are monomials.

An expression formed with the combination of a finite number of constants and variables through the operations (+), (.), (′) is called a Boolean function. If $a.b.c$ are Boolean constants and x, y are Boolean variables, then $f(x,y) = ax + by + c'$ is a Boolean function in x, y.

A Boolean function will assume 0 or 1 when the variables involved in it assume 0 and 1. These values are called truth values. Since each independent variable can assume 0 or 1, a set of n variables will have 2^n possible combination of 0 and 1. A table showing the truth values of a Boolean function for all possible combinations of the truth values of the independent variables is called truth table for the function. Truth table for the Boolean function $f = xy'z' + x'yz' + x'y'z$ is given below.

Table: Truth table for the Boolean function $f = xy'z' + x'yz' + x'y'z$

x	y	z	x'	y'	z'	$xy'z'$	$x'yz'$	$x'y'z$	f
0	0	0	1	1	1	0	0	0	0
0	0	1	1	1	0	0	0	1	1
0	1	0	1	0	1	0	1	0	1
0	1	1	1	0	0	0	0	0	0
1	0	0	0	1	1	1	0	0	1
1	0	1	0	1	0	0	0	0	0
1	1	0	0	0	1	0	0	0	0
1	1	1	0	0	0	0	0	0	0

15.5 DISJUNCTIVE NORMAL FORM

A Boolean function is said to be in disjunctive normal form in $n(>0)$ variables x_1, x_2, $\cdots$, x_n if each term of the function is of the form $f_1(x_1) f_2(x_2) \cdots f_n(x_n)$, where $f_k(x_k) = x_k$ or x_k' for $k = 1,2, \cdots, n$. and no two terms are equal. For x,y,z in B, $xy' + x'y$, $xy'z' + x'yz' + x'y'z$ are the disjunctive normal form of the Boolean function in two and three variables respectively.

It is to be noted that with the basic laws of algebra, viz.,

$$(x+y)' = x'y',\ (xy)' = x' + y',\ x' + x = 1,\ xx' = 0,\ x + x = x,\ xx = x$$

a function having no constant term in Boolean algebra can be expressed in the disjunctive normal form. Consider the function $xyz + (xy)'(xz)'$ which has no constant term. Now

$$\begin{aligned} & xyz + (xy)'(xz)' \\ &= xyz + (x' + y')(x' + z') \\ &= xyz + x' + x'z' + y'x' + y'z' \\ &= xyz + x'(y' + y)(z' + z) + x'z'(y' + y) + y'x'(z' + z) + y'z'(x' + x) \\ &= xyz + xyz' + x'yz + xy'z + xy'z' + x'yz' + x'y'z + x'y'z'. \end{aligned}$$

This is the disjunctive normal form of the given function.

Note:

(i) A Boolean function can be reduced to disjunctive normal form in more than one ways, but when it is expressed in minimum number of variables then the representation is unique.

(ii) Two Boolean functions are said to be equal iff their disjunctive normal forms contain same terms.

(iii) The disjunctive normal form of a Boolean function in n variables is said to be complete if it contains 2^n terms. Example: $xy + xy' + x'y + x'y'$ is the complete disjunctive normal form of a Boolean function in two variables.

Theorem: A complete disjunctive normal form is identically 1.

Proof: In a Boolean function in complete disjunctive normal form the coefficient of any variable and its complement are of the form $y + y'$ and this is equal to 1. So using distributive laws and $y + y' = 1$ we can reduce the number of variables by one. Proceeding in same way n times, we can eliminate all the variables and hence it will be reduced to 1.

15.6 CONJUNCTIVE NORMAL FORM

A Boolean function is said to be in conjunctive normal form in $n(>0)$ variables x_1,x_2, $\cdots$, x_n if the function is a product of distinct linear factors of the form $f_1(x_1) + f_2(x_2) + \cdots + f_n(x_n)$ where $f_k(x_k) = x_k$ or x_k' for $k = 1, 2, \cdots, n$. For x, y, z in B, $(x + y')(x + y)$, $(x + y + z)(x + y' + z)(x + y' + z')$ are the conjunctive normal form of the Boolean function in two and three variables respectively.

Using the basic laws and distributive laws of Boolean algebra, it can be easily shown that a Boolean function having no constant term can be expressed in conjunctive normal form. Consider the function $xy' + yz'$, which has no constant term. Now

$$
\begin{aligned}
& xy' + yz' \\
&= (xy' + y)(xy' + z') \\
&= (x + y)(y' + y)(x + z')(y' + z') \\
&= (x + y + zz')(x + z' + yy')(y' + z' + xx') \\
&= (x + y + z)(x + y + z')(x + z' + y)(x + z' + y')(y' + z' + x)(y' + z' + x') \\
&= (x + y + z)(x + y + z')(x + z' + y')(y' + z' + x').
\end{aligned}
$$

This is the conjunctive normal form of the given Boolean function.

Note:

(i) Two Boolean functions are said to be equal if their conjunctive normal forms contain same factors.

(ii) The conjunctive normal form of a Boolean function in n variables is said to be complete if it contains all factors formed by the sums of all possible combinations of the variables and their complements. Example: $(x + y)(x + y')(x' + y)(x' + y')$ is the complete conjunctive normal form of a Boolean function in two variables.

Theorem: A complete conjunctive normal form is identically 0.

Proof: In a Boolean function in complete conjunctive normal form a pair of terms like $(x_1 + x_2 + \cdots + x_n)(x_1 + x_2 + \cdots + x_n')$ is equal to $(x_1 + x_2 + \cdots + x_{n-1})$. Thus choosing such pairs we can reduce the number of terms one by one. By repeated application of it, we finally get $xx' = 0$.

15.7 CONVERSION

We can convert any one of the above normal forms of the Boolean function to other using the result $(f')' = f$. Since each variable in each of the 2^n terms of a Boolean function of n variables in complete disjunctive normal form appears as x or x', the complement of the disjunctive normal form of a function will contain those terms of the complete disjunctive normal form which are absent in the given function. This is called method of complete DNF. Similarly, the complement of the conjunctive normal form of a function will contain those terms of the complete conjunctive normal form which are absent in the given function. This is called method of complete CNF.

15.8 SWITCHING CIRCUIT

Application of Boolean algebra to switching circuits helped a lot in designing electrical networks and that is why Boolean algebra is called backbone of computer circuit analysis. Switching algebra considers two elements 0 and 1 which correspond the states open (current is not flowing) and closed (current is flowing). Switches used in our houses are the model of such two state switches.

A network of switches is said to be basic parallel network if the switches x and y are arranged in parallel. The parallel network of two switches x and y is closed if at least one of the switches is closed, see following figure and table.

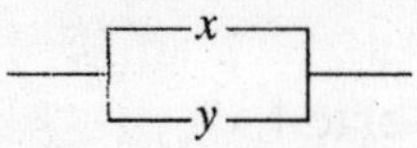

Fig. Parallel Network

Table: Closure table for parallel network

x	y	$x+y$
0	0	0
1	0	1
0	1	1
1	1	1

A network of switches is said to be basic series network if the switches x and y are arranged in series. The series network of two switches x and y is closed if both of the switches are closed, see following figure and table.

Fig. Series Network

Table: Closure table for series network

x	y	$x+y$
0	0	0
1	0	0
0	1	0
1	1	1

15.9 SWITCHING AND BOOLEAN FUNCTION

A switch is an electrical circuit of two states, so it can be represented by a Boolean variable x. If two states are in opposite states, then one is said to be complement of other. If we denote the closed switches as x, then open one is denoted by x'. In switching algebra, each electric circuit corresponds an algebraic expression involving the operations (+) (logical OR), (.) (logical AND) and (′) (logical NOT) and conversely. Such algebraic expression is a Boolean function, called closure function of the circuit. For example, consider the Boolean function

$$f = (x + y)\,(y + z)\,(z + x),$$

whose corresponding circuit is

Now we see that

$$\begin{aligned}
& (x+y)\,(y+z)\,(z+x) \\
= \; & (x+y)\,(z+y)\,(z+x) \\
= \; & (x+y)(z+xy) \\
= \; & (x+y)z+(x+y)xy \\
= \; & xz+yz+xy+xy \\
= \; & xz+yz+xy
\end{aligned}$$

and the corresponding circuit is

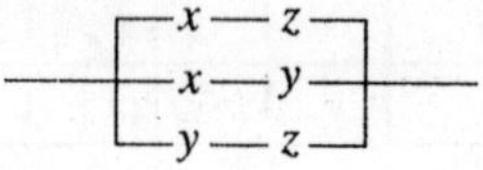

This is shown that a complicated series can be simplified to circuit with smaller number of switches having same effect.

15.10 LOGIC GATES

Gate is a logic circuit which is designed to implement given logic statement represented by a Boolean equation. A logic gate receives one or more binary information, called signals, as its inputs and produces a signal as output after manipulating the inputs in its internal circuits. The output can be predicted as a gate works according to some definite logic.

There are two types of gates: (1) Basic gates, (ii) Universal gates.

Basic Gate: The gates which do not depends on any other gate are called basic gates. The gates OR,AND, NOT gates are the examples of basic gates.

Universal Gate: There are some other gates which are a combination of the basic gates, these gates are called universal gates. Example: NAND (combination of NOT and AND gates), NOR (combination of NOT and OR gates).

OR gate: OR gate has two or more inputs and a single output. An OR gate yields the output as 1 if atleast one of the inputs is 1 (see the truth table), i.e., an OR gate is a logical function equivalent to parallel network of switches.

The algebraic form of the OR gate is: $f = x + y$,

where the symbol (+) represents the logical OR function (see figure).

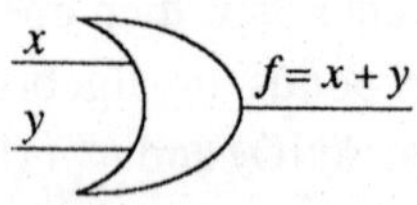

Fig. OR gate

Table: Truth table for OR gate

x	y	$f = x + y$
0	0	0
1	0	1
0	1	1
1	1	1

AND gate: AND gate has two or more inputs and a single output. An AND gate yields the output as 1 if all the inputs are 1 (see the truth table), i.e., an AND gate is a logical function equivalent to series network of switches.

The algebraic form of the AND gate is : $f = x.y$,
where the symbol . represents the logical AND function (see figure).

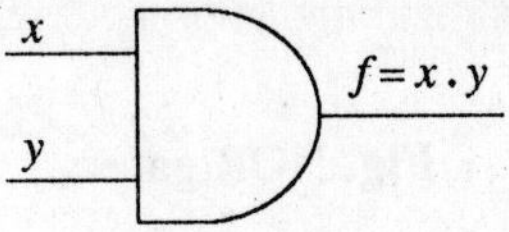

Fig. AND gate

Table: Truth table for AND gate

x	y	$x.y$
0	0	0
1	0	0
0	1	0
1	1	1

NOT gate: A NOT gate has only one input and one output. NOT gate yields the output as 1 if the input is 0 and vice versa., i.e., NOT gate yields the output as the complement of input. The algebraic function of NOT gate is: $f = x'$, where $'$ is the logical NOT function (see figure).

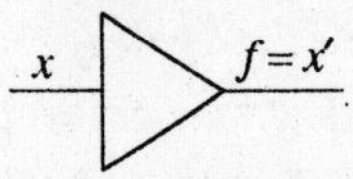

Fig. NOT gate

Table: Truth table for NOT gate

x	$f = x'$
0	1
1	0

Boolean laws for basic gates are as follows:

OR gate
$x + y = y + x$
$x + x = x$
$x + 1 = 1$
$x + 0 = x$

AND gate
$x.y = y.x$
$x.x = x$
$x.1 = x$
$x.0 = 0$

NOT gate
$(x')' = x$

NOR gate: NOR gate has two or more inputs and a single output. A NOR gate is a combination of NOT and OR gates and yields the output as 1, when all the inputs are 0, and 0 if any one of the input is 1.

The algebraic function of NOR gate is: $f = (x + y)'$,

where (′) is the logical NOT function and the symbol (+) denotes the logical OR function (see figure).

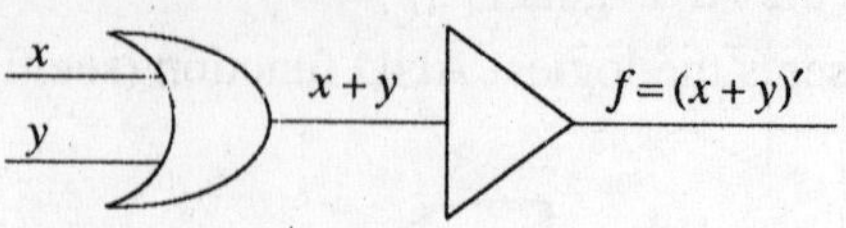

Fig. NOR gate

Table: Truth table for NOR gate

x	y	$f = (x + y)'$
0	0	1
1	0	0
0	1	0
1	1	0

NAND gate: NAND gate has two or more inputs and a single output. A NAND gate is a combination of NOT and AND gates and yields the output as 1, if any one of the inputs is 0 and an output 0, if all the inputs are 1.

The algebraic function of NOR gate is: $f = (x.y)'$,
where (′) is the logical NOT function and the symbol (.) denotes the logical AND function (see figure).

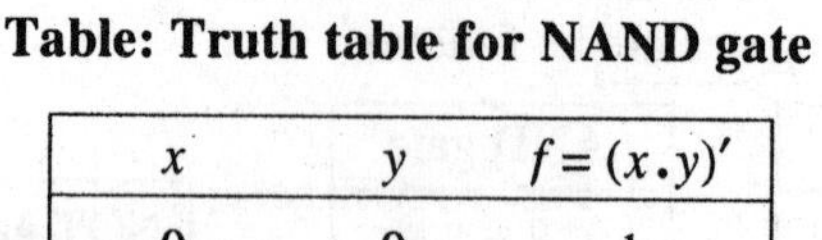

Fig. NAND gate

Table: Truth table for NAND gate

x	y	$f = (x . y)'$
0	0	1
1	0	1
0	1	1
1	1	0

XOR gate (Exclusive OR gate): An XOR gate has two or more inputs and a single output. It is a logical operation which gives an out put 1 if an odd number of 1's appeared as input. XOR gate is a combination of all the basic gates as shown in the figure.

The algebraic function of XOR gate is: $f = x \oplus y = x.y' + x'.y$

where ($'$) is the logical NOT function, the symbol (+) denotes the logical OR function and the symbol (.) denotes the logical AND function (see figure)

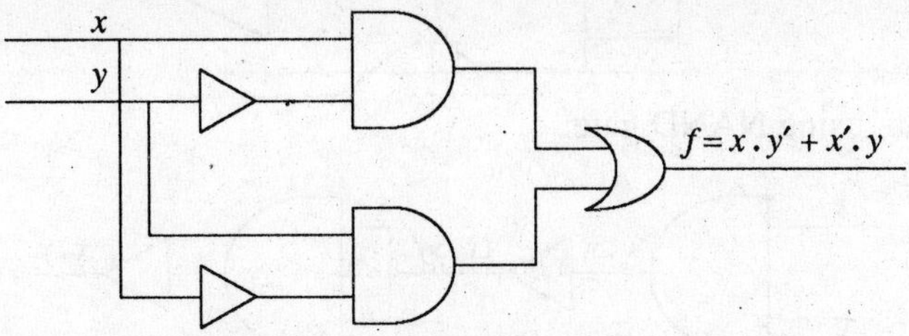

Fig. XOR gate

Table: Truth table for XOR gate

x	y	$f = x.y' + x.y'$
0	0	0
1	0	1
0	1	1
1	1	0

XNOR gate (Exclusive NOR gate): An XNOR gate has two or more inputs and a single output. It is a logical operation which gives an output 0, if an odd number of 0's appeared as input. Like XOR gate, XNOR gate is also a combination of all the basic gates as shown in the figure.

The algebraic function of XNOR gate is: $f = (x \oplus y)' = x.y + x'.y'$,

where ($'$) is the logical NOT function, the symbol (+) denotes the logical OR function and the symbol (.) denotes the logical AND function (see figure)

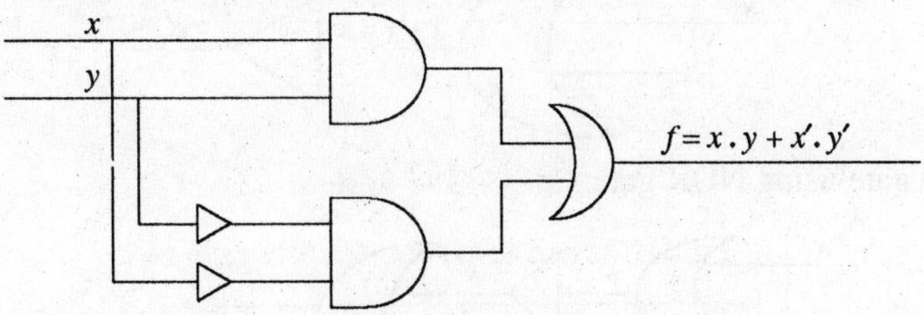

Fig. XNOR gate

Table: Truth table for XNOR gate

x	y	$f = x.y + x'.y'$
0	0	1
1	0	0
0	1	0
1	1	1

15.11 TO EXPRESS BASIC GATES USING UNIVERSAL GATES

(a) NOT gate using NAND gate:

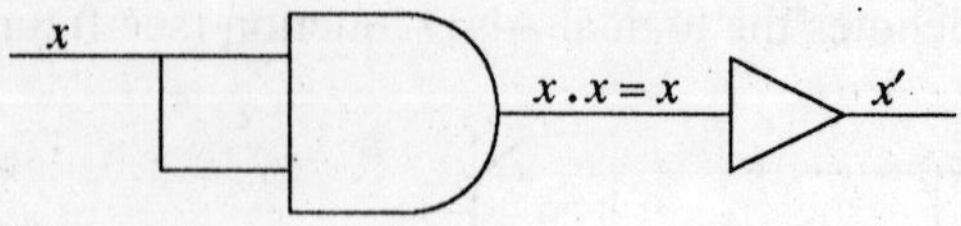

(b) AND gate using NAND gate:

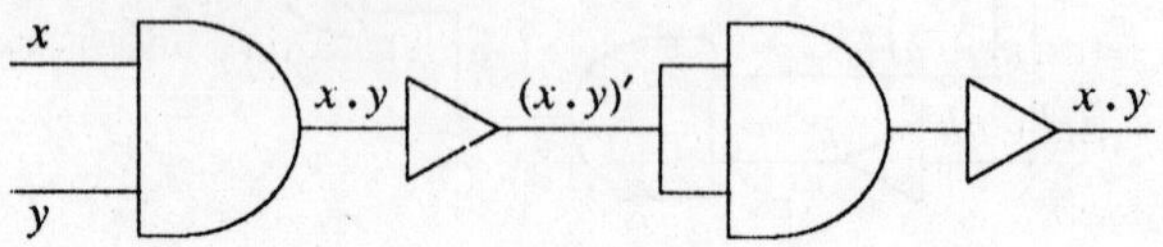

(c) OR gate using NAND gate:

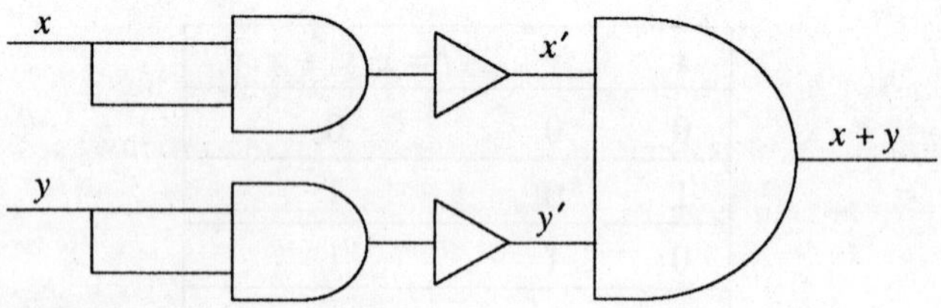

(d) OR gate using NOR gate:

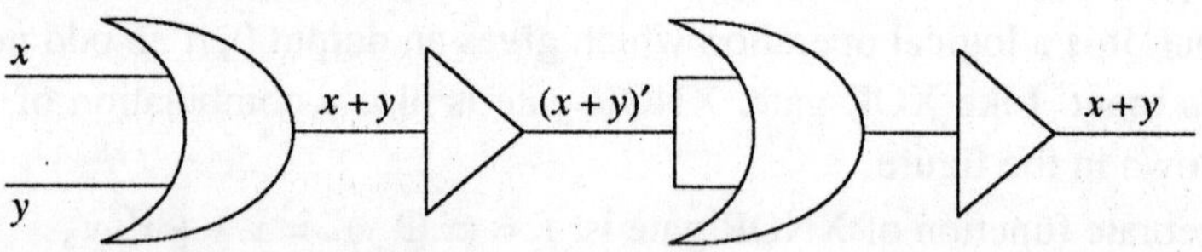

(e) NOT gate using NOR gate:

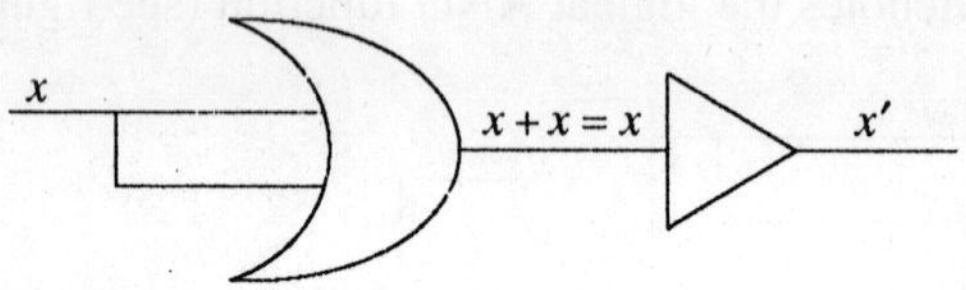

(f) AND gate using NOR gate:

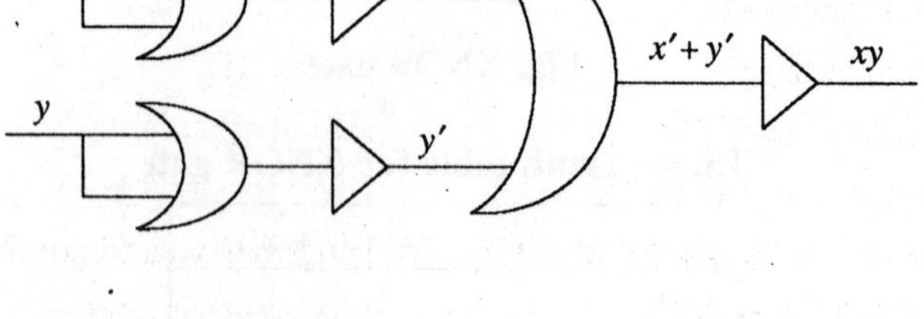

Examples

1. In a Boolean algebra B, $a, b, c \in B$, show that

(a) $(a+b)\,(b+c)\,(c+a) = ab + bc + ca$,

(b) $ab + a'\,b' + bc = ab + a'\,b' + a'c$

(c) $(a+b)' + (a+b')' = a'$

(d) $ab + a(ab)' = a$

(e) $a\{b + c(ab+ac)'\} = ab$

❑ (a) $(a+b)(b+c)(c+a)$

$= (a+b)(c+b)(c+a)$	[commutative law]
$= (a+b)(c+ab)$	[distributive law]
$= (a+b)c + (a+b)ab$	[distributive law]
$= ac + bc + (aa)b + a(bb)$	[distributive and associative laws]
$= ac + bc + ab + ab$	
$= ac + bc + ab$	
$= ab + bc + ca$ **(Proved).**	[commutative law]

❑ (b) $ab + a'b' + bc$

$= ab + a'b' + bc(a+a')$	
$= ab + a'b' + bca + bca'$	[distributive law]
$= ab + abc + a'b' + a'bc$	[commutative law]
$= ab(1+c) + a'(b' + bc)$	[distributive law]
$= ab + a'(b'+b)(b'+c)$	[distributive law]
$= ab + a'(b'+c)$	
$= ab + a'b' + a'c$ **(Proved).**	[distributive law]

❑ (c) $(a+b)' + (a+b')'$

$= a'b' + a'b$	[De Morgan's law]
$= a'(b'+b)$	[distributive law]
$= a'.1$	
$= a'$ **(Proved).**	

❑ (d) $ab + a(a.b)'$

$= ab + a(a'+b')$	[De Morgan's law]
$= ab + aa' + ab'$	[distributive law]
$= ab + 0 + ab'$	
$= ab + ab'$	
$= a(b+b')$	
$= a.1 = a$ **(Proved).**	

❑ (e) $a\{b + c(ab+ac)'\}$

$= a\{b + c(ab)'.(ac)'\}$	[De Morgan's law]
$= a\{b + c(a'+b')(a'+c')\}$	[De Morgan's law]
$= a\{b + ca'(a'+b') + cc'(a'+b')\}$	[distributive and commutative laws]
$= a\{b + ca'a' + ca'b'\}$	[distributive law]
$= ab + caa' + caa'b$	[distributive and commutative laws]
$= ab + 0 + 0$	
$= ab$ **(Proved).**	

2. In a Boolean algebra B, $a,b,c \in B$, find the complements of

(a) $ab' + c$, (b) $ab + ac + b'c$, (c) $(abc' + b')(a + b'c)$.

❑ (a) $(ab' + c)' = (ab')'c' = (a' + b)c'$ [De Morgan's law]

(b) $(ab + ac + b'c)' = (ab)'(ac)'(b'c)'$ [De Morgan's law]

$= (a'+b')(a'+c')(b + c')$ [De Morgan's law]

(c) $[(abc' + b')(a + b'c)]'$

$= (abc' + b')' + (a + b'c)'$ [De Morgan's law]

$= (abc')'\, b + a'\,(b'c)'$ [De Morgan's law]

$= (a' + b' + c)\, b + a'(b + c')$ [De Morgan's law]

$= a'b + b'b + cb + a'b + a'c'$ [distributive law]

$= a'b + 0 + a'b + cb + a'c'$ [commutative law]

$= a'b + cb + a'c'$.

3. Write separately and simplify the three functions f_1, f_2 and f_3 specified by the table:

x	y	z	f_1	f_2	f_3
1	1	1	0	0	1
1	1	0	1	1	1
1	0	1	0	1	0
1	0	0	1	0	0
0	1	1	0	0	0
0	1	0	0	1	0
0	0	1	0	1	1
0	0	0	0	0	1

The values of f_1 is 1 in the 2nd and 4th rows and its value is 0 in all other rows. The disjunctive normal form consists of these terms and each term will be 1. So

$f_1 = xyz' + xy'z'$

$= xz'y + xz'y'$ [commutative law]

$= xz'(y + y')$ [distributive law].

$= xz'.1$

$= xz'$

Proceeding in the similar ways we get

$f_2 = xyz' + xy'z + x'yz' + x'y'z$

$= x(yz' + y'z) + x'(yz' + y'z)$ [distributive law]

$= (x+x')(yz' + y'z)$ [distributive law]

$= 1.(yz' + y'z)$

$= yz' + y'z$

and

$f_3 = xyz + xyz' + x'y'z + x'y'z'$

$= xy(z + z') + x'y'(z + z')$ [distributive law]

$= xy.1 + x'y'.1$

$= xy + x'y'$.

4. Factorize: (i) $xy' + x'y$, (ii) $xy + yz + xz$.

❑ (i) $xy' + x'y = xy' + xx' + y'y + x'y$

$= x(y' + x') + (y' + x')y$ [distributive laws]

$= x(y' + x') + y(y' + x')$ [commutative law]

$= (x + y)(y' + x')$. [distributive law]

(ii) $xy + yz + xz = x(y + z) + yz$ [distributive law]

$= yz + x(y + z)$ [commutative law]

$= (yz + x)(yz + y + z)$ [distributive law]

$= (y + x)(z + x)(y + z)$. [distributive and absorption laws]

5. Express the following Boolean functions in disjunctive normal form in the smallest possible number of variables:

(i) $xy' + x'y + xz$

(ii) $(xy' + xz)' + x'$

(iii) $(x + y + z)(xy + x'z)'$

(iv) $[x'y + (xz)'](x + yz)'$

❑ (i) $xy' + x'y + xz$

$= xy'(z + z') + x'y(z + z') + xz(y + y')$

$= xy'z + xy'z' + x'yz + x'yz' + xzy + xzy'$ [distributive law]

$= xy'z + xy'z' + x'yz + x'yz' + xyz + xy'z$ [commutative law]

$= xy'z + xy'z' + x'yz + x'yz' + xyz$.

(ii) $(xy' + xz)' + x'$

$= (xy')'(xz)' + x'$ [De Morgan's law]

$= (x' + y)(x' + z') + x'$ [De Morgan's law]

$= x' + yz' + x'$ [distributive law]

$= x' + x' + yz'$ [commutative law]

$= x' + yz'$

$= x'(y + y')(z + z') + yz'(x + x')$

$= x'(yz + y'z + yz' + y'z') + yz'x + yz'x'$ [distributive law]

$= x'yz + x'y'z + x'yz' + x'y'z' + yz'x + yz'x'$ [distributive law]

$= x'yz + x'y'z + x'yz' + xyz' + x'yz' + x'y'z'$ [commutative law]

$= x'yz + x'y'z + x'yz' + xyz' + x'y'z'$.

(iii) $(x + y + z)(xy + x'z)'$

$= (x + y + z)(xy)'(x'z)'$ [De Morgan's law]

$= (x + y + z)(x' + y')(x + z')$ [De Morgan's law]

$= (x + y + z)(x'x + x'z' + y'x + y'z')$ [distributive laws]

$= (x + y + z)(x'z' + y'x + y'z')$

$= xx'z' + xy'x + xy'z' + yx'z' + yy'x + yy'z' + zx'z' + zy'x + zy'z'$ [distributive law]

$= xx'z' + xxy' + xy'z' + x'yz' + yy'x + yy'z' + x'zz' + xy'z + y'zz'$ [commutative law]

$= xy' + xy'z' + x'yz' + xy'z$

$= xy'(z + z') + xy'z' + x'yz' + xy'z$

$= xy'z + xy'z' + xy'z' + x'yz' + xy'z$ [distributive law]

$= xy'z + xy'z + xy'z' + x'yz'$ [commutative law]

$= xy'z + xy'z' + x'yz'.$

(iv) $[x'y + (xz)'](x+yz)'$

$= [x'y + x' + z']x'(y' + z')$ [De Morgan's law]

$= [x'(y + 1) + z']x'(y' + z')$

$= x'(z' + x')(z' + y')$ [commutative law]

$= x'(z' + x'y')$ [distributive law]

$= x'z' + x'y'$ [distributive law]

$= x'z'(y + y') + x'y'(z + z')$

$= x'z'y + x'z'y' + x'y'z + x'y'z'$ [distributive law]

$= x'yz' + x'y'z' + x'y'z.$ [commutative law]

6. Express the following Boolean functions in conjunctive normal form in the smallest possible number of variables

(i) $(a+b)(a'+b)(a+c')$

(ii) $abc + (a + b)(a + c)$

(iii) $(a+b+c)\{(ab)'(a'c)'\}$

(iv) $(a+b')(a'+c)(b+c')$

❑ (i) $(a+b)(a'+b)(a+c')$

$= (a+b+cc')(a'+b+cc')(a+c'+bb')$

$= (a+b+c)(a+b+c')(a'+b+c)(a'+b+c')(a+c'+b)(a+c'+b')$ [distributive law]

$= (a+b+c)(a+b+c')(a'+b+c)(a'+b+c')(a+b+c')(a+b'+c')$ [commutative law]

$= (a+b+c)(a+b+c')(a'+b+c)(a'+b+c')(a+b'+c').$

(ii) $abc + (a+b)(a+c)$

$= (abc+a+b)(abc+a+c)$ [distributive law]

$= (a+a+b)(b+a+b)(c+a+b)(a+a+c)(b+a+c)(c+a+c)$ [distributive law]

$= (a+b)(a+b)(c+a+b)(a+c)(a+c)$

$= (a+b)(c+a+b)(a+c)$

$= (a+b+cc')(c+a+b)(a+c+bb')$

$= (a+b+c)(a+b+c')(a+b+c)(a+c+b)(a+c+b')$ [distributive and commutative laws]

$= (a+b+c')(a+b+c)(a+b+c)(a+b+c)(a+b'+c)$ [commutative law]

$= (a+b+c')(a+b+c)(a+b'+c).$

(iii) $(a+b+c)\{(ab)'(a'c)'\}$

$= (a+b+c)(a'+b')(a+c')$ [De Morgan's law]

$= (a+b+c)(a'+b'+cc')(a+c'+bb')$

$= (a+b+c)(a'+b'+c)(a'+b'+c')(a+c'+b)(a+c'+b')$ [distributive law]

$= (a+b+c)(a'+b'+c)(a'+b'+c')(a+b+c')(a+b'+c').$ [commutative law]

(iv) $(a+b')(a'+c)(b+c')$

$= (a+b'+cc')(a'+c+bb')(b+c'+aa')$

$= (a+b'+c)(a+b'+c')(a'+c+b)(a'+c+b')(b+c'+a)(b+c'+a')$ [distributive law]

$= (a+b'+c)(a+b'+c')(a'+b+c)(a'+b'+c)(a+b+c')(a'+b+c').$ [commutative law]

7. The values of a function $f(a,b,c)$ are given in the following table for different values of the variables:

a	b	c	$f(a, b, c)$
0	0	0	0
0	0	1	1
0	1	0	0
0	1	1	0
1	0	0	1
1	0	1	1
1	1	0	1
1	1	1	0

❑ The values of f is 1 in the 2nd, 5th, 6th and 7th rows and its value is 0 in all other rows. We ignore the combinations of the variables of the rows with zero values of f. The disjunctive normal form consists of these terms and each term will be 1. So

$$f(a,b,c) = a'b'c + ab'c' + ab'c + abc'.$$

For the conjunctive normal form of the function f we consider the combinations of the variables of the rows in which f is zero. The conjunctive normal form consists of these terms and each term will be 0. So

$$f(a, b, c) = (a + b + c)\,(a + b' + c)\,(a + b' + c')\,(a' + b' + c').$$

8. Change the following functions from the disjunctive normal form to conjunctive normal form

(i) $abc + a'bc + ab'c' + a'bc'$,

(ii) $abc + a'bc' + ab'c' + a'b'c'$.

❑ (i) $abc + a'bc + ab'c' + a'bc'$

$= [(abc + a'bc + ab'c' + a'bc')']'$

$= [(abc)'\,(a'bc)'\,(ab'c')'\,(a'bc')']'$ [De Morgan's law]

$= [(a' + b' + c')(a + b' + c')(a' + b + c)(a + b' + c)]'$ [De Morgan's law]

$= (a + b + c)(a + b + c')(a' + b' + c)(a' + b + c')$, by the method of complete CNF.

(ii) $abc + a'bc' + ab'c' + a'b'c'$

$= [(abc + a'bc' + ab'c' + a'b'c')']'$

$= [(abc)'\,(a'bc')'\,(ab'c')'\,(a'b'c')']'$ [De Morgan's law]

$= [(a' + b' + c')(a + b' + c)(a' + b + c)(a + b + c)]'$ [De Morgan's law]

$= (a'+b+c')(a'+b'+c)(a+b'+c')(a + b + c')$, by the method of complete CNF.

9. Change the following functions from the conjunctive normal form to disjunctive normal form:

(i) $(a + b')(a' + b)(a' + b')$,

(ii) $(a + b' + c')(a' + b + c)(a' + b + c')(a + b' + c)$.

(i) $(a + b')(a' + b)(a' + b')$

$= [\{(a + b')(a' + b)(a' + b')\}']'$

$= [(a + b')' + (a' + b)' + (a' + b')']'$ [De Morgan's law]
$= [a' b + ab' + ab]'$ [De Morgan's law]
$= a'b'$, by the method of complete DNF

(ii) $(a+b'+c')(a'+b+c)(a'+b+c')(a+b'+c)$
$= [\{(a+b'+c')(a'+b+c)(a'+b+c')(a+b'+c)\}']'$
$= [(a+b'+c')'+(a'+b+c)'+(a'+b+c')'+(a+b'+c)']'$ [De Morgan's law]
$= [a'bc + ab'c' + ab'c + a'bc']'$ [De Morgan's law]
$= abc + abc' + a'b'c + a'b'c'$, by the method of complete DNF.

10. If $f(x, y, z) = (x' + y' + z)(x + y' + z')(x' + y + z')$, express $f'(x, y, z)$ in disjunctive normal form.

$f'(x,y,z) = [(x' + y' + z)(x + y' + z')(x' + y + z')]'$
$= (x' + y' + z)' + (x + y' + z')' + (x' + y + z')'$ [De Morgan's law]
$= xyz' + x'yz + xy'z.$

This is the disjunctive normal of the function $f'(x,y,z)$.

11. If $a + b = a + c$ and $a.b = a.c$, then show that $b = c$.

$a + b = a + c$
$\Rightarrow a'.(a + b) = a'.(a + c)$
$\Rightarrow a'.a + a'.b = a'.a + a'.c$ [distributive law]
$\Rightarrow 0 + a'.b = 0 + a'.c$
$\Rightarrow a'.b = a'.c.$

Again, $a.b = a.c$

Adding, we get $a'.b + a.b = a'.c + a.c$
$\Rightarrow (a' + a).b = (a' + a).c$ [distributive law]
$\Rightarrow 1.b = 1.c$
$\Rightarrow b = c.$

12. If x, y, z be three switches, then draw the switching circuit representing $xyz' + x'(y + z')$.

❑

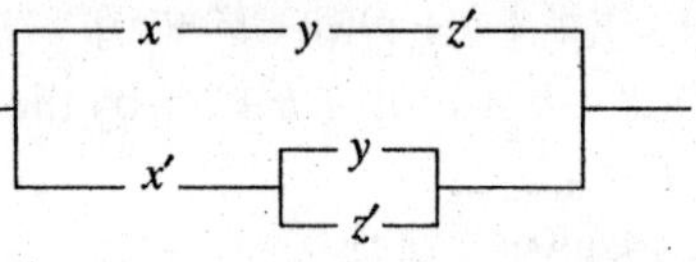

13. If x,y,z be three switches, then simplify the circuit represented by the Boolean function $xy' + yz + x'y'$ and draw the simplified circuit.

❑ $xy' + yz + x'y'$
$= yz + xy' + x'y'$ [commutative law]
$= yz + (x + x')y'$ [distributive law]
$= yz + y'.$

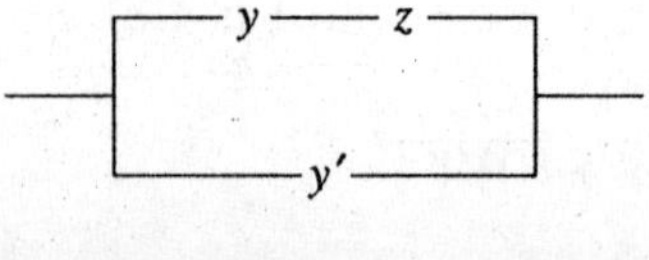

14. If x, y, z be three switches, then simplify the circuit represented by the Boolean function $(x' + y)(z' + y)(x' + z')(x + y + z')$ and draw the simplified circuit.

❑ $(x' + y)(z' + y)(x' + z')(x + y + z')$

$= (x'z' + y)\{x'(x+y)+z'\}$ [distributive law]

$= (x'z' + y)(x'x + x'y + z')$ [distributive law]

$= (x'z' + y)(x'y + z')$

$= x'z'(x'y + z') + y(x'y + z')$ [distributive law]

$= x'z'y + x'z' + yx' + yz'$

$= x'yz' + yz' + x'z' + x'y$ [commutative law]

$= (x' + 1)yz' + x'(z' + y)$ [distributive law]

$= yz' + x'(z' + y).$

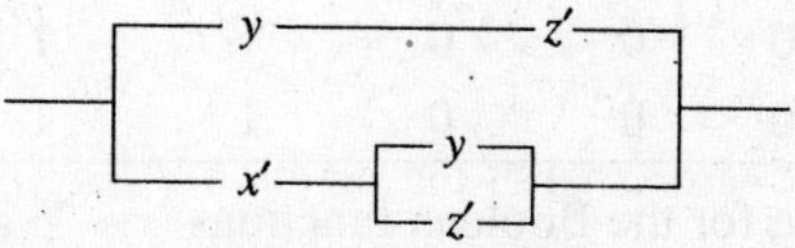

15. Show that current will flow through the network given by the function

$$[ab(a'b+ab')]'$$

irrespective of the switches a and/or b are closed or open.

❑ $[ab(a'b+ab')]'$

$= [aba'b + abab']'$ [distributive law]

$= [aa'bb+aabb']'$ [commutative law]

$= [0.b + a.0]'$

$= 0' = 1.$

Hence, whether the switches a and/or b are closed or open, current will flow through the network represented by the given function.

16. Construct the closure table of the Boolean function $x + x'y$ and draw its simplified circuit.

❑ $x + x'y$

$= (x + x')(x + y)$ [distributive law]

$= 1.(x + y)$

$= x + y.$

Closure table

x	x'	y	$x + x'y$	$x + y$
1	0	1	1	1
0	1	1	1	1
1	0	0	1	1
0	1	0	0	0

Thus the circuits

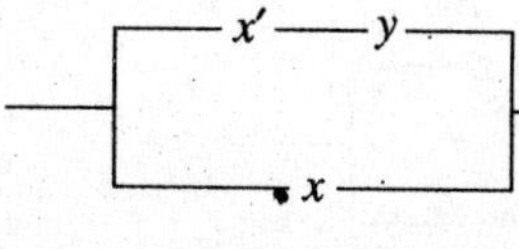

and

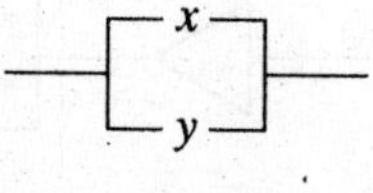

are equivalent.

17. By constructing the closure tables of the Boolean functions $xy + z$ and $xy + z(x' + y')$, show that they are equal.

❑ We construct the closure tables for both the functions in the table given below:

x	y	z	x'	y'	$x' + y'$	xy	$xy + z$	$xy + z(x' + y')$
0	0	0	1	1	1	0	0	0
0	0	1	1	1	1	0	1	1
0	1	0	1	0	1	0	0	0
0	1	1	1	0	1	0	1	1
1	0	0	0	1	1	0	0	0
1	0	1	0	1	1	0	1	1
1	1	0	0	0	0	1	1	1
1	1	1	0	0	0	1	1	1

The above closure table for the Boolean functions $xy + z$ and $xy + z(x' + y')$ shows that they are equal.

18. Draw the logic diagram for the following Boolean expressions:

(i) $L = AB'(B + C)$

(ii) $L = A + BC'$

(iii) $L = A'B + B'A$

(iv) $L = (A + B)CD'$

(v) $L = A'B + C'A + B'C$

(vi) $L = ABC' + AB'C + A'BC$

(vii) $L = (A' + B')(A + C)(B + C')$

(i)

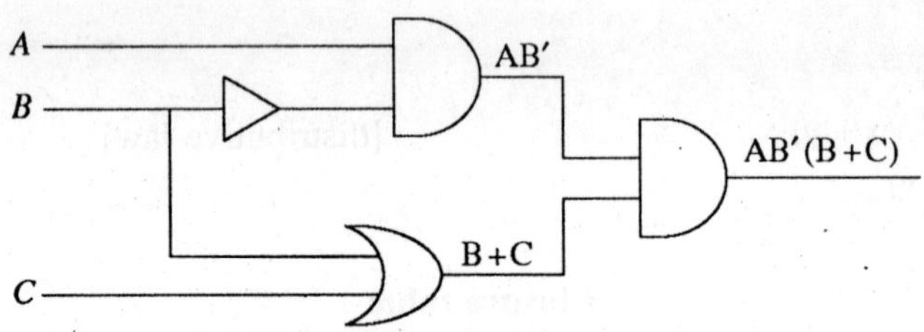

(ii)

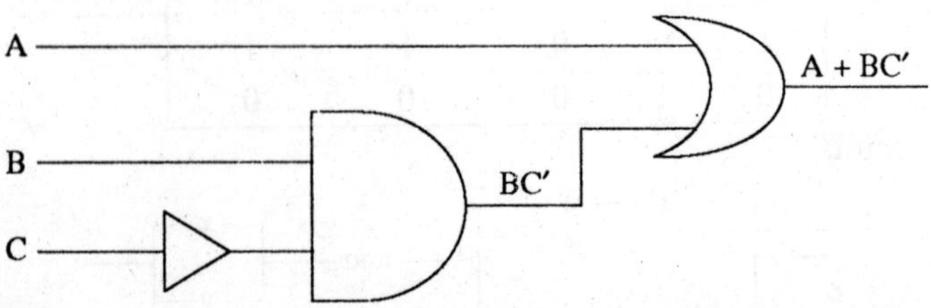

(iii)

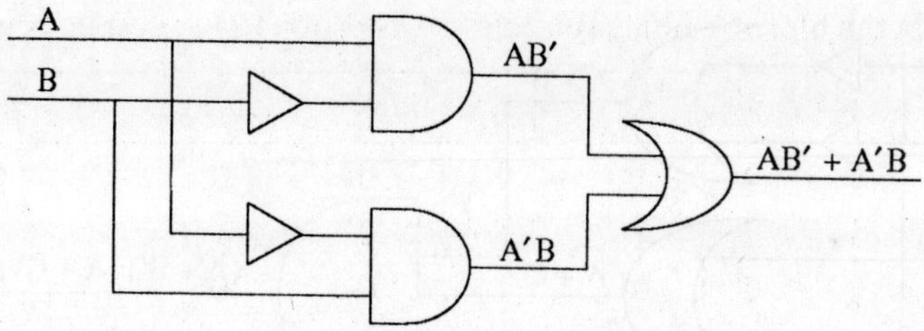
A
B
AB′
A′B
AB′ + A′B

(iv)

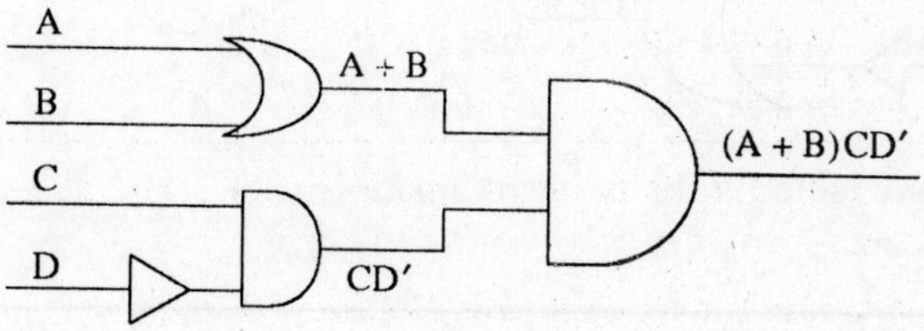
A
B
C
D
A + B
CD′
(A + B)CD′

(v)

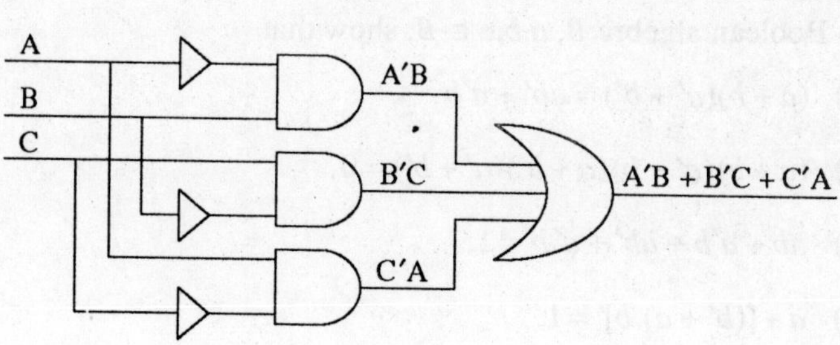
A
B
C
A′B
B′C
C′A
A′B + B′C + C′A

(vi)

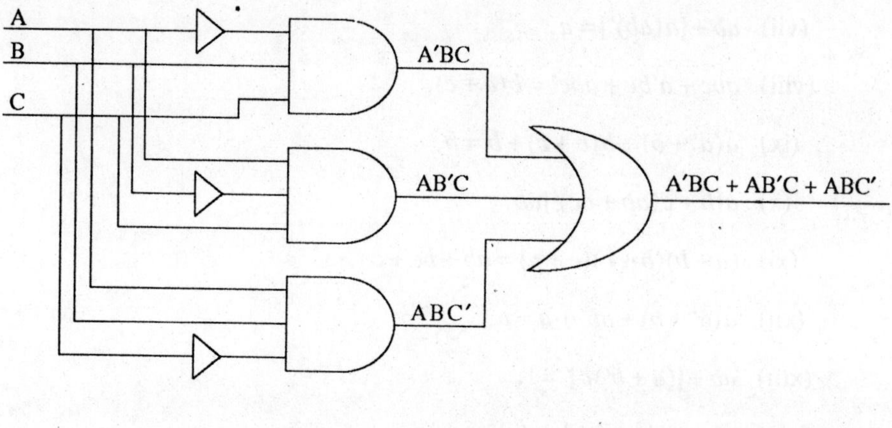
A
B
C
A′BC
AB′C
ABC′
A′BC + AB′C + ABC′

(vii)

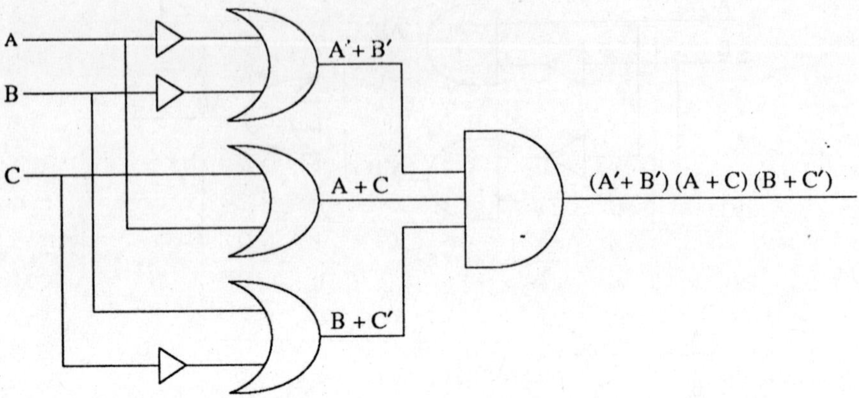

EXERCISE

1. (a) State the postulates of Boolean algebra. Explain the difference between Boolean algebra and the algebra of real numbers.

   (b) In a Boolean algebra B, $a,b,c \in B$, show that

   (i) $(a+b)(a'+b') = ab' + a'b$,

   (ii) $(a+b)(a'+b)(a+b')(a'+b') = 0$,

   (iii) $ab + a'b + ab' + a'b' = 1$,

   (iv) $a + [(b'+a).b]' = 1$,

   (v) $a[b'a+b]' = 0$,

   (vi) $[(a'+b)(b'+c)](a'+c)' = 0$,

   (vii) $ab + [a(ab)'] = a$,

   (viii) $abc + a'bc + abc' = b(a+c)$,

   (ix) $a(a'+b) + b(b+c) + b = b$,

   (x) $a[b + c(ab+ac)']ab$,

   (xi) $(a+b)(b+c)(c+a) = ab + bc + ca$,

   (xii) $a(a'+b) + ab' + a = a$,

   (xiii) $ab + [(a+b')b]' = 1$,

   (xiv) $ab + ab' + a' + b = 1$.

2. Write separately and simplify (if possible) the three functions f_1, f_2 and f_3 specified by the table:

x	y	z	f_1	f_2	f_3
1	1	1	0	0	1
1	1	0	1	0	0
1	0	1	0	1	0
1	0	0	1	0	0
0	1	1	0	0	1
0	1	0	0	1	0
0	0	1	1	0	0
0	0	0	0	0	1

3. Factorize the following expressions:

(i) $xz + x'y$, (ii) $w + xyz$, (iii) $xz + uv$, (iv) $xz + x'y + xy$.

4. Find the complements of

(i) $ab + a'c' + bc'$,

(ii) $(abc + a')(ab + c')$,

(iii) $(a + b)(c' + b)$,

(iv) $(abc + a'b')' + bc$

(v) $ab + a'b'$.

5. Obtain the truth tables for the following expressions and find the truth values of the following functions in which the variables have only two values 0 and 1.

(i) $f = abc + a'bc' + a'b'c'$,

(ii) $f = (a + b)(b + c)(c + a)$,

(iii) $f = abc' + a'b'$,

(iv) $f = ab'c + a'(b + c')$,

(v) $f = ab' + a'b$.

6. Express the following Boolean functions in disjunctive normal form in the smallest number of variables:

(i) $(a + b')(b + c')(c + a')(a' + b')$,

(ii) $(a + b + c)(ab + ac)$,

(iii) $(a + b)(a + b')(a' + c)$,

(iv) $a + bc'$,

(v) $(a + bc)(b + c'a)$,

(vi) $abc + (a + b)(a + c)$,

(vii) $ab' + c + a'b$,

(viii) $ac + b'$.

7. Express the Boolean functions in disjunctive normal form in three variables a,b,c :

(i) $a + b'$, (ii) $ab + c'$, (iii) $ac + c'$.

8. Express the following Boolean functions in conjunctive normal form in the smallest number of variables:

(i) $ab + c'$, (ii) $abc + (a + b)(a + c)$, (iii) $abc + a'bc' + ab'c$, (iv) $(a + b + c)(ab + a'c)'$, (v) $abc + (ab)'(ac)'$, (vi) $(a'b + bc)' + b'$, (vii) $ab' + bc' + ca'$.

9. Change the following functions from the disjunctive normal form to conjunctive normal form:

(i) $ab + a'b$,

(ii) $abc + a'bc' + ab'c'$,

(iii) $abc + a'bc' + ab'c' + a'b'c + a'b'c'$.

10. Change the following functions from the conjunctive normal form to disjunctive normal form:

(i) $(a + b')(a' + b)$,

(ii) $(a + b' + c')(a' + b + c)(a' + b + c')$,

(iii) $(a + b' + c')(a' + b + c)(a' + b + c')(a + b' + c)(a' + b' + c')$.

11. If a, b, c be three switches, then draw a switching circuit representing

(i) $(a + b' + c)(a + bc')$, (ii) $a(b + c) + bc$, (iii) $bc' + a(b + c')$.

12. If a, b, c be three switches, then draw a switching circuit representing $a'b' + ab' + bc$ and simplify it, if possible.

13. Write short notes on: (i) Basic Gates, (ii) Universal gates.

14. Explain (i) NAND, (ii) NOR, (iii) XOR, (iv) XNOR gates with block diagram and truth table.

15. Design all basic gates using NAND gates.

16. Design all basic gates using NOR gates.

17. Draw a gate network for $A'BC' + A'BC + AB'C$; taking A, B, C as inputs, not their complements.

18. Draw a gate network for $A'B + AB'$; taking A, B as inputs not their complements.

19. Draw logical diagrams for the following expressions using three basic gates only:

(i) $a + bc'$, (ii) $ab + a'b + ab'$, (iii) $a'b + ab'$, (iv) $(a + b')cd'$.

ANSWERS

3. (i) $(x + y)(x' + z)(y + z)$, (ii) $(w + x)(w + y)(w + z)$, (iii) $(x + u)(x + v)(z + u)(z + v)$, (iv) $(x + y)(y + z)$.

4. (i) $(a' + b')(a + c)(b' + c)$, (ii) $(a' + b' + c')a + (a' + b')c$, (iii) $a'b' + cb'$, (iv) $(abc + a'b')(b' + c')$, (v) $(a' + b')(a + b)$.

6. (i) $abc + abc' + ab'c + ab'c' + a'bc + abc$, (ii) $abc + abc' + ab'c$, (iii) $abc + ab'c$, (iv) $abc + abc' + ab'c + ab'c' + a'bc + a'b'c$, (v) $abc + abc' + ab'c + ab'c'$, (vi) $abc + abc' + ab'c + ab'c' + a'bc$, (vii) $a'bc + abc' + ab'c + ab'c' + a'bc' + a'b'c$, (viii) $abc + ab'c + a'b'c + ab'c' + a'b'c'$.

7. (i) $abc + ab'c + abc' + a'b'c + ab'c' + a'b'c'$, (ii) $abc + abc' + ab'c' + a'bc' + a'b'c'$, (iii) $abc + ab'c + abc' + a'b'c + ab'c' + a'b'c'$.

8. (i) $(a+b+c')(a+b'+c')(a'+b+c)(a'+b+c')$,
(ii) $(a+b+c)(a+b+c')(a+b'+c)$,
(iii) $(a+b+c)(a+b+c')(a+b'+c')(a'+b+c)(a'+b'+c)$,
(iv) $(a+b+c)(a+b'+c')(a+b+c')(a'+b'+c)(a'+b'+c')$,
(v) $(a'+b+c)(a'+b+c')(a'+b'+c)$, (vi) $(a+b'+c)(a+b'+c')(a'+b'+c')$,
(vii) $(a+b+c)(a'+b'+c')$

9. (i) $(a+b')(a'+b')$, (ii) $(a+b+c)(a+b+c')(a'+b'+c)(a+b'+c')(a'+b+c')$,
(iii) $(a'+b'+c)(a+b'+c')(a'+b+c')$

10. (i) $ab+a'b'$, (ii) $abc+abc'+a'bc'+a'b'c+a'b'c'$, (iii) $abc'+a'b'c+a'b'c'$.

INDEX

UNIVERSITY QUESTION PAPERS

BURDWAN UNIVERSITY

2001

Fourth Paper

(Numerical Analysis & Computer Part)

I. (g) Find the relative error of the number 8.6, if both of its digits are correct.

(h) Find the value of $(1+\Delta)(1-\nabla)$ on $f(x)$, where the symbols have their usual meaning and h is the space length.

[Hints: See Ch. 10.]

(i) Write the Euler's formula for the first order ordinary differential equation.

[Hints: See article 9.5.]

(j) Why is an operating system necessary for a computer?

[Hints: See $ 11.8.]

(k) Find the values of (I*2)/2 and (I/2)*2 for I = 11 in FORTRAN.

[Hints: 11 and 10.]

(l) Write the full form of 'CPU'

(m) Write the names of three computer high level language.

[Hints: See article 11.9.]

II. (f) Write down Newton-Raphson iterative formula for the computation of the real simple root of $3x + \sin x - e^x = 0$.

(g) Write the Lagrange's interpolating polynomial for two points (x_0, y_0), (x_1, y_1).

[Hints: See Ch.10.]

(h) Calculate $\Delta(\sin 2x)$ with space length h.

[Hints: $\Delta(\sin 2x) = \sin 2(x + h) - \sin 2x = 2\cos(2x + h)\sin h$]

(i) What do you mean by the order of convergence of an iteration?

[Hints: See Ch.10.]

(j) State the hierarchy of arithmetic operations in FORTRAN.

(k) Write the equivalent FORTRAN expression for the following mathematical expression:

$$\log_e \left| \frac{x + \sqrt{x^2 + 1}}{x - \sqrt{x^2 + 1}} \right|$$

[Hints: ALOG ((X + SQRT (X**2 + 1)) / (X – SQRT (X**2 + 1)))]

(l) What is the difference between 'STOP' and 'END' statements in FORTRAN?

[Hints: See articles 14.8.11 and 14.8.12.]

(m) Draw the flowchart for functioning of the logical IF statement in FORTRAN.

(n) State the five functional units of a computer.

[Hints: See article 11.7.]

III. (e) Derive the fixed point iteration method for computing a real root of $f(x) = 0$. Mention the condition of convergence of the method. When does the method fail?

[Hints: See article 5.7.]

(f) For any positive integer k, show that

$$\nabla^k y_n = \sum_{i=0}^{n} (-1)^n \binom{k}{i} y_{n-i},\ \nabla \text{ being the backward difference operator.}$$

[Hints: See Ex. 2 of Ch. 2.]

(g) Apply the Gauss-Seidel iteration method (4 steps) to the following system, starting from (0,0,0):

$$10x + y + z = 6$$

$$x + 10y + z = 6$$

$$x + y + 10z = 6.$$

(h) Write short notes on (i) Compiler, (ii) Algorithm, (iii) DIMENSION statement.

[Hints: See article 11.9, article 13.1 and article 14.8.9.]

(i) Write a FORTRAN program to find the real roots of an equation $Ax^2 + Bx + C = 0$ with real non-zero coefficients A, B, C.

[Hints: See Ex. 8 of Ch. 14.]

(j) (i) Write a note on Arithmetic IF statement in FORTRAN.

[Hints: Article 14.8.7.]

(ii) Write the output of the following program segment

```
      IT = 0
      DO 10 I = 1.5
      IT = IT + 1
10    CONTINUE
      IT = IT + 1
      WRITE(*,*) 'IT ='IT
      STOP
      END
```

[Hints: IT = 6.]

4. (a) Solve the equation $\dfrac{dy}{dx} = x + y$ by Euler's method with the initial condition $y(0) = 0$ and obtain y for $x = 0.4$ and 0.8, choosing the step length $h = 0.2$.

(b) Derive Simpson's $\dfrac{1}{3}$rd rule for numerical integration from Newton-Cote's formula. State the error in the Simpson's $\dfrac{1}{3}$rd formula.

[Hints: See article 7.1(b).]

5. (a) Using Lagrange's interpolation formula, find a cubic polynomial from the following table:

x:	-2	-1	2	3
y:	-12	-8	3	5

(b) If $f(x)$ is a function such that $f(0) + f(6) = -107$, $f(1) + f(5) = -36$ and $f(2) + f(4) = -3$, find $f(3)$.

6. (a) Write the arithmetic expressions corresponding to the following FORTRAN statements:

(i) A*B/(C + D**F/G + H) + E, (ii) (A*B) + C/(D**F + G) + A/B + C.

(b) Write the output of the following FORTRAN program:

```
      IT1 = 1
      IT2 = 1
      DO 20 I = 1,5
      ITP = IT1 + IT2
      IT1 = IT2
      IT2 = ITP
      WRITE(*,*) ITP
20    CONTINUE
      STOP
      END
```

[Hints: 2,3,5,8,13]

7. (a) Draw a flow chart to find a real, simple root of an equation $f(x) = 0$ using Newton-Raphson method.

[Hints: See Ex. 17 of Ch. 13.]

(b) Write short notes on (i) Assignment statement, (ii) Computed GO TO statement.

[Hints: See article 14.4.4 and article 14.8.7.]

2002

Fourth Paper

(Numerical Analysis & Computer Part)

1. (e) State the condition of convergence for solving a linear system of equations by Gauss-Seidel method.

(f) Find the value of $\Delta^2 f(x)$, where $f(x) = .39x^2 - 12x + 4$ and the interval of differencing is unity.

(g) Round off the number 0.0001021 to three significant figures.

[Hints: 0.000102]

(h) What is an assembler?

[Hints: See article 11.9.]

(i) Define a source program.

[**Hints:** See article 11.9]

(j) Write the full-form of MS-DOS

[**Hints:** Microsoft Disk Operating System]

2. (e) What is the geometrical significance of Simpson's $\frac{1}{3}$rd rule for numerical integration?

(f) When should we use Newton's forward interpolation formula?

(g) Take $\pi = 3.14$ instead of $\pi = 22/7$. Calculate the absolute and relative errors.

(h) Write the names of four FORTRAN integer variables.

(i) Why are binary numbers used in computer design?

(j) Draw the block diagram of a computer.

7. Define n-th order forward and backward differences for a function tabulated at equally spaced points. Show that $\Delta^n\{15e^{cx}\} = 15\left(e^{ch} - 1\right)^n e^{cx}$, c being a constant and the interval of differencing is h.

[**Hints:** $\Delta^n\{15e^{cx}\} = 15(e^{c(x+h)} - e^{cx}) = 15e^{cx}(e^{ch} - 1) \Rightarrow \Delta^2\{15e^{cx}\}$

$$= 15(e^{c(x+h)} - e^{cx})(e^{ch} - 1) = 15e^{cx}(e^{ch} - 1)^2.$$

Proceeding in this way we get $\Delta^n\{15e^{cx}\} = 15(e^{ch} - 1)^n e^{cx}$]

8. Establish the trapezoidal rule for numerical integration. Write the expression of error. What does the formula represent graphically?

[**Hints:** See article 7.4.]

9. Use Newton-Raphson method to deduce the iterative scheme $x_{n+1} = \frac{1}{2}\left(x_n + \frac{a}{x_n}\right)$ for evaluating $\sqrt{a}$. Compute $\sqrt{2}$ using the above iterative scheme corrected up to four significant figures taking the initial approximation, $x_0 = 1.4$.

[**Hints:** See Ch. 10.]

10. Write a FORTRAN program to compute the integration $\int_0^5 \log_e(1 + x^2)\,dx$, numerically by Simpson's $\frac{1}{3}$rd rule with 10- subintervals.

11. Write short notes on:

(i) Arithmetic IF statement, (ii) Computer high level language.

12. Write FORTRAN equivalent of the following arithmetic expressions and indicate the order of execution: (i) $\dfrac{N(ab - cd)^2}{(a + b)(c + d)(a + c)(b + d)}$, (ii) $\sin\{\log_e(x^2 - 36)\}$.

16. (a) Solve the following system by Gauss-elimination method:

$$x + 2y + 3z = 10$$
$$x + 3y - 2z = 7$$
$$2x - y + z = 5.$$

(b) Solve the differential equation $\frac{dy}{dx} = x + y$ with the initial condition $y(0) = 1$ by Picard's method of successive approximation to compute $y(0.1)$.

17. (a) Establish the Lagrange's interpolation formula for n distinct points (x_0, y_0), $(x_1, y_1), \ldots, (x_n, y_n)$.

[Hints: See article 3.10.]

(b) Construct the divided difference table for a function given by the following table:

x:	0	2	4	6	8
y:	–18	6	54	174	414

18. (a) Write the output of the following segment of the FORTRAN program:

```
      INTEGER SUM
      SUM = 100
      DO 30 I = 5, 14, 2
      SUM = SUM + I
      WRITE(*,*)'SUM=',SUM
30    CONTINUE
      STOP
      END
```

[Hints: SUM = 105, SUM = 112, SUM = 121, SUM = 132, SUM = 145]

(b) Write short note on the DO-statement.

19. (a) Draw the flow chart for finding the largest of given numbers.

(b) Write a FORTRAN program for finding the L.C.M. of two integers.

[Hints: See Ex. 7 of Ch. 14.]

2003

Fourth Paper

(Numerical Analysis & Computer Part)

1. (g) Write the number of significant digits of the number 0.020530.

(h) What will be the second order difference for a linear function?

(i) When does Newton-Raphson method fail to find a real root of the equation $f(x) = 0$?

(j) Write the condition of convergence for the fixed point iteration.

(k) What is the full-form of FORTRAN?

(l) Name one secondary storage device in a computer.

(m) Find the value of J*4 + K/N*L when J = 3, K = 2, N = 5, L = 4 in FORTRAN.

(n) What is the function of a compiler?

[Hints: See article 11.9.]

2. (g) Round off the number 48.2461 correct to 4 significant digits.

[Hints: 48.25]

(h) Write an interpolating polynomial considering two points (x_0, y_0) and (x_1, y_1).

(i) Why does the approximate value of $\int_0^1 x dx$ calculated by trapezoidal rule is free from inherent error?

[Hints: Error term in trapezoidal rule contains $f''(\xi)$, which is equal to zero in this case.]

(j) State the advantages of using the Lagrange's interpolation formula over other interpolation formula.

(k) What are the differences between the executable and non-executable statements in FORTRAN?

[Hints: See article 14.8.]

(l) Write the five functional value of a computer.

(m) Write general form for writing ARITHMETIC statement in FORTRAN.

[Hints: See article 14.8.4.]

(n) What will be the output for the following two FORTRAN statements:

L = I + J/K, X = L, if I = 2, J = 3 and K = 4?

5. Answer any three of the following:

(a) For any positive integer n, show that

$$\Delta^n y_0 = \sum_{i=0}^{n} (-1)^n \binom{n}{i} y_{n-i}, \ \Delta \text{ being the forward difference operator.}$$

(b) What do you mean by interpolation? Establish Newton's backward interpolation formula when the function $f(x)$ is known to have $(n + 1)$ distinct values corresponding to equispaced points.

(c) Find by Lagrange's formula, the interpolation polynomial which corresponds to

x:	–1	0	2
y:	9	5	3

(d) Derive Simpson's $\frac{1}{3}$rd rule for numerical integration $\int_a^b f(x)dx$. State its geometrical significance.

[Hints: See article 7.5.]

(e) Explain Newton-Raphson's method to determine approximately one simple real root of the equation $f(x) = 0$, Derive the condition of convergence.

[Hints: See article 5.9.]

6. Answer any one of the following:

(a) Solve the following system of equations by Gauss-Seidel iteration method (4 steps), starting from (0, 0, 0).

$$27x + 6y - z = 85$$

$$6x + 15y + 2z = 72$$

$$x + y + 54z = 110.$$

(b) Deduce Euler's method for numerical solution of the differential equation $\frac{dy}{dx} = f(x, y)$ subjected to the condition $y = y_0$ at $x = x_0$.

[Hints: See article 9.5.]

7. Answer any three from the following:

(a) Write short notes on any two: (i) Algorithm and Flow chart, (ii) Hardware and Software, (iii) DO –statement.

(b) Write a FORTRAN expression corresponding to the following mathematical expressions:

(i) $\left(a^b + b^a\right)\left(a + b + \frac{ab}{(a+b)^2}\right)$, (ii) $\sin\left(x + \sqrt{x^2+1}\right) + \frac{x^3}{3}\log_e|x - y|$

(c) For I = 2, J = 3, K = 4, A = 2.0, B = 5.0, find the value of the expression A + J*K**I/J/B showing the hierarchy of the arithmetic operation.

(d) Define numeric variables in FORTRAN. State their classifications with examples.

(e) Draw a flowchart to find the roots of quadratic equation having real non-zero coefficients.

[Hints: See Ex.8 of Ch.13.]

8. Answer any one of the following:

(a) Write a FORTRAN program to compute the integration $\int_1^2 \log_e x dx$ numerically by trapezoidal rule with 10 equal subintervals.

(b) Find out error, if any

(i) IF(X.GE.Y) GO TO STOP, (ii)5A = A + 4.*A, (iii) FORCE = MASS* ACCELERATION, (iv) I = J = 5, (v) GO TO STEP 10 , (vi) X + 5. = 5. + X

[Hints (i) either STOP or some level after GO TO is allowed , (ii) 5A is not a variable name, (iii) ACCELERATION contains more than 6 characters, (iv) either I = J or I = 5 or J = 5,(v) some level after GO TO is allowed, (vi) X+5. is not a variable name.]

2004

Fourth Paper

(Numerical Analysis & Computer Part)

1. (h) Find $\Delta^3(2x^2 - 2x + 1)$.

[Hints: As $\Delta^n x^k = 0$ for $n > k$, the result is zero.]

(i) Write an interpolating polynomial considering two points (x_0, y_0) and (x_1, y_1).

(j) Compute $\int_0^1 (1 + x)dx$ by Simpson's $\frac{1}{3}$rd rule with space length 0.5.

(k) Define source program.

(l) Write the full-form CPU

(m) Name three of the computer high level languages.

(n) Find the value of (X + Y)**2/J where X = 4.0, Y = 2.0 J = 5.

2. (g) Is the form $x = x^2 - 0.2$ valid for finding the real simple root of $x^2 - x - 0.2 = 0$ between 1 and 2 by fixed point iteration method?

[**Hints**: $2x > 1$ for $x > 1$, so the given form is invalid.]

(h) Examine whether Gauss-Seidel iteration method is applicable for solving:

$$2x + 6y - z = 6$$
$$x + y + 3z = 3$$
$$4x + 3y - 2z = -7.$$

(i) When should we generally use Newton's forward interpolation formula?

(j) Find $y(0.2)$ by Euler's method in solving the differential equation $\frac{dy}{dx} = xy$ with $y(0) = 1$.

(k) Name two numeric FORTRAN constants.

(l) What is the difference between STOP and END statements in FORTRAN?

(m) What are the FORTRAN representation of the functions $\log_{10} x, \sin^{-1} x$?

(n) What are the symbols used to denote the following instructions of a flow chart:

Terminals, I/O, Connectors, Decision?

5. Answer any three of the following:

(a) Establish the Trapezoidal rule for numerical integration. Give its geometrical representation and account for the name "Trapezoidal".

[**Hints**: See article 7.4.]

(b) Describe Regula-Falsi method for finding a simple real root of the equation $f(x) = 0$ and interpret it geometrically.

[**Hints**: See article 5.5.]

(c) Find the polynomial which takes the following values. Hence obtain $f(0.5)$.

x:	0	1	2	3	4	5
$f(x)$:	1	4	9	16	25	36.

(d) Let $f(x) = \frac{1}{x^2}$. Obtain the divided difference $f(a, b), f(a, b, c)$ and $f(a, b, c, d)$.

(e) Let the curve y = f(x) pass through the points (x_i, y_i), $i = 0(1)4$. Find the polynomial that approximate $f(x)$ by Newton's forward interpolation formula.

6. Answer any one of the following:

(a) Solve the following equations by Gauss's successive elimination method:

$$x + 5y - 3z = 30$$
$$2x + y - 8z = -40$$
$$x - y + 5z = 10.$$

(b) Solve $\dfrac{dy}{dx} = \dfrac{x^2}{1+y^2}$ with the initial condition $y(0) = 0$, by Picard's method to obtain y for $x = 0.15$, correct up to 3 decimal places.

7. Answer any three of the following:

(a) Write short notes on any one of the following:

(i) Assigned GO TO in FORTRAN, (ii) Logical IF statement in FORTRAN.

(b) Give algebraic expressions corresponding to the following FORTRAN expressions:

(i) SQRT(ALOG(X + 2.0*Y)**3.0) + ALOG10(2.0*A–3.0*B)**2

(ii) SIN (X + SQRT(X*X + 1.0)) + 0.5*X*X*ALOG(ABS(X–Y))

[**Hints**: (i) $\sqrt{\log_e(x+2y)^3} + \log_{10}(2a-3b)^2$,

(ii) $\sin(x+\sqrt{x^2+1}) + \frac{1}{2}x^2 \log_e|x-y|$.]

(c) Draw a flowchart for finding H.C.F and L.C.M. of two positive integers.

(d) Explain why the following real constants in FORTRAN are invalid:

(i) 3521, (ii) –3,245.21 (iii) 100/6 (iv) 23.21 E123 (v) 0.21E1.75

[**Hints**: (i) comma is not allowed,(ii) same as (i),(iii); mathematical operation is not allowed within the name, (iv) & (v) after E at most two integers are allowed.]

(e) Write short notes on any two of the following:

(i) Compiler, (ii) Operating system, (iii) Algorithm.

8. Answer any one of the following:

(a) What will be the output of the following programming segment:

```
      K = 10
      I = 1
      S = 0.0
5     WRITE(*,*) K,S
      S = S + 1
      K= K – I
      I = I + 2
      IF(I – 9)5,5,6
6     STOP
      END
```

(b) Write the FORTRAN program to find a real simple root of an equation by Newton-Raphson method imposing the condition of convergence of the method.

2005

Fourth Paper

(Numerical Analysis & Computer Part)

1. (ix) Write down the number of significant figures in (a) 2.00125, (b) 0.012500, (c) 2.3×10^3, (d) 21340.

[**Hints**: (a) 6, (b) 3, (c) 4, (d) 5]

(x) Calculate the relative error in computation of $(x - y)$ for $x = 3.21$ and $y = 2.12$ having absolute errors $\Delta x = 0.003$ and $\Delta y = 0.001$.

(xi) Find the function f(x) whose first difference is e^x.

(xii) Find the value of $(1 + \Delta)(1 - \nabla)$ on $f(x)$ where the symbols have usual meanings.

(xiii) Write down the Newton-Raphson iteration formula relating the nth and $(n + 1)$th approximation to a real root of $x^2 - 5x + 2 = 0$.

(xiv) Find the value of $\int_0^4 x^2 dx$ by trapezoidal rule with space length unity.

(xv) Why are binary numbers used in computer design?

(xvi) Draw the block diagram of a computer.

(xvii) Convert the following integer variable names into the real variable names:

(a) LENGTH , (b) MONTH , (c) NEXT, (d) JUMP.

(xviii) What is ALU? State its two main functions.

(xix) Indicate mistakes in the following statements:

(a) (X + Y)*Z = A, (b) 8.7 = D + C

(xx) Find the value of K of the following program segment:

```
J = 5
K = 5 + J
L = J
J = K
K = L
```

[**Hints**: K = 5]

4. Answer any three of the following questions:

(a) Derive Simpson's $\frac{1}{3}$rd rule for numerical integration $\int_a^b f(x)dx$. State its geometrical significance.

[**Hints**: See article 7.5.]

(b) Describe "Fixed point iteration method". Mention the condition for convergence. When does the method fail?

[Hints: See article 5.7]

(c) Write down the quadratic polynomial by applying Lagrangian interpolation, which takes same value as $f(x)$ at $x = -1,0,1$ and obtain the integration rule

$$\int_{-1}^{1} f(x)\,dx = \frac{1}{3}\left[f(-1) + 4f(0) + f(1)\right].$$

(d) Given $x_1 = x_0 + h, x_2 = x_0 + 2h, x_3 = x_0 + 3h (h > 0)$

and $u_3(x) = (x - x_0)(x - x_1)(x - x_2)(x - x_3)$

Prove that $\Delta^3 u_3(x) = 4! h^3 (x - x_0)$.

(e) Compute $f(3.8)$ by Newton's backward interpolation formula from the following data:

x:	0	1	2	3	4
y:	1	1.5	2.2	3.1	4.2

5. Answer any one of the following questions:

(a) Apply Gauss-Seidel iteration method (4 steps) for solving following system of equations, with initial value (0,0,0) of x, y, z:

$$10x + y + z = 6,$$
$$x + 10y + z = 6,$$
$$x + y + 10z = 6.$$

(b) Given $\frac{dy}{dx} = x^2 + y$ with $y(0) = 1$. Determine by Euler's method $y(0.02)$ with step length $h = 0.01$.

6. Answer any three of the following questions:

(a) Write a FORTRAN program to find the real roots of a quadratic equation $AX^2 + BX + C = 0$ with real (non-zero) coefficients A, B and C.

(b) Draw the flowchart for finding the largest of a given set of numbers:

$$x_1, x_2, \ldots, x_n.$$

(c) Write short note on any one of the following:

(i) Compiler and Interpreter, (ii) Hardware and Software.

(d) Explain any one of the following with example:

(i) Arithmetic IF statement, (ii) READ and WRITE statements.

(e) Write down the FORTRAN expressions for the following algebraic expressions:

(i) $\frac{a+b}{c} - b(d-e)(a \times d)^2$, (ii) $\sqrt{a^2 + b^2 - 2ab\cos(x)}$

7. Answer any one of the following questions:

(a) Point out the error of any four in the following FORTRAN statements:
(i) IF X.EQ.Y GO TO 10, (ii) GO TO STEP 10, (iii) READ(*,*)5 (iv) PRINT* A,B,5, (v) FORCE = MASS*ACCELERATION, (vi) DO I = 1,8,3

[**Hints**: (i) Logical expression must be given within parentheses, (ii) Step 10 is not a level number. (iii) 5 is not a variable name. (iv) same as (iii). (v) ACCELERATION contains more than 6 characters. (vi) After DO level number is missing.]

(b) Let the initial value of N and J be 5 and 10 respectively. What would be the result of the following program segment?

```
IF(2*J.LE.3*N) GO TO 10
```

```
      N = N+1
      GO TO 20
10    N = J
20    N = N+J
      WRITE(*,*)N
      STOP
      END
```

[**Hints**: Since 20 > 15, control will be transferred to next executable statement where N is increased by 1 and then control will be transferred to level number 20, where N will take the value 16.]

2006

Fourth Paper

(Numerical Analysis & Computer Part)

1. (ix) For a given step length h, write expressions for the first order forward and backward differences of a function $f(x)$ and give relation between them.

(x) If 1/3 is approximated by 0.333, find absolute and relative errors.

(xi) Compute $\int_{-1}^{1} |x| dx$ by Simpson's $\frac{1}{3}$rd rule with space length unity.

(xii) Find the value of y at $x = 2$ by Lagrange's interpolation formula from the table:

x:	1	3
y:	3	12

(xiii) Find the second order approximation to the real root of $f(x) = x^3 - 4x - 1 = 0$ lying between 2 and 3 by Regula Falsi method.

(xiv) Round off the following numbers as instructed:

(a) 2.45 and 1.55 each to 1 decimal place, (b) 1.2535 to 3 and 2 decimal places.

(xv) If I = 2, J = 5, K = 1 and N = 6 ; evaluate I*J+(5+K)/J +N/J.

(xvi) Name one executable and one non-executable FORTRAN statements.

(xvii) Discuss the validity of the following variable names:

READ, 4TEMP,MASS,COMMISSION.

[**Hints**: (a) Being a part of FORTRAN language it cannot be variable name, (b) variable name cannot start with digit, (c) valid, (d) contains more than 6 characters.]

(xviii) What are the full forms of ALU and FORTRAN?

(xix) What is an operating system of a computer? Name two computer operating systems.

(xx) What are the FORTRAN representations of the functions: $\log_{10}x$ and $\sin^{-1}x$?

4. Answer any three of the following questions:

(a) The operator E is defined by $E\{f(x)\} = f(x+h)$. Find the relation between E and Δ and prove that for any positive integer K, $y_K = \sum_{i=0}^{K} {}^K C_i \Delta^i y_0$, $y_0, y_1, \cdots, y_K$ are the values of y at $x_0, x_0 + h, \cdots, x_0 + Kh$.

[**Hints**: See article Ex.3 of Ch 2.]

(b) What do you mean by interpolation? Obtain Newton's forward interpolating polynomial to represent the function $f(x)$ which is known to have $(n+1)$ distinct values corresponding to equispaced points.

[**Hints**: See article 3.1 and article 3.4.]

(c) Establish the trapezoidal rule for numerical integration. Give its geometrical interpretation.

[**Hints**: See article 7.4.]

(d) Explain Newton-Raphson's method to determine approximately one simple root of $f(x) = 0$. Mention the condition of convergence.

[**Hints**: See article 5.9.]

(e) Find the first and second orders divided differences of $f(x) = \sin x$ for the points $\frac{\pi}{2}, \frac{\pi}{3}, \frac{\pi}{4}$ and hence show that divided difference is symmetric.

5. Answer any one of the following questions:

(a) Obtain $y^{(3)}(x)$ by Picard's method for solving the equation $\frac{dy}{dx} = xy + 1$ with the condition $y(0) = 1$.

(b) Solve the following system of equations by Gauss elimination method(row operation in the augmented matrix can be applied):

$$x_1 + 2x_2 + x_3 = 0,$$
$$2x_1 + 2x_2 + 3x_3 = 3,$$
$$-x_1 - 3x_2 = 2.$$

6. Answer any three of the following questions:

(a) Draw a flow chart for finding $\int_a^b x^2 dx$ by trapezoidal rule taking 10 equal sub-intervals.

(b) Write a FORTRAN program for finding H.C.F. and L.C.M. of two positive integers.

(c) Write short note on any one of the following:
(i) Bits, Bytes and Word lengths, (ii) Computer program and programming language.

(d) Explain any one of the following with example:
(i) Unconditional GO TO and Computed GO TO statements, (ii) STOP and END statements.

(e) (i) Write the mathematical expression for the following FORTRAN expression

(((3.*A-2.)*B**5) + 3.6*(C**(2*K+1)))/5*D

(ii) Rewrite the following using arithmetic IF statement:

```
IF(COKE.LE.PEPSI)GO TO 77
GO TO 777
```

[Hints: IF(COKE-PEPSI)77,77,777.]

7. Answer any one of the following:

(a) Point out error, if any, in each of the following statements with proper justification:

(i) WORK = FORCE*DISPLACEMENT, (ii) A**2 – B**2 = (A + B)*(A – B) (iii) IF(X + Y)GO TO 50, (iv) IF(X**2.GT.5.0)5,6,7, (v) 5A = A + 4.*A.

[Hints: (ii) A**2 – B**2 is not a variable name, (iii) X+Y is not a logical expression (V) 5A is a variable name.]

(b) If the initial values of N and J are assumed to be N = 5 and J = 10; what would be the final value of N at the end of the following program segment?

```
      IF(2*J.LE.3*N) GO TO 10
      N = N + 1
      GO TO 20
10    N = J
20    N = N + J
      WRITE(*,*)N
      STOP
      END
```

2007

Fourth Paper

(Numerical Analysis & Computer Part)

1. (x) Show that $\Delta \log f(x) = \log\left\{1 + \frac{\Delta f(x)}{f(x)}\right\}$.

[Hints: See Ch.10.]

(xi) Evaluate $\Delta^3(1-ax)(1-bx^2)$.

(xii) Compute $\int_0^1 (1+x)\,dx$ by Simpson's $\frac{1}{3}$rd rule with space length 0.5.

(xiii) Approximate the following numbers correct up to 4 significant figures:

(a) 5.2056, (B) 0.0056812, (c) 1.9998, (d) 1.03578.

[Hints: 5.206,0.005681,2.000,1.036]

(xiv) Write down Newton's backward interpolation formula.

(xv) For a given step length h, prove that $E\nabla = \Delta$, symbols having usual meanings.

(xvi) Write the full forms of VDU and CAD.

(xvii) Draw the block diagram of a computer.

(xviii) Give example of a real and integer variables in FORTRAN.

(xix) Indicate mistakes in the following statements:

(a) 5A = A + 4.*A, (ii) X + 5. = 5. + X

[**Hints:** Neither 5A nor X + 5. is a variable name.]

(xx) What are the Logical constants?

4. Answer any three of the following questions:

(a) Derive Simpson's $\frac{1}{3}$rd rule for numerical integration $\int_a^b f(x)\,dx$. State its geometrical significance.

[**Hints:** See article 7.5.]

(b) Describe Regula-Falsi method for finding a simple real root of the equation $f(x) = 0$ and interpret it geometrically.

[**Hints:** See article 5.5.]

(c) If x_0, x_1, x_2, x_3 be equi-spaced arguments with space length h and y_0, y_1, y_2, y_3 be the corresponding values of $y = f(x)$, then show that $f[x_0, x_1, x_2, x_3] = \dfrac{\Delta^3 y_0}{3!h^3}$.

[**Hints:** See article 4.4.]

(d) Find the polynomial $p(x)$ to approximate $f(x)$ by Lagrange's interpolation formula from the following table:

x:	–1	0	2
y:	9	5	3

(e) Which one of the arrangements (i) $x = (9x-1)^{1/3}$, (ii) $x = \dfrac{1}{9-x^2}$, (iii) $x = \sqrt{9 - 1/x}$ will give the root between 0 and 1 by fixed point –iteration method of the equation $x^3 - 9x - 1 = 0$? Give reasons.

[**Hints:** (ii) As $\dfrac{2x}{(9-x^2)^2} < 1$ for $0 < x < 1$.]

5. Answer any one of the following questions:

(a) Obtain solution of the following equations up to 3 significant figures by Gauss-Seidel iteration method (3 steps):

$$8x_1 - 3x_2 + 2x_3 = 20,$$
$$4x_1 + 11x_2 - x_3 = 33,$$
$$x_1 + x_2 + 4x_3 = 9.$$

(b) Given: $\dfrac{dy}{dx} = x + y$ with y(0) = 0 and h = 0.2. Determine by Euler's method $y(1.0)$.

6. Answer any three of the following questions:

(a) Write a FORTRAN program to find the largest and smallest of three given positive numbers.

(b) Write short note on any one of the following:

(i) Algorithm, (ii) Memory Unit and External Storage Device.

(c) (i) Assuming X = 2.5, Y = 3.7, M = 2 and N = –3, find the value of A = X + M*N**2 + Y – 24.2.

(d) Translate the following into FORTRAN equivalent form: $x^2 + \frac{1}{x^3 + x^4}$.

(e) Explain any one of the following:

(i) READ and WRITE statements, (ii) Logical and Arithmetic IF statements.

[**Hints**: See article 14.8.5 and article 14.8.7.]

7. Answer any one of the following questions:

(a) Point out the error, if any, in each of the following statements with proper reasons:

(i) GO TO STEP 10, (ii) PRINT*,A,B,5, (iii) X(K) = X*K + 1.0 (iv) M + 1 = (2*M + 2)/2.

[Hints: (i) Step 10 is not a level number. (ii) 5 is not a variable. (ii) If K is defined before the statement, it has no error. (iv) M + 1 is not a variable name.]

(b) Find the final value of M given by the following FORTRAN program segment:

```
      M = 2
      DO 20 I = 3,8,2
20    CONTINUE
      STOP
      END
```

[**Hints**: During execution of the statements the value of M is not changed, so its final value is 2.]

2008

Fourth Paper

(Numerical Analysis & Computer Part)

1. (g) Rounded off the given numbers as instructed:

(i) 0.00493 and 2.1345 (correct to 3 decimal places)

[**Hints:** 0.005, 2.134]

(i) 1.0352 and 0.005736 (correct to three significant figures.)

[**Hints:** 1.04, 0.00574]

(h) Prove that $E = 1 + \Delta$, where E, Δ have their usual meanings,

[**Hints:** See article 2.3.]

(i) If $f(x) = \dfrac{1}{x}$ and a, b, c are three distinct values of x, then show that $f(a, b, c) = \dfrac{1}{abc}$, where $f(a, b, c)$ denotes second order divided difference operator.

[**Hints:** $f(a, b) = \dfrac{1}{b-a}\left(\dfrac{1}{b} - \dfrac{1}{a}\right) = \dfrac{1}{ab}$. Similarly, $f(b, c) = -\dfrac{1}{bc}$

Now $f(a, b, c) = \dfrac{1}{c-a}\left(-\dfrac{1}{bc} + \dfrac{1}{ab}\right) = \dfrac{1}{abc}$.]

(j) Evaluate $\int_{-1}^{3} |x|\, dx$ by trapezoidal rule using four equal subintervals.

(k) Convert the following algebraic expressions into equivalent FORTRAN expressions:

(a) $\dfrac{c(a+b)}{d} - \cos x$, (b) $e^{xy} + |x^2 + y^2|$.

[**Hints:** (a) $c*(a+b)/d - \cos x$, (b) $\exp(x * y) + abs(x** 2 + y ** 2)$]

(l) Explain the following FORTRAN statement: STOP and END.
[**Hints:** See articles 14.8. 11-12.]

(m) Indicate the mistakes in the following statements:
(a) $(X + Y) * X = A - B$, (b) $8.7 = D + C$.
[**Hints:** None of the (X + Y) * X and 8.7 can be variable name]

(n) What is the full form of FORTRAN? What is the application area of FORTRAN?

3. Answer any one question from (a) and (b):

(a) (i) What do you mean by an interpolation polynomial for a function $y = f(x)$ when its values are given at $(n + 1)$ distinct points x_i $(i = 0, 1, 2, \ldots, n)$?
[**Hints:** See article 3.2.]

(ii) Establish Newton's forward interpolation formula when the function $f(x)$ is known at $(n + 1)$ equispaced points.
[**Hints:** See article 3.4.]

(iii) Deduce trapezoidal rule (without error) for the numerical integration $\int_{a=x_0}^{b=x_n} f(x)\, dx$.
[**Hints:** See article 7.4.]

(b) (i) Show that the equation $x + \log x = 2$ has a real root between 1 and 2.

[**Hints:** Taking $f(x) = x + \log x - 2$ we note that $f(1)\, f(2) < 0$ and $f'(x) = 1 + \dfrac{1}{x} > 0$ $\forall x \in [1, 2]$.]

(ii) Find the nth iteration formula by Newton Raphson's method for determination of a simple real root of an equation $f(x) = 0$. When does the method fails?

[Hints: See article 5.9. The method fails if $f'(x)$ becomes closed to zero.]

(iii) If $f(x) = x^2 - a^2 = 0$, show the iteration formula for determination of the root of the equation is $x_{n+1} = \frac{1}{2}\left[x_n + \frac{a^2}{x_n}\right] \forall n.$

[Hints: See problem no 41 of chapter 10.]

4. Answer any one question from (a) and (b):

(a) (i) Define Algorithm.

(ii) Prepare an algorithm to find the average value of three given numbers.

(iii) Draw a flow chart to obtain the sum of the finite series: $1 + x + x^2 + \cdots + x^N$.

[Hints: See example 10 of chapter 13.]

(b) (i) If $I = 2$, $J = 5$, $K = 1$ and $N = 6$; evaluate $I * J + (5 + K)/J + N/J$.

[Hints: 10 + 1 + 1 = 12]

(ii) Explain DO statement. How many will the loop defined by the following FORTRAN77 DO statements be executed? (i) *DO* 10 L = 2, 8, 2, (ii) *DO* 100 I = 1, 6, 1.

[Hints: Once in each case]

(iii) Write down a programme in FORTRAN77 that will help to obtain GCD of two non zero integers.

[Hints: See programme No. 6 of chapter 14.]

6 Answer any two questions from (a) to (c):

(a) Construct a backward difference table in respect of a function $f(x)$ using the following data:

x:	2	4	6	8
$f(x)$:	15	28	56	89

Hence find $f(x)$ when $x = 7$.

(b) Obtain $y^2(x)$ by Picard's method for solving the following differential equation: $\frac{dy}{dx} = 2x - y$ with the initial condition $y = 3$ when $x = 1$. Also find y when $x = 2$ (correct to decimal places)

(c) Solve by Gauss elimination method: $2x + 4y + z = 3$; $3x + 2y - 2z = -2$; $x + y + z = 4$. When does the method fails?

7. Answer any two questions from (a) to (c):

(a) Write short notes on: CPU, I/O devices, Library function in FORTRAN.

(b) (i) Discuss the validity given variable names: READ, 4 TEMP, MASS, COMMISSION.

[Hints: Invalid being a statement, Invalid as it cannot start with digit, Valid integer variable, Invalid as it consist of more than 8 alphabets/digits]

(ii) Point out the error and write the FORTRAN statement correctly: IF X.EQ.Y, GO TO N.

[Hints: Condition must be given within a parenthesis, no comma is allowed after the parenthesis and N should be prescribed. So the correct form is IF (X.EQ.Y) GO TO 10.]

(iii) Find the value of : 2.5 + 3.0 ** 2/3.0.

(c) What will be the outcome of the following programme:

```
     K = 10
     I = 1
     S = 0.0
5    WRITE (*,*) K,S
     S = S + 1
     K = K – 1
     I = I + 2
     IF (I – 9) 5, 5, 6
6    STOP
     END
```

[Hints: 10, 0, 9, 1, 8, 2, 7, 3, 6, 4]

2009

Fourth Paper

(Numerical Analysis & Computer Part)

1. (g) Draw the following symbols of a flow chart: (i) Terminal, (ii) Input/Output, (iii) Decision, (iv) Connector.

(h) Represent the equivalent FORTRAN expressions for the following algebraic expressions

(i) $x+\dfrac{1}{y+\dfrac{1}{z+\dfrac{1}{w}}}$, (ii) $\sqrt{u^2+2fs}$.

[Hints: (i) $x+1/(y+1/(z+1/w))$, (ii) $sqrt\,(u**2+2*f*s)$]

(i) Define subscripted variable with example.

(j) Given that A = 2, 1, B = 0.1, J = 3, K = 2, find the value of I stored in the following statement:

$$I = A + J * K ** 2 + B.$$

[Hints: 14.2]

(k) State the rule of rounding off a number to n significant digits.

[Hints: See article 1.3.]

(l) Show that the absolute relative error of $u = \frac{x}{y}$ is less than or equal to the sum of the absolute relative errors of x and y.

(m) Define shift operator E and prove that $E = e^{hD}$, $D = \frac{d}{dx}$ and h = step length.

[Hints: See example 7 of chapter 2.]

(n) Interpret geometrically the trapezoidal rule used for the numerical integration $\int_a^b f(x)\,dx$.

[Hints: See article 7.4.]

3. Answer any one question from (a) and (b):

(a) (i) Show that Gauss-Seidel iteration method is applicable to solve the following system of equations and hence solve the equations (upto 2 step's approximation):

$$3x + y + z = 3;\ 2x + y + 5z = 5;\ x + 4y + z = 2.$$

[Hints: Rewriting the system of equations as $3x + y + z = 3$; $x + 4y + z = 2$; $2x + y + 5z = 5$ we see that the coefficient matrix is diagonally dominant.]

(ii) Describe the method of fixed point iteration for determination of a real root of the equation $f(x) = 0$. State the condition of convergence of this method.

[Hints: See article 5.7.]

(b) (i) Write down the Lagrange's interpolation formula for a function $f(x)$ given at $(n + 1)$ distinct arguments.

[Hints: See article 3.10.]

(ii) State the Newton Cote's formula for the numerical integration $\int_a^b f(x)\,dx$ taking two subintervals. Hence using Simpson's rule establish the formula given below $\int_{-1}^{1} f(x)\,dx = \frac{1}{3}[f(-1) + 4f(0) + f(1)]$.

[Hints: See article 7.1.]

4. Answer any one question from (a) and (b):

(a) (i) Give the general forms of the Logical IF and Arithmetic IF statements in FORTRAN77 programme also distinct clearly between them.

[Hints: See (a), (d) of article 14.8.7.]

(ii) Write a programme to find the real roots of a quadratic equation $ax^2 + bx + c = 0$ where a, b, c are not zero and real. What will be the output of the programme when $a = b = 1.0$ and $c = -2.0$.

[Hints: See programme no. 8 of chapter 14.]

(b) (i) Prepare a FORTRAN programme to find the values of $f(x) = x(x + 1)$ at the equidistance arguments $x = 0, 0.2, 0.4, 0.6, 0.8, 1.0$.

(ii) From the given programme segment obtain the final value N if the initial values of N and J are 5 and 10 respectively.

```
      IF (2 * J.LE.3 * N) GO TO 10
      N = N + 1
      GO TO 20
   10 N = J
   20 N = N + J
      WRITE (*,*) N
      STOP
      END
```

[**Hints:** 16]

6. Answer any two questions from (a) to (c):

(a) Show that the equation $x^3 - x - 11 = 0$ has a real root between 2 and 3. Find the root correct two decimal places by regula falsi method.

(b) State Newton's Forward interpolation formula with reminder term. Find $f(0.5)$ from the given tabulated values.

x:	0	1	2	3
$f(x)$:	1	2	11	34

(c) Explain Euler's method to solve following differential equation $\frac{dy}{dx} = f(x)$ given $y = y_0$ when $x = x_0$. Why Euler's method is called a single step method of order 1?

[**Hints:** See articles 9.3, 9.1.]

7. Answer any two questions from (a) to (c):

(a) Name the principal components of the block diagram of a computer. Describe briefly the functions of each of the components.

[**Hints:** See article 11.7.]

(b) Write short notes on (any two)

(i) Integer variable and real variable, (ii) System software and application software, (iii) Machine language and Assembly language.

[**Hints:** See articles: 14.3, 11.6, 11.9.]

(c) Define the term algorithm. Write an algorithm to evaluate the definite integral $\int_a^b f(x)\,dx$ by trapezoidal rule taking N equal subintervals.

[**Hints:** See article 13.1 and example no. 19 of chapter 13.]

2010

Fourth Paper

(Numerical Analysis & Computer Part)

1. (g) What is ALU? State its two main functions.

(h) What do you mean by High Level Language used in computer programming?

(i) Translate the following algebraic expression into FORTRAN equivalent form

$$= \log \frac{x + \sqrt{x^2+1}}{x - \sqrt{x^2-1}}.$$

[Hints: $= \log((x + sqrt(x ** 2+1)) / (x - sqrt(x **2-1)))$]

(j) Why is the FORTRAN END statement used in where it is placed in a computer programming?

(k) When should we generally use Newton's forward interpolation formula?

[Hints: See example 52 of chapter 10.]

(l) Find the value of (1 + Δ) (1 – ∇) on $f(x)$ where the symbols have their usual meanings.

[Hints: See example 7 of chapter 10.]

(m) If $(x) = \frac{1}{x}$, find $f(2, 3, 5)$, where $f(2, 3, 5)$ denotes second order divided difference operator.

(n) Write down fixed point iterative formula for the equation $f(x) = 0$ and mention its condition of convergence.

[Hints: See article 5.7.]

3. Answer any one question from (a) and (b):

(a) (i) Derive Newton-Raphson formula to find a real root of the equation $f(x) = 0$. What is the order of convergence of Newton-Raphson formula and when does the method fails?

(ii) Solve by Gauss elimination method:

$$x_1 + x_2 + x_3 = 3\,;\, 2x_1 + 3x_2 + x_3 = 6\,;\, x_1 - x_2 - x_3 = -3.$$

(b) (i) Establish Newton's forward interpolation formula. Write down the error term also.

(ii) Given the values of $f(x)$ for x in the following table, show that $f(x)$ will be a polynomial of degree atmost three. Hence find $f(x)$.

x:	0	1	2	3	4
$f(x)$:	3	6	11	18	27

4. Answer any one question from (a) and (b):

(a) (i) Explain by examples the following FORTRAN statement:

Computed GO TO and Unconditional GO TO.

[Hints: See (c), (b) of article 14.8.7.]

(ii) What will be the values of *I, J* and *K* after execution of the programme segment given below?

```
        I = 1
        J = 1
        K = 1
15      IF (J – 10) 20, 25, 25
20      I = I + J
        J = J + 1
        GO TO 15
25      STOP
        END
```

[**Hints:** 46, 10, 1]

(b) (i) Write a FORTRAN programme to find the highest common factor of two given positive integers *K* and *L*. What will be the output of the programme when $K = 12$ and $L = 30$?

(ii) Determine the output of the following segment of the FORTRAN programme and state how many times does the DO-loop execute the programme segment.

```
      INTEGER TERM1, TERM2, TERMP, SUM
      SUM = 10
      TERM1 = 1
      TERM2 = 2
      DO 100 I = 1,7
      TERMP = TERM1 + TERM2
      TERM1 = TERM2
      TERM2 = TERMP
      SUM = SUM + TERMP
      WRITE (*,*)' 'TERM = ', TERMP
100   CONTINUE
      WRITE (*,*)' SUM OF THE TERMS = ', SUM
      STOP
      END
```

[**Hints:** TERM = 3, TERM = 5, TERM = 8, TERM = 13, TERM = 21, TERM = 34, TERM = 55, SUM OF THE TERMS = 149]

6. Answer any two questions from (a) to (c):

(a) Write short notes on (i) Rounding off numbers, (ii) Difference table.

(b) State geometrical significance of the trapezoidal rule for numerical integration. Use trapezoidal rule with five equal subintervals to evaluate the following (correct to two decimal places): $\int_1^3 \frac{dx}{1+x}$.

(c) Establish Picard's iteration formula for solution of the differential equation $\frac{dy}{dx} = f(x, y)$, given the initial condition $y(x_0) = y_0$.

[**Hints:** See article 9.2.]

7. Answer any two questions from (a) to (c):

(a) How does a flow chart help for solving a problem by computer? Draw a flow chart to find the largest of three given numbers.

(b) (i) Define subscripted variable with example

(ii) Write short notes on: DIMENSION- statement.

[**Hints:** See article 14.8.9.]

(c) (i) Point out the error, if any, in each of the following statements with proper justifications.

X(K) = X * K + 1.0

I – 5 = (5 * I – 25)/5

(ii) If 1.0, $B = 8.0$, $C = 3.0$ and $D = 2.0$, then find the value of the expression

$$A + C/D * B + SQRT\ (B/D).$$

[**Hints:** 15.]

2011

Fourth Paper

(Numerical Analysis & Computer Part)

1. (g) Write down different components of the CPU of a computer.

(h) If I = 2, J = 3, K = 4 and C = 5.0, then evaluate the Fortran expression: I/J * (C + 1.5 * K).

(i) (a) Translate the following algebraic expression into FORTRAN equivalent form:

$$a = p\left(1+\frac{r}{100}\right)^n.$$

[**Hints:** $a = p * (1 + r/100) ** n$]

(b) Translate the following FORTRAN expression into equivalent algebraic expression:

A * B/(C + D ** F/G + H) + E

(j) Define an assignment statement and state its action.

(k) What do you mean by significant figure of a number.

(l) Show that the absolute relative error of the product of two approximate numbers is less than or equal to the sum of the absolute relative errors of the numbers.

(m) Explain geometrical significance of an interpolation method.

(n) Examine whether Gauss Seidel iteration method is applicable to solve the following system of equation: $3x + y + z = 31$; $2x + y + 5z = 5$; $x + 4y + z = 2$.

3. Answer any one question from (a) and (b):

(a) (i) Describe the method of false position to find the real root of the equation $f(x) = 0$. State the order of convergence of the method.

[Hints: See article 5.5 and example 38 of chapter 10.]

(ii) Show that the Newton-Raphson's iteration formula for determination of the root of the equation $x^2 - m = 0$ is $x_{n+1} = \frac{1}{2}\left[x_n + \frac{m}{x_n}\right] \forall n$. Using that formula find the value of $\sqrt{3}$ (correct to two decimal places.)

(b) (i) For the numerical integration $\int_a^b f(x)\,dx$, using Newton's forward interpolation polynomial in place of $f(x)$ deduce Simpson's one-third rule in composite form.

[Hints: See article 7.5.]

(ii) Find the value of π correct up to two places of decimals from the formula $\frac{\pi}{4} = \int_0^1 \frac{dx}{1+x^2}$ by numerical integration taking 4 equal subdivisions of the interval [0, 1].

4. Answer any one question from (a) and (b):

(a) (i) What do you mean by software and hardware of a computer?

(ii) State which of the followings are softwares and hardwares: WINDOWS, VDU, COMPILER, RAM.

(iii) Define the term "System software".

(iv) The formula for the sum of squares of the first n natural numbers is given by

$$SUM = n(n + 1)(n + 2)/6.$$

Develop a FORTRAN programme to read n and print the value of the SUM.

[Hints:
```
READ (*,*) N
SUM = N * (N + 1) * (N + 2)/6
WRITE (*,*) SUM
STOP
END]
```

(b) (i) Differentiate between the terms given below: Object programme and Source programme.

(ii) Explain the following FORTRAN statements: STOP and END.

(iii) What will be the value of X and SUM after the STOP statement in the given programme is executed:

```
      X = 1
      SUM = 0
    3 IF (10.0 – X) 8,8,7
    7 SUM = SUM + 1.0/X
      X = X + 1
```

```
      GO TO 3
    8 STOP
      END
```

6. Answer any two questions from (a) to (c):

(a) If $\phi(x)$ be a polynomial of degree n then prove that the nth difference of $\phi(x)$ is constant.

(b) The values of a function $f(x)$ at some points x are given below. Construct the divided difference table. Using suitable interpolation formula find the value of $f(0)$.

x:	−1	−2	2	4
$f(x)$:	1	−9	11	69

(c) Obtain the solution of the following differential equation over the interval [0, 1] with step size 0.25 by Euler's method correct to two decimal places.

$$\frac{dy}{dx} = xy,\ y(0) = 1.$$

7. Answer any two questions from (a) to (c):

(a) What is a flowchart? Why it is so called? Draw any four symbols used in a flowchart and state with examples what does each symbol represent?

(b) Write short notes on: (i) Executable and Non-executable FORTRAN statements, (ii) Library function in FORTRAN, (iii) State whether CONTINUE is an executable statement.

(c) (i) Find the final value of K after the execution of the following programme segment.

```
      K = 2
      DO 10 N = 3, 8, 2
      K = K + N
      IF (K.GT.6) GO TO 20
   10 CONTINUE
   20 K = 2 * K
```

[**Hints:** 34]

(ii) Write a programme in FORTRAN language to find the largest number from three given numbers A, B, C.

VIDYASAGAR UNIVERSITY

2001

Fourth Paper

Group A

(Numerical Analysis)

Answer any three questions

1. (a) Define the following terms:

(i) Truncation Error, (ii) Round off error.

(b) Let x_T and x_A denote respectively the true approximate values of a number. Find the absolute error, relative error and percentage error in x_A, when $x_T = \frac{1}{3}$ and $x_A = 0.333$.

(c) It is required to compute $\tan^{-1}x$, accurate to six decimal places, for $x = 0.5$. Find the required number of terms of the series, where

$$\tan^{-1} x = x - \frac{x^3}{3} + \frac{x^5}{5} - \ldots.$$

(d) Evaluate $\left(\frac{\Delta^2}{E}\right)x^3$, symbols have usual meanings.

2. (a) Deduce Newton's backward difference interpolation formula.

(b) Use Newton's divided difference interpolation formula to find the interpolating polynomial for the function $y = f(x)$ given by the table.

x:	-1	1	4	6
y:	1	-3	21	127

(c) Find a polynomial for $f(x)$ for the following function values using Lagrange's method

$$f(0) = 1, f(1) = 3 \text{ and } f(3) = 55.$$

3. (a) Derive Newton-Cote's integration formula.

(b) Evaluate the integral $\int_4^{5.2} \ln x dx$ using Simpson's one-third rule with $h = 0.1$ and compare the result with exact value.

(c) Find the function whose first difference is $9x^2 + 11x + 5$.

4. (a) Describe Gauss's method of elimination for numerical solution of a system of linear equations.

(b) Prove that

$$\Delta^k f(x) = \sum_{i=0}^{k} (-1)^i \binom{k}{i} f\{x + (k-i)h\},$$

where h is the step length.

(c) Use modified Euler's method to compute the solution of the following problem in the interval [0,0.1] with $h = 0.02$.

$$y' = x + y, \ y(0) = 1.$$

5. Describe Newton-Rapshon method for computing a simple real root of an equation $f(x) = 0$ (with the condition of convergence). Give a geometrical interpretation of the method. Write the advantages and the disadvantages of this method.

Group B

(Elements of Computer Science)

Answer Q. No. 6 and any four from the rest

6. Answer any five questions:

(i) Calculate the value of any one of the following expressions:

(a) 3**2/2 + 4/5*2 (b) 12.0/3*2.0 + 2**4/2.

(ii) Write FORTRAN expressions equivalent to any one of the following mathematical expressions:

(a) $\cos(\log_{10}(a + 3b))$, (b) $e^{x+y} - \sin(x + 4y)$.

(iii) Suppose A, B have the following values:

$A = 3.5$, $B = 1.2$. Find the vales of A and B as a result of the list of any one of the following:

(a) $A = B$
$B = A$

(b) $T = A$
$A = B$
$B = T$

(iv) Suppose J contains 5 and K contains 10. Find the final value of J after each FORTRAN program segment (answer any one):

(a)
```
      IF(4*4-2*K)10,20,20
   10 J = K
   20 J = J + 1
```

(b)
```
      IF (K .GT. J)GO TO 50
      J = J + 1
      GO TO 60
   50 J = K
   60 J = J + K
```

(v) N = 17 and VAL = 1985.95. It is desired to get the printout of the form

THE VALUE AFTER 17 YEARS IS 1985.95

Write the output statement.

(vi) Write the FORTRAN IF-THEN statement for the following flow chart

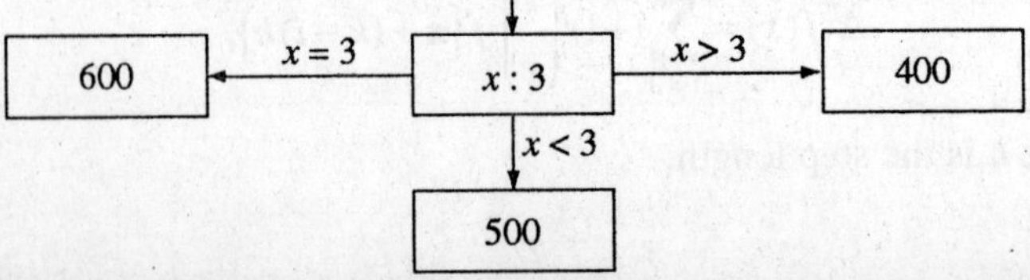

(vii) Define: (i) Source program and (ii) Object program.

(viii) Given that $A = 123.78$, $B = -84 \times 10^{-23}$, $I = -857$. Write the appropriate READ statements with FORMAT statement to read A, B, I.

7. (a) Explain with example the difference between implied DO statement and DO statement.

(b) Draw a flow chart to calculate the largest among the three numbers A, B, C.

8. (a) (i) Convert the decimal number $(61)_{10}$ to binary number. (ii) Find the decimal number equivalent of the binary number $(11.00011)_2$. (iii) Subtract $(101111)_2$ from $(110101)_2$.

(b) Write a FORTRAN program that reads an integer N and prints the sum of the following:

(i) $2 + 4 + 6 + 8 + \cdots + 2N$, (ii) $1 + 3 + 5 + 7 + \cdots + (2N - 1)$.

9. (a) Explain Arithmetic IF statement with example.

(b) Write a program to calculate $\cos x$ using the following series:

$$\cos x = 1 - \frac{x^2}{2!} + \frac{x^4}{4!} - \cdots.$$

Take x as input.

10. (a) Write an algorithm to evaluate $I = \int_0^1 f(t)\,dt$, where $f(t) = \frac{1}{1+t}$ with equally spaced intervals of length $h = 0.1$ by Simpson's $\frac{1}{3}$rd rule.

(b) Write a FORTRAN program to arrange 20 numbers in an array in ascending order.

11. (a) State the postulates of Boolean algebra. Explain the difference between Boolean algebra and algebra of real numbers.

(b) For all $x, y, z \in B$ (a Boolean algebra) prove, not using truth table, that

$$x + xyz + xy'z' + yz + y'z = x + z.$$

12. (a) Explain NOR gate with block diagram and truth tables.

(b) Define disjunctive normal form of a Boolean function and express the following function in SOP form in smallest number of variables

$$x'yz + xy'z' + x'y'z + x'yz' + xy'z + x'y'z'.$$

13. Write short notes on the following:

(i) Subscripted variables

(ii) Hierarchy of operations in evaluating an arithmetic expression and the rule of evaluation in case of equal hierarchy in FORTRAN 77.

(iii) Basic theorems of Boolean algebra.

2002

Fourth Paper

Group A

(Numerical Analysis)

Answer any three questions

1. (a) Define the terms: Significant figures and relative error, in connection with numerical computation. Give an illustrative example with each definition.

(b) Solve the following system of linear equations by Gauss-Seidel method:

$$27x+6y-z=85;\ 6x+15y+2z=72;\ x+y+54z=110.$$

2. (a) Obtain the Lagrange's interpolation for (n+1) distinct and finite interpolating points x_i, $i = 1, 2, \ldots, n$ (error is not required).

(b) Using the given table of values of Bessel function $y = J_0(x)$, find the roots of the equation $J_0(x) = 0$ lying in (2.4,2.6) correct up to three significant digits.

x:	2.4	2.5	2.6
y:	0.0025	–0.0484	–0.0968

3. (a) Explain how an algebraic equation can be solved by the method of iteration and hence deduce the condition of convergence.

(b) Describe Regula-Falsi method for finding a real root of the equation $f(x) = 0$. Give its geometrical interpretation.

4. (a) A curve is drawn to pass through the points given by the following table:

x:	1.0	1.5	2.0	2.5	3.0	3.5	4.0
y:	2.0	2.4	2.7	2.8	3.0	3.6	2.4

Use Weddle's rule to estimate the area bounded by the curve and the lines $x = 1$, $x = 4$ and x-axis.

(b) Find the value of $\int_0^1 \frac{dx}{1+x^2}$ taking 5 sub-intervals, by Trapezoidal rule, correct to five significant figures. Also find the error by comparing with the exact value.

5. (a) Describe Euler's method for solving the differential equation

$$\frac{dy}{dx} = f(x, y)$$

in a finite interval $[a,b]$ assuming that $y(a)$ has a known value y_0.

(b) Define kth order difference of a function. Prove that

$$\Delta^k f(x) = \sum_{i=0}^{k} (-1)^{i\,k}C_i f\{x+(k-i)h\}.$$

where h is the step length.

Group B

(Elements of Computer Science)

Answer Q. No. 6 and any four from the rest

6. Answer any five questions from the following:

(i) Calculate the value of any one of the following:

(a) SQRT(4.0**4) + ABS(4.0–8.0) + 41/2,

(b) (2**3)**2 + (2**3**2) + 15/4.

(ii) Write FORTRAN expression equivalent to any one of the following:

(a) $\dfrac{pe^q}{2q} + \log p + \dfrac{|x|}{\sin y^2}$

(b) $\sqrt{a^2 + b^2}\,\tan(a/b) + \log_{10} c + \cos^2 x.$

(iii) If A = 2.7, B = 3.5, I = 3 and K = −2, find L = ABS(A–J*B) +7./4. +11/5.

(iv) What is the value of C in the following segment, if A = 8, B = 8?

```
IF(A.GE.B) THEN
C =10
ELSE
C = A+B
ENDIF
```

(v) (a) For what purpose, in a FORTRAN coding sheet, the sixth column is used?

(b) Fill in the gap: FORTRAN 77 is a .... language.

(vi) Perform the following operation on the following binary numbers:

$$(110110.01)_2 - (101101.11)_2$$

(vii) Write FORTRAN IF-THEN-ELSE statement for the following flow chart:

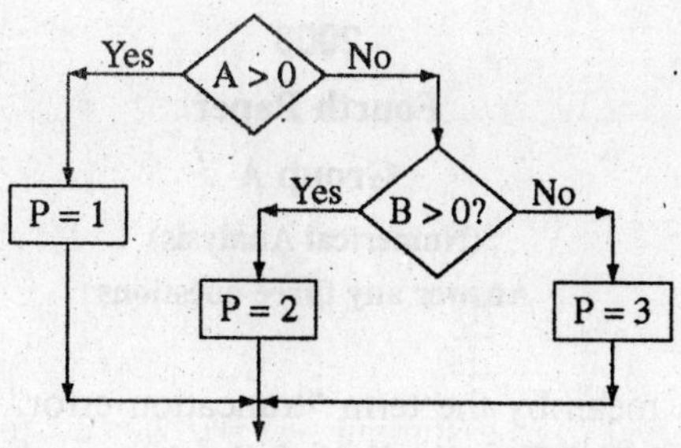

(viii) Write the De Morgan's law for three Boolean variables.

7. (a) Write a FORTRAN program segment to evaluate and print values of $f(x)$ for $x = 0.1$ to 1.0 with spacing $h = 0.1$, where

$$f(x) = \begin{cases} 4x^2 + 1 \text{ for } 0 \le x < 0.5 \\ 4x^2 - 1 \text{ for } 0.5 \le x < 1.0 \end{cases}$$

(b) Draw a flow chart to integrate a function $f(x)$ in $[a, b]$ by trapezoidal rule.

8. (a) The values A = 29.3682, B = 1.2432 ×10⁻⁶, I = 486, J = 19, K = −15 are stored, indicate the output after the execution of

```
    WRITE(6,15) A,B,I,J,K
15  FORMAT(3X,F10.5,E16.5,3I5)
```

(b) Write a FORTRAN program to evaluate the infinite series

$$x-\frac{x^3}{3!}+\frac{x^5}{5!}-\frac{x^7}{7!}\cdots,$$

correct to six decimal places for $x = 0.68, 0.72, 0.76, 0.80$ in an execution.

9. (a) Write an algorithm to find the real roots of a quadratic equation with proper messages indicating the character of the roots.

(b) The values A = 1234.3056, B = 3.28×10^{-8}, I = −123, J = 25 are to be read by the computer. Write a READ statement and the necessary statement to do this using F format for A, E format for B and I format for I, J.

10. (a) Explain with example, the difference between unconditional GO TO statement and computed GO TO statement in FORTRAN.

(b) Write a FORTRAN program that reads n numbers and prints their mean = $\frac{\Sigma x}{n}$

and standard deviation = $\sqrt{\frac{\Sigma x^2}{n}-\left(\frac{\Sigma x}{n}\right)^2}$.

11. (a) Determine the conjunctive normal form of

$$x'y'z' + x'y'z + x'yz + xyz + xy'z + xy'z'.$$

(b) Find the disjunctive normal form of $xy + x'z$.

12. (a) Explain the NAND gate with block diagram and truth table.

(b) Draw a gate network to implement F = (A+B′)(CD+E) using NOR gate only.

13. Write short notes on (i) Principle of duality. (ii) Executable and non-executable statements. (iii) Different components of a computer with block diagram.

2003

Fourth Paper

Group A

(Numerical Analysis)

Answer any three questions

1. (a) What do you mean by the term "truncation error" and "round off" error in numerical analysis? Round off the following numbers to five significant figures:

562.55, 17896.6, 0.0036945, 3.470003, 24.39578, 67.0085.

(b) The number 1/3 is approximated by 0.33. Find the absolute, relative and percentage errors in approximation.

(c) Define divided difference of various order (up to 3rd order) and show that the divided differences are symmetric functions of their arguments.

2. (a) Given a table of values of $f(x)$ for $x = x_0 + ih(h > 0)$, $r = 0, 1, 2, \cdots, n$, find the formula for numerical computation of $f'(x)$ and $f''(x)$, when x is closed to x_n. Also find the formula of $f'(x)$ and $f''(x)$ for $x = x_0$.

(b) Solve the following system of linear equations by Gauss elimination method:

$$x_1 - x_2 - x_3 = 1$$
$$2x_1 - 3x_2 + x_3 = 1$$
$$3x_1 - x_2 - x_3 = 2.$$

3. (a) Explain Newton-Raphson method for finding a real root of the equation $f(x) = 0$. Give its geometrical interpretation.

(b) Show that Newton-Raphson method is a special case of the method of fixed point iteration. Find the sufficient condition of convergence of the Newton – Raphson method.

4. (a) Find the value of $\int_0^1 (4x - 3x^2)\,dx$ by Simpson's $\frac{1}{3}$rd rule taking $n = 10$. Compute the exact integral and find the relative error in your result.

(b) Derive Newton-Cotes integration formula (error is not required) and deduce the particular formula with two subintervals.

5. (a) Describe Euler's method for solving the differential equation

$$\frac{dy}{dx} = f(x, y)$$

in a finite interval $[a, b]$ assuming that $y(a)$ has a known value y_0.

(b) State Euler-Maclaurin sum formula and use it to sum the following series:

$$\frac{1}{11^2} + \frac{1}{12^2} + \cdots + \frac{1}{50^2}.$$

Group B

(Elements of Computer Science)

Answer Q. No. 6 and any four from the rest

6. Answer any five questions from the followings:

(i) Evaluate the following expression for I = 4, J = 5,K = 3, L= 2.

M = J + L*1/3 +J/I –K**I/L

(ii) Indicate whether the following statements are valid in FORTRAN and, if not, why ?

(a) Y = B(A+D) – C/3 +A/–3

(b) V = –C/2 + A*5–SQRT(–9)+Y**-3.

(iii) Write FORTRAN expression corresponding to the following algebraic expression:

$$2.5\log_{10}x + \cos 32° + x^2 - y^2 + \sqrt{(2xy)}$$

(iv) A ______ program is transformed into an ________ program by an assembler or a compiler.

(v) Write the mathematical expression corresponding to the following FORTRAN expression:

(a) −1/(2.*X) + A**2/(4.*X**2)

(b) (1/E06 + A*X) **2./3

(vi) Perform the following operation:

$$(11001.01)_2 - (10011.11)_2$$

(vii) Write READ and FORMAT statements to read the following data:

2.2bb450bbb29.05

7. (a) Explain with example, the difference between Arithmetic IF and logical IF statements in FORTRAN.

(b) Write the algorithm to arrange the given numbers $a_1, a_2, \cdots, a_n$ in the descending order.

8. (a) Write a FORTRAN program for evaluating the integral by Simpson's $\frac{1}{3}$rd rule.

$$\int_1^2 y dx, \text{ where } y = \frac{x^{1/2}(1+\tan x)}{e^x + x^{5/3}}.$$

(b) Give a print statement in FORTRAN of $-75{,}123.35, .732\times10^{-06}$ with a gap of 3 spaces between them using I,F and E formats respectively. Give also another print statement to get a printout of the above numbers in three different lines.

9. (a) Write statements to compute the function, using Arithmetic IF statement, $y(x)$ is defined as

$$y = 4x + 6, \text{ if } x \le 3.5$$
$$y = 6x + 1.6, \text{ if } x > 3.5.$$

(b) Write the equivalent logical statement for the following computed GO TO statement:

GO TO (99,55,77),L

(c) Draw a flow-chart to find the real roots of a quadratic equation.

10. (a) Write a FORTRAN program to evaluate the infinite series:

$$1 - \frac{x^2}{2!} + \frac{x^4}{4!} - \frac{x^6}{6!} \cdots,$$

correct to six decimal places taking $x = 0.1$ as the input.

(b) Draw a flow chart to find the largest value amongst the given three values.

11. (a) Simplify the following Boolean function:

$$abc + \bar{a}bc + a\bar{b}c + ab\bar{c} + a\bar{b}\bar{c} + \bar{a}b\bar{c} + \bar{a}\bar{b}\bar{c}, \text{ not using truth table.}$$

(b) Explain the NOR gate with block diagram and truth table. Design a two input XOR gate ($f = x'y + xy'$) using NAND gates.

12. (a) Find the following:

(i) Decimal equivalent of $(11010.101)_2$,(ii) Binary equivalent of $(11.25)_{10}$, (iii) $(1101)_2 \times (10.1)_2$, (iv) $(100011)_2 \div (101)_2$.

(b) Write down the postulates of Huntington for Boolean algebra.

13. Write short notes on any two of the following:

(i) Subscripted variable, (ii) ALU and Control Unit of a Computer, (iii) Format specifications I,E and A formats for input data in FORTRAN.

2004
Fourth Paper
Group A
(Numerical Analysis)

Answer any three questions

1. (a) Define absolute, relative and percentage error. Find the relative and percentage error in the computation of $x - y$ for $x = 9.05$ and $y = 6.56$ having absolute errors $\Delta x = 0.001$ and $\Delta y = 0.003$ respectively.

(b) Find the number of correct significant figures in $x_A = 0.3941$, given its absolute error as 0.25×10^{-2}.

(c) For any positive integer n, show that $\Delta^n y_0 = \sum_{i=0}^{n} (-1)^i \, {}^nC_i y_{n-i}$.

2. (a) Explain the method of iteration for the numerical solution of an equation $f(x) = 0$ by resetting in the form $x = \phi(x)$. State the condition of convergence of this method. Is it possible to formulate more than one iteration scheme for an equation? Formulate a convergent iteration scheme for solving the equation $2x - \sin x - 1 = 0$.

(b) For Lagrangian interpolation, show that the sum of the Lagrangian coefficients is unity.

3. (a) Derive the numerical integration formula from Newton's forward interpolation formula (error is not required). Then deduce Simpson's $\frac{1}{3}$rd rule.

(b) Solve the following system of equations by Gauss-Seidel iteration method (3 steps) starting from (0,0,0):

$$27x + 6y - z = 85$$
$$6x + 15y + 2z = 72$$
$$x + y + 54z = 110.$$

4. (a) Using Euler's method , solve the following differential equation:

$$\frac{dy}{dx} + 2y = 0,\ y(0) = 1.$$

Take $h = 0.1$ and obtain $y(0.3)$.

(b) Taking equally spaced interpolating points, establish Newton's backward polynomial interpolation formula.

5. (a) Describe Regula Falsi method for finding a real root of the equation $f(x) = 0$.

(b) For finding the square root of $a(a > 0)$, derive the iteration formula:

$$x_{n+1} = \frac{1}{2}\left(x_n + \frac{a}{x_n}\right), n = 0,1,2,\cdots,$$

where x_0 (> 0) is any initial approximation and x_n is the n-th approximation. Apply this formula to find $\sqrt{2}$ correct up to three significant figures.

Group B

(Elements of Computer Science)

Answer Q. No. 6 and any four from the rest

6. Answer any five questions from the following:

(i) Find the value of I from the following (when A = 1.5, J = 2 and B = 3.0):

I = B/2 + B*4/A – 8 + A**3 – J**3/8

(ii) Write FORTRAN expression corresponding to the following algebraic expressions:

$$\sin\left(x + \sqrt{x^2 + 1}\right) + x^3 \log_e |x - y| / 3.$$

(iii) Find out if the following statements have any error :

(a) A – B + C/–5+D(2*E-F),

(b) X**2 – SQRT(81.0) + ASIN(–1.35).

(iv) Write mathematical expressions for the following FORTRAN expressions:

(a) T = TANH(SQRT((X – Y)/(X + Y))),

(b) COS(2*X*Y) + ABS(X**2 + Y**2) + EXP(X*Y).

(v) Write down the equivalent statement of the following single statement using nested DO loops.

WRITE(*,5) ((C(I,J),J = 3,5);I = 10,15)

(vi) Indicate the value of A that will be stored after execution the statement:

A = 3.E02/4.0**0.5*2.0 + C/8.0*3.0 +3.0**2/2,

where C = (0.8, –2.4).

(vii) Draw a logical gate network of $x'y + xy'$ using three basic gates only.

(viii) Write down the dual of the statements:

(a) $(a + b).(b + c) = a.c + b$ (b) $(a + 1).(a + \theta) = a$.

(ix) Find the decimal equivalent of $(1110.10)_2$ and binary equivalent of $(27)_{10}$.

(x) Perform the following operation:

$$(1110010)_2 - (10110)_2.$$

7. (a) Write a FORTRAN program for solving the following equation:

$$x^3 + 2x^2 + 2.2x + 4 = 0,$$

using Newton-Raphson method.

(b) Explain with example, the difference between implied DO loop and DO loop.

8. (a) Draw a flow-chart to evaluate the following series:

$$x-\frac{x^3}{3!}+\frac{x^5}{5!}+\frac{x^7}{7!}\cdots.$$

(b) The values A = −34.638, B = 3.4187 × 10^{-9}, I = 786, J = −5, K = 259 are stored, indicate the output after the execution of

```
      WRITE(1,10) A,B,I,J,K
10    FORMAT(5X,F10.6,E16.5,3I5)
```

9. (a) Write an algorithm to find the value of the roots of a quadratic equation with appropriate messages giving the nature of roots.

(b) Write a FORTRAN program for summing up the series

$$\sum_{m=1}^{50}\sum_{n=1}^{100}\frac{\sin mx\cos nx}{m^2+n^2}\quad \text{for } x = 0.2.$$

10. (a) Find how many times the statement numbered 10 will be executed in the following program segment:

```
      K = 0
      DO 9 I = 2,10,2
      K = K + 2
      DO 9 J = 1,3
      IF(K.GT.3) GO TO 9
10    X = X + 1
9     CONTINUE
```

(b) Correct the mistake(s) in the following program segment:

```
      KOUNT = 0
40    DO 50 I = 1,100
      READ(*,1)A
      IF(A.LT.45.0)GO TO 40
50    KOUNT = KOUNT + 1
```

(c) Write a FORTRAN program to compute the integral $\int_1^2 \log e^x dx$ numerically by trapezoidal rule with 10 equal intervals.

11. (a) Express $(x+y)(x+y+z')(y+z)$ as a function in conjunctive normal form.

(b) Explain NAND gate with block diagram and truth table. Design all basic gates using NAND gates.

12. (a) Find the binary difference $(111000)_2-(11001)_2$ and check by converting the binary numbers to decimal.

(b) Write an algorithm to compute the integral $\int_0^2 \frac{dx}{\sqrt{1+x^2+x^4}}$ taking 24 equal intervals.

13. Write short notes on any two of the following:

(a) Algorithm and flow chart, (b) GO TO statement, (c) Machine, assembly and high level language.

2005

Fourth Paper

Group A

(Numerical Analysis)

Answer any three questions

1. (a) Let x_A and x_T denote the true and approximate values of a number respectively.
   (i) For the following numbers x_A and x_T, determine how many significant figures x_A has with respect to x_T:
   (1) $x_A = -0.054113$, $x_T = -0.05418$
   (2) $x_A = 39.4213$, $x_T = 39.4604$.
   (ii) Round off the following numbers to three significant figures:
   (1) 0.01201, (2) −239.85, (3) 0.004923, (4) 63.8543.
   (iii) Find the number of significant figures in (1) $x_A = 1.8921$, given its relative error as 0.1×10^{-2}, (2) $x_A = 0.3941$, given its absolute error as 0.1×10^{-2}.

   (b) Solve the following system by the method of elimination:
   $$x_1 + 2x_2 + x_3 = 0$$
   $$2x_1 + 2x_2 + x_3 = 3$$
   $$-x_1 - 3x_2 = 2.$$

2. (a) Define divided difference of various order (up to 3rd order) and show that the divided differences are symmetric functions of their arguments.
   (b) Given a table of values of $f(x)$ for $x = x_0 + ih (h > 0), r = 0, 1, 2, \cdots, n$, find the formula for numerical computation of $f'(x)$ and $f''(x)$, when x is closed to x_n. Also find the formula of $f'(x)$ and $f''(x)$ for $x = x_0$.

3. (a) Calculate by Simpson's $\frac{1}{3}$rd rule the value of $\int_{1.2}^{1.6}\left(x + \frac{1}{x}\right)dx$ correct up to two significant figures, taking four intervals.
   (b) Explain the Gauss-Seidel iterative method for solving a system of equations. Give sufficient conditions for the convergence of the process.

4. (a) Describe modified Euler's method for solving the differential equation $\frac{dy}{dx} = f(x, y)$ in a finite interval $[a, b]$ assuming that $y(a)$ has a known value y_0.
   (b) Derive Newton-Cote's integration formula (error is not required) and deduce the particular formula with two sub-intervals.

5. (a) Find by Lagrange's formula, the interpolation polynomial which corresponds to

x:	−1	0	2
$f(x)$:	9	5	3

   (b) Explain Newton-Raphson method for finding a real root of the equation $f(x) = 0$. Give its geometrical significance.

Group B

(Elements of Computer Science)

Answer Q. No. 6 and any four from the rest

6. Answer any five questions from the following:

(i) Fill in the blanks:-

FORTRAN program is written in ______ language which can be translated into ______ language by a ______. This translated program is called ______ program.

(ii) Find the value of K1 = J/3 + J*L/5 + K**I/L + L/J, where I = 4, J = 5, K = 3, L = 2.

(iii) Write the FORTRAN expression for the following:

$$g = e^{-x^2} \sin^2(x + |x|) + x \log(y^2)/\cos^{-1}\left(\sqrt{x+y}\right)$$

(iv) Write READ and FORMAT statements to read the following data:

7.59bb455bbbb34.792

(v) Perform the following operation on the following binary numbers:

$$(110110.01)_2 - (101101.11)_2$$

(vi) Write down the De Morgan's law for three Boolean variables.

(vii) Indicate whether the following statements are valid in FORTRAN and, if not, why?

(a) X = A(B + C) – D/5 + A/–3

(b) V = –P/2 + X*5 – SQRT(–9) + A**–3.

(viii) Write the mathematical expression corresponding to the following FORTRAN expressions:

(a) X**5/(3.*A*4) + SQRT(X*X + 5)

(b) (1.0E – 06+X*Y)**5./4.

7. (a) Write an algorithm to find the real roots of $ax^2 + bx + c = 0$, where a, b are constants. With appropriate messages indicating to the character of the roots.

(b) Write a short note on BLOCK IF statement with an example and flow-chart.

8. (a) Draw a flow chart to find the largest value amongst the three given values.

(b) Write a FORTRAN program that reads three positive numbers A,B,C and determine whether A,B,C can form the sides of a triangle and, if so, determine whether the triangle will be a right angled triangle.

9. (a) Write a FORTRAN program to evaluate the infinite series $x - \frac{x^3}{3!} + \frac{x^5}{5!} - \frac{x^7}{7!} \cdots$ correct up to six decimal places taking $x = 0.3$ as input.

(b) Give a printout statement in FORTRAN of $-28, 111.01, 0.5 \times 10^{-2}$ with a gap of 3 spaces between them, using I,F and E formats. Give also a statement to get a printout of the above numbers in three different lines.

10. (a) Define numeric variables in FORTRAN. State their classifications with examples.

(b) The values A = 1235.2783, B = 3.27×10^{-8}, I = −123, J = 25 are to be read by the computer. Write a READ statement and necessary statement to do this using F format for A, E format for B and I format for I, J.

11. (a) Write a FORTRAN program to compute the integration $\int_1^2 \log_e x\,dx$ numerically by trapezoidal rule with 10 sub-intervals.

(b) If A,B,C are inputs, obtain the circuit using any gate to get the output of the function F, where

$$F = AC' + A'(BC' + B'C).$$

12. (a) Find the binary multiplication of the following:

$$(1101.101)_2 \times (101.1)_2$$

and check by converting the binary numbers to decimal.

(b) Simplify the following Boolean expression:

$$X = A'B(B+C) + BC(B' + A').$$

13. Write short notes on any two of the following:

(a) AND-gate and OR-gate, (b) Arithmetic IF statement, (c) FORMAT statement.

2006

Fourth Paper

Group A

(Numerical Analysis)

Answer any three questions

1. (a) Define the following terms with suitable examples:

(i) Truncation error, (ii) Round off error.

(b) If we take p =3.14 in place of 3.14156, find the absolute error and relative error.

(c) Solve the following system of equations by Gauss-Seidel method:

$$2x + 10y + z = 13$$
$$10x + y + z = 12$$
$$2x + 2y + 10z = 14.$$

2. (a) Obtain the Lagrange's interpolation polynomial for $(n+1)$ distinct and finite interpolating points (error is not required).

(b) Determine the polynomial which interpolates the following points:

$$(1, -1), (2, -1), (3, 1), (4, 5).$$

3. (a) Derive trapezoidal formula by integrating Newton's forward difference interpolation formula.

(b) Using Euler-Maclaurin sum formula find the following sum correct to four significant figures:

$$\frac{1}{11^3}+\frac{1}{12^3}+\cdots+\frac{1}{50^3}.$$

4. (a) Taking equally spaced interpolating points, establish Newton's forward interpolation formula.

(b) For finding the square root of $a(a > 0)$, derive the iteration formula

$$x_{n+1}=\frac{1}{2}\left(x_n+\frac{a}{x_n}\right), n=0,1,2,\cdots,$$

where $x_0(> 0)$ is any initial approximation and x_n is the n-th approximation. Apply this formula to find $\sqrt{2}$, correct up to three significant figure.

5. (a) Describe Gauss-elimination method for the solution of a system of linear equations.

(b) Solve: $\frac{dy}{dx}=1+xy$, where $y(0) = 1$, correct up to three significant figures using Euler's method.

Group B

(Elements of Computer Science)

Answer Q. No. 6 and any four from the rest

6. Answer any five:

(a) Find the value of KX, if KX = A + J *K**I/J/B, where I = 2, J = 3, K = 4, A = 2.0, B = 5.0.

(b) Write the FORTRAN expression of the following:

$$f=e^{x^2}+\log_{10}(x^2+1)/\log_e(x^2+\tan^{-1}x).$$

(c) Draw a logical diagram for $(a+bc')$ using three basic gates only.

(d) Find the value of the variable ROOT after completion of the following program segment.

```
      ROOT = 0
      DO 20 I =1,10
      ROOT = ROOT + I*0.1
20    CONTINUE
```

(e) Subtract $(101011)_2$ from $(111110)_2$.

(f) Convert 23.25 to binary.

(g) Write truth table $AB' + A'B + AB$.

(h) State the principle of duality in Boolean algebra.

7. (a) Write a FORTRAN program that reads n numbers and prints their mean = $\frac{\Sigma x}{n}$ and standard deviation = $\sqrt{\frac{\Sigma x^2}{n} - \left(\frac{\Sigma x}{n}\right)^2}$.

(b) Explain with example, the difference between Arithmetic IF and logical IF statements in FORTRAN.

8. (a) Write a FORTRAN program to find the roots of $ax^2 + bx + c = 0$ $(a,b,c$ are constants) with appropriate comment indicating the type of the roots.

(b) Write a short notes on DIMENSION statement and when it is used.

9. (a) Draw a flow chart to find the root of $f(x) = 0$ by Newton-Rapshon method.

(b) What is SUBROUTINE in FORTRAN? Explain the advantage of using this.

10. (a) Write a FORTRAN program to read a matrix A of order $m \times n$ and to print its transpose.

(b) Write Read statement and necessary statements to read A = 1234.5678, B = 5.67×10^8, I = −539, J = 23 by the computer using F format for A, E format for B and I format for I, J. Give also statements to get a printout in the said form but in three different lines.

11. (a) State the postulates of Boolean algebra. Explain the difference between Boolean algebra and the algebra of real numbers.

(b) Simplify the Boolean expression

$x'yz + xy'z' + x'y'z + x'yz' + xy'z + x'y'z'$, not using the truth table.

12. (a) Construct the truth table for the Boolean expression: $xy' + y(z + x')$ and hence write it in C.N.F.

(b) Give a functional description of different components of a computer.

13. (a) Draw a gate network for $A'BC' + A'BC + AB'C$; using NAND gates only and taking A, B, C as inputs, not their complements.

(b) What are logic gates? Write down truth table and logic symbol for AND and OR gates each with three inputs.

2007

Fourth Paper

Group A

(Numerical Analysis)

Answer any three questions

1. (a) Define the terms 'absolute error', 'relative error' and 'percentage error' as used in numerical computations. The number 1/3 is approximated by 0.33. Find the absolute, relative and percentage errors in approximation.

(b) The following numbers are all approximate and correct as far as their last digits only. Find their sum: 32.537, 7.3, 71.45, 125.2357, 22.3451.

(c) Prove that the divided difference $f(x_0, x_1)$ is independent of x_0 and x_1, if $f(x)$ is linear.

2. (a) Define divided differences of various orders and show that the divided differences are symmetric functions of their arguments.

(b) Explain regula falsi method to compute a real root of the equation $f(x) = 0$.

3. (a) Describe Gauss-Seidel iteration method for the solution of a system of equations.

(b) Explain the principle of numerical differentiation. Deduce numerical differentiation formula from Newton's forward interpolation formula.

4. (a) Deduce Newton's forward interpolation formula.

(b) Describe Euler's method for solving a differential equation of the form

$$\frac{dy}{dx} = f(x, y), \text{ where } y(x_0) = y_0.$$

5. (a) Using Newton-Raphson method to find a real positive root of the equation $x^3 - 2x - 5 = 0$, correct up to three significant figures by choosing the initial approximation $x_0 = 2$.

(b) Deduce trapezoidal rule for integrating a function $f(x)$ between a and b.

Group B

(Elements of Computer Science)

Answer Q. No. 6 and any four from the rest

6. Answer any five:

(a) Define disjunctive normal form of a Boolean function.

(b) Design OR gate using NAND gates only.

(c) Find the output of the following logical diagram:

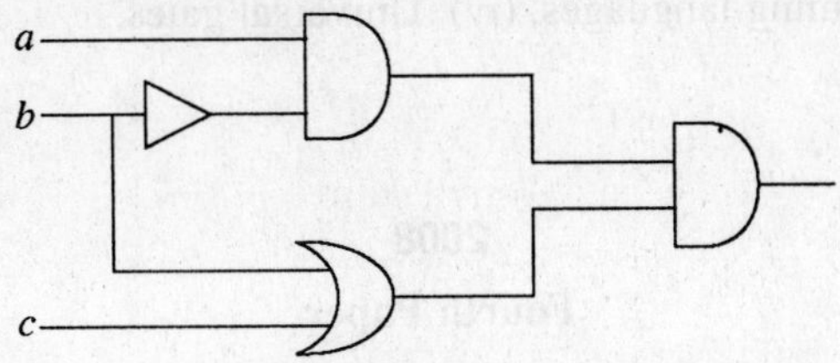

(d) Find the remainder and quotient when $(101101.1010)_2$ is divided by $(110)_2$.

(e) What do you mean by source program and object program?

(f) Compare compiler and interpreter.

(g) State De Morgan's laws in Boolean algebra.

(h) Probe that in a Boolean algebra $a + a.b = a$.

7. (a) Write a FORTRAN program that reads n values of x and prints the corresponding values of $f(x)$, where $f(x)$ is given by

$$f(x) = \begin{cases} xe^x + \sin^{-1} x \text{ for } x \le 1 \\ \log_{10} x + \tan^{-1} \text{ for } x > 1 \end{cases}$$

(b) Write short notes on DO loop in FORTRAN.

8. (a) Write a flow chart to find the maximum and minimum among n numbers.

(b) Write the FORTRAN expression of $\tan^{-1}(e^{x^2}) + \log_{10}(\sin x + 2x)/\cos^{-1} x$.

(c) Find the value of IA if IA = X*M+Y**I/J/K in FORTRAN, where X = 2.0, M = 2, Y = 4.0, I = 2, J = 4, K = 2.

9. (a) Write an algorithm to find the numerical value of $\int_a^b f(x)\,dx$ using Simpson's $\frac{1}{3}$rd rule.

(b) State Huntington's postulates of Boolean algebra.

(c) Write the following Boolean expression in two normal forms

$$x'y + y'z + zx.$$

10. (a) Write a FORTRAN program to arrange a set of numbers in ascending order.

(b) Define the truth table. Construct truth table of the following Boolean expression:

$$x'yz' + xy' + zy'.$$

11. (a) Compare logical IF and arithmetic IF statements in FORTRAN.

(b) Write a FORTRAN program to read two $m \times n$ matrices and find their sum.

12. (a) Give a brief history of development of computers.

(b) Compare the working process of a clerk and a computer.

(c) Simplify the following Boolean expression, not using truth table

$$abc + abc' + a'bc + a'bc' + a'b'c + a'b'c'.$$

13. (a) Draw a gate network for $A'B + AB'$ using NOR gates only and taking A, B as inputs, not their complements.

(b) Write short notes on any two: (i) FORMAT statement, (ii) Flow-chart, (iii) Programming languages, (iv) Universal gates.

2008

Fourth Paper

Group A

(Numerical Analysis)

Answer any three questions

1. (a) If the first significant figure of a number is k and the number is correct up to n significant figures, prove that the relative error is less than $\dfrac{1}{k \times 10^{n-1}}$.

(b) Evaluate: $\left(\frac{\Delta^2}{E}\right)x^3$.

(c) Find the number of correct significant figures in (i) $x_A = 0.3941$ given its absolute error is 0.25×10^{-2}, (ii) $x_A = 1.8921$ given its relative error is 0.1×10^{-2}.
[**Hints:** See example 8 of chapter 10.]

2. (a) Write the expression of Lagrangian interpolation. For Lagrangian interpolation, show that the sum of the Lagrangian coefficients is unity.

(b) Deduce Simpson's one-third rule of numerical integration for a function $f(x)$ between a and b.

3. (a) Find a real root of the equation $\cos x = 2x - 3$ correct up to three decimal places by using iteration method.

(b) Give geometrical meaning and condition for convergence of Newton-Raphson method.

4. (a) Describe modified Euler's method for solving a differential equation of the form

$$\frac{dy}{dx} = f(x, y) \text{ with } y(x_0) = y_0.$$

(b) State Euler-Maclaurin's sum formula and use it to find the sum of the following series:

$$\frac{1}{21^2} + \frac{1}{23^2} + \frac{1}{25^2} + \ldots + \frac{1}{35^2}.$$

5. (a) Define second order divided difference of $f(x)$ for nodes x_0, x_1, x_2. Prove that $f(x_0, x_1, x_2) = 1$ if $f(x) = x^2$ where x_0, x_1, x_2 are distinct.
[**Hints:** See article 4.2]

(b) State the condition of convergence for the solution of a system of linear algebraic equations with three variables by using Gauss-Seidel iteration method.
[**Hints:** The coefficient matrix should be diagonally dominant.]

(c) Prove that $\Sigma_{i=0}^{n} c_i = 1$ where c_i $(i = 0, 1, 2, \ldots, n)$ are Cote's coefficients of Newton-Cote's closed type formula.
[**Hints:** See P1 of article 7.1.]

Group B

(Elements of Computer Science)

Answer Q. No. 6 and any four from the rest

6. Answer any five questions from the following:

(a) Write FORTRAN expression equivalent to the following mathematical expression.

$$b = \frac{1}{2x^2}\log\left|a + \sqrt{b}\right| + \tan^{-1}(ae^x + \cos x).$$

(b) Explain the terms "hardware" and "software" of a computer.

(c) Suppose *J* contains 5 and *K* contains 10. Find the final value of *J* after each FORTRAN programme statement:

```
20    IF (2 * K.GE.3 * J) GO TO 70
      J = J + 1
60    K = K + J
      GO TO 20
70    J = J + 2
      STOP
```

[Hints: 7]

(d) Design OR-gate using NOR-gates only.

[Hints: See (d) of article 15.11.]

(e) Write FORTRAN IF-THEN-ELSE statement for the following flowchart

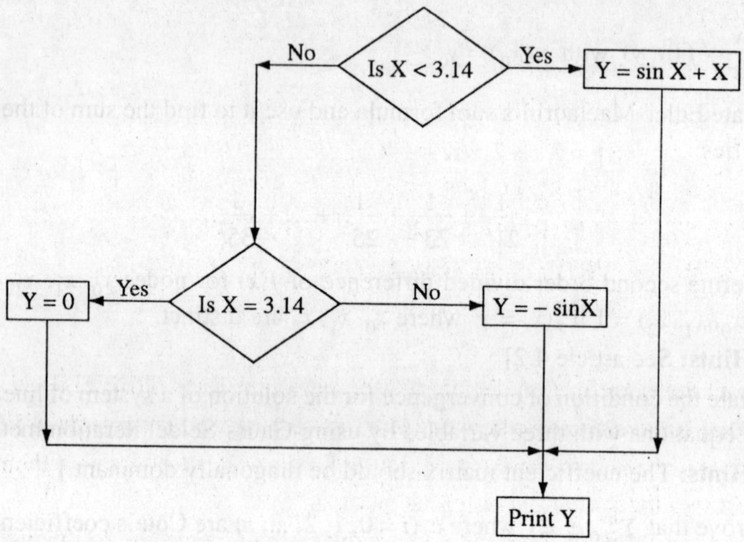

[Hints:
```
IF (X.LT.3.14) THEN
Y = SIN(X) + X
ELSE IF (X.EQ.3.14) THEN
Y = 0
ELSE
Y = –SIN(X)
END IF
WRITE (*,*) Y]
```

(f) State the principle of duality in Boolean algebra.

(g) Find decimal equivalent of $(110001.01)_2$ and binary equivalent form of $(67)_{10}$.

(h) What do you mean by algorithm and flowchart?

7. **(a) Simplify the following Boolean expression without using truth table:**

$$\bar{a}bc+\bar{a}\bar{b}c+abc+a\bar{b}c.$$

[Hints: $\bar{a}bc+\bar{a}\bar{b}c+abc+a\bar{b}c=\bar{a}c(b+\bar{b})+ac(b+\bar{b})=\bar{a}c+ac=(\bar{a}+a)c=c$**]**

(b) Write a FORTRAN program that reads three positive numbers *A*, *B* and *C* and determine whether *A*, *B* and *C* can form the sides of a triangle and if, so determine the nature of the triangle.

[Hints: See programme No. 16 of chapter 14.]

8. **(a) Write an algorithm to find the solution of a transcendental equation by iteration method.**

(b) Explain NOR gate with block diagram and truth table. Design a two-input XOR gate ($f=x'y+xy'$) by using NAND gate.

9. **(a) Write a programme that read *N* integers and first find sum of the even and odd integers separately and also find the sum of all integers.**

[Hints:

```
      READ (*,*) N
      S1 = 0
      S2 = 0
      NN = N/2
      DO 5K = 1, NN
      S1 = S1 + 2 * K – 1
    5 S2 = S2 + 2 * K
      SUM = S1 + S2
      WRITE (*,*) S1, S2, SUM
      STOP
      END]
```

(b) State the rules used to declare one-dimensional array in FORTRAN.

(c) What is the difference between DO and Implied DO statements in FORTRAN?

10. **(a) What is meant by input and output units of a computer? Explain their functions. List two input and two output devices.**

(b) Find the Boolean expression in product of sum from the following truth table and draw the corresponding logic diagram.

Inputs			Outputs
x	y	z	f
0	0	0	0
1	0	0	0
0	1	0	0
1	1	0	1
0	0	1	0
1	0	1	0
0	1	1	1
1	1	1	0

11.(a) Write a FORTRAN programme that read $2n$ real numbers $x_1, x_2, \ldots, x_n$ and $y_1, y_2, \ldots, y_n$ and print out correlation coefficient $r = \dfrac{\Sigma(x_i - \bar{x})(y_i - \bar{y})}{\sqrt{\Sigma(x_i - \bar{x})^2 \Sigma(y_i - \bar{y})^2}}$.

(b) Give a print statement in FORTRAN of –75, 123.05 and 0.732×10^{-6} with a gap of three spaces between them using *I*, *F* and *E* formats. Give also another print statement to get a print-out of above numbers in three different lines.

12.(a) Write a flowchart to arrange a set of integers in ascending order.

(b) Write short notes on control statements in FORTRAN.

13.(a) Write a FORTRAN program to read an $m \times n$ matrix and find the transpose of this matrix.

(b) Write short notes on any two: (i) CONTINUE statement, (ii) Library functions, (iii) Compiler, (iv) High level language.

2009

Fourth Paper

Group A

(Elements of Computer Science)

Compulsory

1. Answer any one questions:

(a) What is binary number system? How can a decimal number be converted into a binary number? What are the rules followed in binary multiplication? What is Boolean algebra? State the basic theorems and properties of Boolean algebra.

(b) Draw a block diagram to illustrate the basic organization of a computer system. Explain the functions of various units.

(c) What do you mean by algorithm and flowchart? Draw a flowchart to compute the roots of the quadratic equation $ax^2 + bx + c = 0$. State different types of variables in FORTRAN77. Define logical operators and discuss its hierarchy.

2. Answer any two questions:

(a) Which of the following real constants are invalid in FORTRAN77 and why?

$$37{,}35{,}0 - 67.4E0.5,\ E5,\ 7.6E,\ 9.5 - E01,\ +.65,\ -25.,\ 0.25E + 04$$

(b) Write a FORTRAN77 programme for evaluating the integral $\int_a^b f(x)\,dx$ by Simpson's one-third rule taking *N* number of intervals where the expression of *f*(*x*) is given. If there is an restriction on *N*, mention it.

(c) Write down the FORTRAN77 expression for the following mathematical expressions:

(i) $\dfrac{10^6 p}{q(d-r)} + \dfrac{1}{2}s^2$, (ii) $(\alpha + \beta)^{\delta} + \left(\dfrac{q}{p}\right)^{p-1}$,

(iii) $\dfrac{-B+\sqrt{B^2-4AC}}{2A}$, (iv) $\log|\sec x+\sin x|$.

(d) What is an array? What do you mean by subscript? State the rules used to declare one dimensional array.

3. Answer any three questions:

(a) Convert the decimal number $(154)_{10}$ to binary number. Add the binary numbers $(100111)_2$ and $(11011)_2$.

(b) Write the truth table for the Boolean function $W = X+\overline{Y}.Z$.

(c) Give the dual of the following Boolean expression.

(i) $\overline{A}.B+A.\overline{B}$, (ii) $A.(B+C)=A.B+A.C$.

(d) Prove the following: (i) $x + x = x$, (ii) $x + 1 = 1$.

(e) If $I = 2$, $J = 3$ and $K = 6$, the what values do the following expressions have?

(i) *.NOT.I.GT.J.AND.NOT.I * J.LE.K*

(ii) *I,GT.J.AND.(I.LE.K.OR.I * J.LE.K).*

(f) Construct a logic circuit for the Boolean expression: $\overline{A.B}+C.D+\overline{E.F}$.

4. Answer any one question:

(a) Find I, if $I = B/2 + B * 4/A - 8 + A ** 3$, given $A = 1.5$, $B = 3.0$.

(b) What do you mean by a bit in computer terminology?

UNIVERSITY OF NORTH BENGAL

2008

Mathematics (General)

Seventh Paper

(Revised New Syllabus)

Group A

Answer Q. No. 1 and any three from the rest

1. (a) Give the definition of Boolean algebra due to Huntington.

(b) In a Boolean algebra B, prove that $a + (a.b) = a$ for all $a, b \in B$.

(c) Convert $(69)_{10}$ into binary form.

2. (a) In a Boolean algebra B, prove that $(x, y') + (x' . y) = (x + y) . (x' + y')\ \forall x, y \in B$.

(b) Express the Boolean function $(x'y + yz)' + y'$ in conjunctive normal form in the smallest possible number of variables.

(c) Construct a truth table for the Boolean expression $x' + yz$.

3. (a) Draw a circuit using NOR gates only that realizes the truth function $(x, y) = x.y$.

(b) Obtain the Boolean expression from the logic diagram:

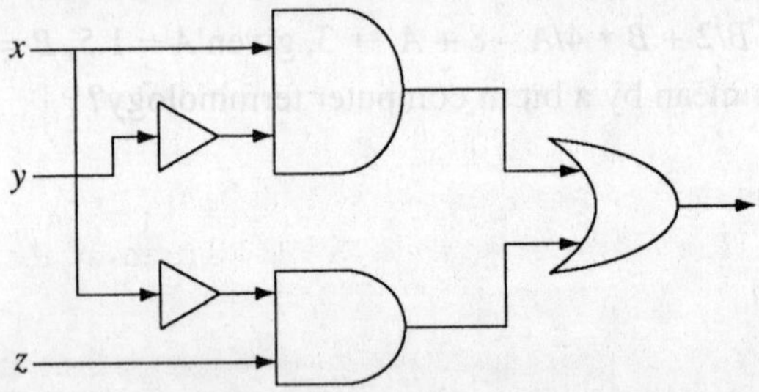

(c) Construct a logic circuit for the Boolean expression $x'y + xy'$.

4. (a) What is the Arithmetic Logic Unit (ALU)? What are its functions? Describe the sequence of Arithmetic logic operations.

(b) What are fifth generation computers? What type of hardware and software will be utilized in it? State the expected advantages compared to the computers of other generations.

(c) Write a short note on Desktop icons.

5. (a) What do you mean by Fixed and floating point representation of real number? What is the normalized form and what is the advantage of using normal form?

(b) Convert the hexadecimal number $2ADAC7B$ to binary and convert the hexadecimal number $4ACEF2B$ to octal.

(c) Using 2's compliment arithmetic, find $1001_2 - 1101_2$.

6. (a) Discuss source program and object program. How source program is converted to object program?

(b) Draw a flowchart to find the sum of first 100 natural numbers.

(c) Write a FORTRAN77 or C program to find the average of 15 given numbers.

2009

1. (a) Show that in a Boolean algebra B, the identity element 0 for addition is unique.
   (b) What is the principle of duality in Boolean algebra?
   (c) Convert $(0.6875)_{10}$ into binary number system.
2. (a) In a Boolean algebra B, show that $(x.y).(x'.z' + z).\{x.(z + y)'\} = 0, x,y,z \in B$.
   (b) Find the disjunctive normal form of $[x'.y + (x.z)'] (x + y.z)'$.
3. (a) Draw a circuit using NAND gate only that realizes the truth function $f(A, B, C) = (A.B.C)'$.
   (b) Construct a switching circuit for the Boolean expression $A.B.C + A'.B.C + A.B'.C + A.B.C'$. Simplify the function and draw the simplified circuit using logic gates.
4. (a) What are the difference between a fourth and fifth generation computers?
   (b) Write a short note on memory unit and software.
   (c) What are assembly level language and high level language?
5. (a) Describe briefly how data are represented in a computer?
   (b) Convert $(3798.3875)_{10}$ to hexadecimal and $(74E.9F)_{16}$ to decimal.
6. (a) Write the FORTRAN expression of $x^{-\frac{5}{2}} - \frac{a}{b}x^2\left(\frac{2}{x^3}+9\right)$.
   (b) A sequence is defined as follows 0, 1, 1, 2, 3, 5, ... ($T_{n+1} = T_n + T_{n-1}, n \geq 1$). Write a FORTRAN77 or C program to find all terms of the sequence which are less than 200.
   (c) Develop an algorithm and draw a flowchart to compute the sum of n given numbers.

2010

1. (a) Give the definition of Boolean algebra due to Huntington.
   (b) In a Boolean algebra B, prove that $(a.b)' = a' + b'$ for all $a, b \in B$.
   (c) Convert $(29)_{10}$ to its equivalent binary form.
2. (a) In a Boolean algebra B, prove that $(x.y + x.y' + x'.y).(x.y + x.y').x = x; x, y \in B$.
   (b) Find the disjunctive normal form of $x + (x'.y' + x'.z)'$.
   (c) Construct the truth table for the Boolean expression $(x + y + z').(x' + y')$.
3. (a) Construct a switching circuit for the Boolean expression $A.B + A'.B + A.B'$. Simplify the function and draw the simplified circuit using logic gates.
   (b) Simplify the following Boolean function using Karnaugh map: $f(x, y, z) = x'yz + xy'z + xyz' + xyz$.
4. (a) Define BIT, BYTE and Word. Discuss briefly the different coding systems.
   (b) What is the arithmetic logic unit? What are its functions? Describe the sequence of arithmetic logic operations.
   (c) Write a short note on ALGOL.

5. (a) Convert $(223.12)_8$ and $(02.5)_8$ to it equivalent decimal form.
   (b) Why are binary numbers used in computer design?
   (c) What do you mean by fixed and floating point representation of real numbers?
6. (a) Using 2's compliment arithmetic, find $10101_2 - 11001_2$.
   (b) Write the FORTRAN expression of $\frac{\tan^{-1} x}{\sqrt{1-x^2}} - (1+x^2)^{\frac{5}{2}}$.
   (c) Write a subroutine subprogram to multiply two matrices A and B of order $x \times y$ and $y \times z$ respectively in FORTRAN or C.
   (d) Draw a flowchart to find the largest number from the numbers: 75, 82, 91, 34, 28, 102, 55.

2011

1. (a) Show that in a Boolean algebra B, the identity element 0 for addition is unique.
   (b) Convert $(0.6875)_{10}$ into binary number system.
   (c) Define Bit and Byte.
2. (a) In a Boolean algebra B, prove that $(x, y') + (x' . y) = (x + y) . (x' + y')\ \forall x, y \in B$.
   (b) Find the disjunctive normal form of $[x'.y + (x.z)'].(x + y.z)'$.
   (c) Construct a truth table for the Boolean expression $x' + y.z$.
3. (a) What do you mean by principle of duality? Write the dual of the statement:
$$(x + y)\ (x + 1) = x + xy + y.$$
   (b) Construct a switching circuit for the Boolean expression $A.B.C + A'.B.C + A.B'.C + A.B.C'$. Simplify the function and draw the simplified circuit using logic gates.
4. (a) What are the differences between fourth and fifth generation computers?
   (b) Write short notes on (i) ASCII code, (ii) Primary and secondary memory.
   (c) Define high level language and machine language. Distinguish between them.
5. (a) Convert $(7F.2C)_{16}$ to equivalent binary form.
   (b) Subtract $(101.001)_2$ from $(1011.11)_2$ using 2's compliment method.
   (c) What are least significant bit and most significant bit? Write a short note on 'Positional Number System'.
6. (a) Draw a flowchart to short the following in descending order: 91, 34, 28, 102, 55, 3, 19, 19, 78, 10, 67.
   (b) Write a short note on LOGICAL-IF statement.
   (c) Write a FORTRAN77 program to calculate the sum of first 100 natural numbers.

2012

1. (a) Show that in a Boolean algebra B, the identity element 0 for addition is unique.
   (b) Convert $(69)_{10}$ into binary form
   (c) Define *AND* gate.
2. (a) In a Boolean algebra B, prove that $(x, y) + (x'.z' + z).\{x.(z + y)'\} = 0,\ x, y, z \in B$.
   [Hints: $(x.y)\,(x'.z' + z).\{x.(z + y)'\} = (x.y)\,(x' + z).\{x.z'.y'\} = x.y.z.x.y'z' = 0$]
   (b) Express the Boolean expression $(xy + yz) + y$ in conjunctive normal form in smallest possible number of variables.
   [Hints: $(x'y + yz)' + y' = (x + y')\,(y + z') + y' = (x + y' + y')\,(y' + z' + y') = (x + y')\,(y' + z') = (x + y' + zz')\,(y' + z' + xx') = (x + y' + z)\,(x + y' + z')\,(y' + z' + x')$]
   (c) Define OR gate. Draw a circuit to realize the two input AND function using only NOR gate.
   [Hints: See (f) of article 15.11.]
3. (a) Draw a circuit using NAND gate only that realizes the truth function $f(A, B, C) = (A.B.C)'$.
   (b) Find the DNF of $(x + y + z).(x.y + x.z)$.
   [Hints: $(x + y + z).(xy + xz) = xy + xz + xy + xyz + xyz + xz = xy + xz + xyz = xy(z + z') + xz(y + y') + xyz = xyz + xyz' + xyz + xy'z + xyz = xyz + xyz' + xy'z$]
   (c) Compute the value of the following FORTRAN expression: (*A*/*B*) ** 3 + *C* + (*A* * *B*/*C*) where $A = 2, B = -1, C = 5$.
   [Hints: $-8 + 5 + 0 = -3$]
4. (a) What is fifth generation computer? What type of hardware and software will be used in it? State the expected advantages compared to the computers of other generations.
   [Hints: See article 11.3.]
   (b) Write a short note on "input" statement and "output" statement.
   (c) Discuss the functions of a Central Processing Unit (CPU). Describe how it helps to execute a programme.
5. (a) Write down the Boolean expression for the following circuit. Draw an equivalent simpler circuit for the expression:

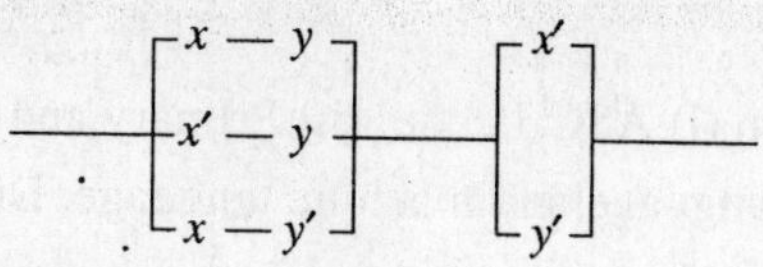

   [Hints: $(xy + x'y + xy')\,(x' + y') = 0 + 0 + x'y + 0 + 0 + xy' = x'y + xy'$]
   (b) Define Bit, Byte and Word.
   (c) Using 2's compliment arithmetic, find $1001_2 - 1101_2$.
6. (a) What are the properties of an algorithm? Write down an algorithm to find the largest of three distinct numbers *X*, *Y* and *Z*.
   (b) Write a FORTRAN/BASIC programme to calculate the perimeter and area of a triangle.